Workbook for

Auto Diagnosis, Service, and Repair

by

Martin W. Stockel

Martin T. Stockel

James E. Duffy

Chris Johanson

Publisher
The Goodheart-Willcox Company, Inc.
Tinley Park, Illinois

Introduction

The **Workbook** for **Auto Diagnosis, Service, and Repair** provides a thorough guide for the **Auto Diagnosis, Service, and Repair** text. It highlights important information, improves understanding, and simplifies the contents of the textbook.

The workbook contains many unique features designed to make learning easier and more interesting. It has three major types of exercises: *Review Questions* (multi-format textbook review), *ASE-Type Questions* (review questions presented in ASE format), and *Jobs* (in-shop activities).

The *Review Questions* are correlated to and serve as a study guide for a textbook chapter. You are lead through the text page by page, making sure you cover the most important material. The Review Questions are organized by subject. A variety of question formats are used, including multiple choice, completion, short answer, matching, and identification.

The ASE-Type Questions are also correlated to a textbook chapter. These questions are presented in the four formats found on the ASE Certification Exam: one-part questions, two-part questions, negative questions, and incomplete sentence questions. These questions are all multiple-choice questions, each with four possible responses.

The workbook jobs supplement the contents of the textbook. The jobs provide easy-to-follow instructions for various hands-on activities. These jobs will help you develop the basic skills needed to service and repair automobiles.

As a student of automotive technology, you will find your workbook to be a valuable learning tool. Its use will make your learning experiences much more enjoyable.

Martin W. Stockel
Martin T. Stockel
James E. Duffy
Chris Johanson

Copyright 2003

by

THE GOODHEART-WILLCOX COMPANY, INC.

Previous Editions Copyright 1996, 1992, 1984

All rights reserved. No part of this book may be reproduced, stored in a retrieval system, or transmitted in any form or by any means, electronic, mechanical, photocopying, recording, or otherwise, without the prior written permission of The Goodheart-Willcox Company, Inc. Manufactured in the United States of America.

International Standard Book Number 1-56637-911-3

1 2 3 4 5 6 7 8 9 0   03   06 05 04 03 02

This book contains the most complete and accurate information that could be obtained from various authoritative sources at the time of publication. Goodheart-Willcox cannot assume responsibility for any changes, errors, or omissions.

The Goodheart-Willcox Company, Inc. and the authors make no warranty or representation whatsoever, either expressed or implied, with respect to the equipment, procedures, and applications described or referred to herein, their quality, performance, merchantability, or fitness for a particular purpose.

Further, the Goodheart-Willcox Company, Inc. and the authors specifically disclaim any liability whatsoever for direct, indirect, special, incidental, or consequential damages arising out of the use or inability to use any of the equipment, procedures, and applications described or referred to herein.

Contents

Instructions for Answering Workbook Questions

Each chapter in this workbook directly correlates to the same chapter in the text. Before answering the questions in the workbook, study the assigned chapter in the text and answer the end-of-chapter review questions while referring to the text. Then, review the objectives at the beginning of each workbook chapter. This will help you review the important concepts presented in the chapter. Try to complete as many workbook questions as possible without referring to the textbook. Then, use the text to complete the remaining questions.

A variety of questions are used in the workbook, including multiple choice, identification, completion, and short answer. These questions should be answered in the following manner.

Multiple Choice

Select the best answer and write the corresponding letter in the blank.

1. Torrington bearings are a type of _____ bearing.
    - (A) needle
    - (B) ball
    - (C) roller
    - (D) None of the above.

1. _____A_____

Identification

Identify the components indicated on the illustration or photograph accompanying the question.

2. Identify the parts of the injector in the following illustration.
    - (A) _____flex hose_____
    - (B) _____filter_____
    - (C) _____winding_____
    - (D) _____needle return spring_____
    - (E) _____needle armature_____
    - (F) _____sealing needle_____

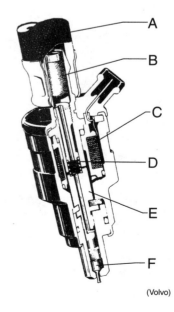

(Volvo)

Completion

In the blank provided, write the word or words that best complete the statement. In some cases you will be asked to draw missing parts or sections on an illustration. Use a ruler for all straight lines. Use colored pencils to help clarify your sketches.

3. On some vehicles with electronic fuel injection, the pulse _____ is increased when the engine is cold.

3. ___width___

Short Answer

Provide complete responses to the statements.

4. What is the purpose of the turbocharger waste gate? ___The turbocharger waste gate prevents the turbocharger from developing too much pressure in the intake manifold.___

Matching

Match the term in the left column with its description in the right column. Place the corresponding letter in the blank.

_C_ 5. Main pressure regulator
_D_ 6. Manual valve
_B_ 7. Shift valves
_F_ 8. Throttle valve
_A_ 9. Governor valve
_H_ 10. Filter
_G_ 11. Accumulator
_I_ 12. Servos

(A) driven by output shaft
(B) redirect pressure to holding members
(C) controls overall system pressure
(D) operated by driver
(E) controls lockup clutch
(F) operated by linkage or modulator
(G) cushions shifts
(H) removes dirt and metal.
(I) apply bands

Other Types of Questions

When other types of workbook questions are presented, follow the specific instructions that accompany the problems.

Instructions for Answering Workbook Jobs

Before starting any job, complete the related textbook and workbook chapters. Read the introduction, objective, and instructions for the job. Ask your instructor for any possible changes in the job procedures and for help as needed. Read and complete each step of the job carefully, while observing all safety rules. Always answer the job questions with complete sentences.

Name _____

Date _____ Period _____

Instructor _____

Score _____

Text pages 15–24

1

Introduction to the Automotive Service and Repair Industry

Objectives: After studying Chapter 1 in the textbook and completing this section of the workbook, you will be able to:

*   Identify the major sources of employment in the automotive industry.
*   List and describe classifications of automotive technicians.
*   Identify different types of automotive service facilities.
*   Identify advancement possibilities for automotive technicians.
*   Identify automotive training opportunities.
*   Explain the benefits of ASE technician certification.

Tech Talk: Career opportunities in the automotive service industry have been very good in the past. In the future, they look even better. This is obvious when looking in the employment sections of most newspapers. Openings for qualified automotive service personnel are almost always listed. If you like work that is mentally and physically demanding, diversified, and financially rewarding, auto service and repair may be for you.

Review Questions

Instructions: Study Chapter 1 of the text and then answer the following questions in the spaces provided.

1. A(n) _____ performs basic maintenance tasks.

2. A(n) _____ can diagnose and repair vehicle systems.

3. Advancement to a supervisory position is dependent on your _____.
    (A) ability to cope with the problems of the position
    (B) knowledge
    (C) skill
    (D) All of the above.

1. _____

2. _____

3. _____

4. A service manager must exercise tact when dealing with _____ problems.

4. _____

5. A sales person will probably *not* be needed in a(n) _____ shop.

5. _____

6. The ultimate endpoint of all parts is the _____.

6. _____

7. Working conditions for the automotive technician are _____ than they were in the past.

7. _____

8. An entrepreneur owns his or her own _____.

8. _____

9. Explain flat-rate pay. _____

_____

_____

10. A shop that repairs brakes and performs no other kinds of work is a _____ shop.

10. _____

11. Name five places where a person can receive training to become an automotive technician.

_____

_____

12. Name three methods by which a technician can keep up with changes in the automotive repair industry. _____

_____

_____

13. Everyone who takes a certain ASE test on the same _____ receives the same test.

13. _____

14. An ASE Master Technician has passed _____ of the tests in a particular area.
    (A) more than one
    (B) three
    (C) five
    (D) all

14. _____

15. The ASE certification program has been extended to all of the following countries, *except:*
    (A) Brazil
    (B) Canada
    (C) Panama
    (D) Mexico

15. _____

ASE-Type Questions

1. Technician A says that a helper will sometimes clean parts for other technicians. Technician B says that taking a job as a helper is a good way to enter the automotive service field. Who is right?
   - (A) A only.
   - (B) B only.
   - (C) Both A and B.
   - (D) Neither A nor B.

1. _____

2. All of the following statements about the technician's work possibilities are true, *except:*
   - (A) it is hard for a good technician to find work.
   - (B) most technicians work in specialty shops.
   - (C) government shops pay a salary.
   - (D) working conditions at independent shops vary widely.

2. _____

3. Technician A says that the shop supervisor usually does the hiring. Technician B says that the shop supervisor does not need to know how to repair vehicles. Who is right?
   - (A) A only.
   - (B) B only.
   - (C) Both A and B.
   - (D) Neither A nor B.

3. _____

4. All of the following are features of working at a dealership, *except:*
   - (A) variety of work.
   - (B) salary guarantees.
   - (C) fast-paced environment.
   - (D) warranty repairs.

4. _____

5. Which statement about the following illustration is true?
   - (A) The illustration shows a common trade magazine.
   - (B) This is the cover of an ASE study guide.
   - (C) This is the cover of an ASE test.
   - (D) The illustration shows an Internet website.

5. _____

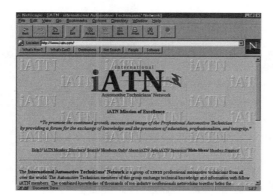

Name _____

Date _____ Period _____

Instructor _____

Score _____

Text pages 25–35

2

Safety and Environmental Protection

Objectives: After studying Chapter 2 in the textbook and completing this section of the workbook, you will be able to:

- Identify the major causes of accidents.
- Explain why accidents must be avoided.
- Recognize unsafe conditions in the shop.
- Give examples of unsafe work procedures.
- Use personal protective equipment.
- Describe types of environmental damage caused by improper auto shop practices.
- Identify ways to prevent environmental damage.

Tech Talk: Do you like working on cars and trucks? Do you want to be able to do it for several more years? Well, the easiest way to end your career as an automobile technician is to have a serious accident. Even minor accidents can cost you your health or your job. Always make sure that you do all repair procedures as safely as possible.

Review Questions

Instructions: Study Chapter 2 of the text and then answer the following questions in the spaces provided.

1. The technician should always try to _____ accidents.

1. _____

2. Some accidents occur when technicians try to take _____.

2. _____

3. The technician must keep _____ in mind at all times.

3. _____

4. List several work procedures that should be followed when working in the automotive shop.

_____

_____

5. Describe the purpose of Material Safety Data Sheets (MSDS).

_____

_____

6. The technician should wear _____ when working on running engines; using drills, grinders, or tire changers; or working around batteries or hot cooling systems.
    (A) respiratory protection
    (B) safety shoes
    (C) eye protection
    (D) All of the above.

6. _____

7. _____ are necessary to protect the technician from the crushing force generated by falling parts, tools, and supplies.

7. _____

8. When should the technician wear rubber gloves?

_____

_____

9. Name the two ways that the automotive technician can cause environmental damage.

_____

_____

10. Automotive shop by-products that can be recycled include _____.
    (A) scrap metal
    (B) tires
    (C) oil and antifreeze
    (D) All of the above.

10. _____

For questions 11–15, describe how you would handle the following shop problems.

11. A radiator hose bursts, covering the shop floor with coolant and water.

_____

_____

12. A can of used carburetor cleaner is discovered behind a workbench.

_____

_____

13. A vehicle is brought into the shop with some of the emission controls disconnected.

_____

_____

14. A vehicle is being operated in the shop when all windows and doors are closed.

_____

_____

15. A fellow technician throws a used alternator into the trash can.

_____

_____

ASE-Type Questions

1. Technician A says that long-term exposure to certain chemicals can cause skin disorders. Technician B says that long-term exposure to some chemicals can cause lung damage. Who is right?
    (A) A only.
    (B) B only.
    (C) Both A and B.
    (D) Neither A nor B.

    1. _____

2. Shopkeeping involves all of the following, _except:_
    (A) keeping the shop premises in order.
    (B) identifying and correcting unsafe conditions.
    (C) keeping tools and equipment in working order.
    (D) adjusting vehicles for lowest emissions.

    2. _____

3. Respiratory protection should be worn whenever you are working with _____.
    (A) starting system components
    (B) clutches
    (C) petroleum products
    (D) engine coolant

    3. _____

4. Technician A says that waste disposal guidelines are established and enforced by the Environmental Protection Agency. Technician B says that the Environmental Protection Agency recycles the hazardous wastes generated in the auto shop. Who is right?
    (A) A only.
    (B) B only.
    (C) Both A and B.
    (D) Neither A nor B.

    4. _____

5. The technician should never make any kind of vehicle adjustment without first determining the effect on _____.
    (A) fuel economy
    (B) emissions
    (C) performance
    (D) durability

    5. _____

Name _____

Date _____ Period _____

Instructor _____

Score _____

Text pages 37–59

3

Basic Tools and Service Information

Objectives: After studying Chapter 3 in the textbook and completing this section of the workbook, you will be able to:

- Identify and describe the proper use of the hand tools commonly used by the automotive technician.
- List the safety rules for hand tools.
- Identify and describe the proper use of the power tools commonly used by the automotive technician.
- List the safety rules for using hand and power tools.
- Identify types of service information and training materials.
- Select the correct tools and service information for a given job.

Tech Talk: Attempting to work on a car without the proper tools is like trying to fight a forest fire with a squirt gun—frustrating if not impossible. In a sense, hand tools serve as extensions of the human body. They increase the physical abilities of a technician's fingers, hands, arms, legs, eyes, ears, and back. Professional auto technicians often spend thousands of dollars on their tools. This is a wise investment. A well selected set of tools increases productivity, precision, and job satisfaction.

Review Questions

Instructions: Study Chapter 3 of the text and then answer the following questions in the spaces provided.

1. The cheaper grades of tools are generally _____.
    (A) made of alloy steel
    (B) cumbersome
    (C) easy to clean
    (D) None of the above.

1. _____

2. Fast, efficient work and _____ cannot exist together.

2. _____

3. The tools in the following illustration are arranged for display. However, when used properly, the cabinet pictured provides proper storage and _____ to tools.

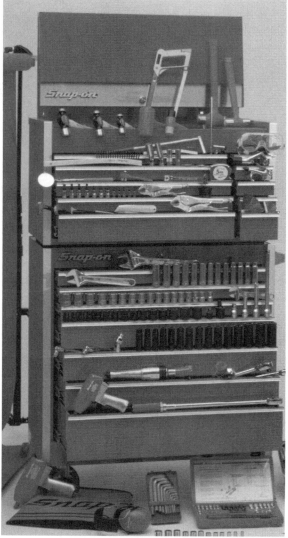

(Snap-on Tools)

4. List the types of screwdrivers that should be owned by the automotive technician.

_____

_____

5. Box-end wrenches are available with _____-point openings.
   (A) 6
   (B) 12
   (C) Both A and B.
   (D) Neither A nor B.

5. _____

6. Double offset wrenches allow more handle _____.

6. _____

7. Sockets are available with _____-point openings.
    (A) 10
    (B) 12
    (C) 18
    (D) All of the above.

7. _____

8. Define drive size.

_____

_____

9. The ratchet handle allows both heavy force and _____.

9. _____

10. Spinner handles _____.
    (A) are used exclusively in areas where limited swing
       is necessary
    (B) are used in the same manner as screwdrivers
    (C) provide heavy turning leverage
    (D) None of the above.

10. _____

11. The reach of a socket set can be increased by using a(n)

_____.
    (A) extension
    (B) impact socket
    (C) T-handle
    (D) None of the above.

11. _____

12. Sockets of one particular drive size can be turned with the
handles of another size by using a(n) _____.
    (A) transition piece
    (B) adapter
    (C) universal fitting
    (D) None of the above.

12. _____

13. When using any type of wrench, it is important to _____
on the handle.

13. _____

14. Pliers are used for _____.
    (A) tightening nuts
    (B) loosening tubing fittings
    (C) crimping
    (D) None of the above.

14. _____

15. Identify the tools shown in the following illustration.

(A) _____

(B) _____

(C) _____

(D) _____

(E) _____

(F) _____

(G) _____

16. Identify the specialty wrenches shown in the following illustration.

(A) _____

(B) _____

(C) _____

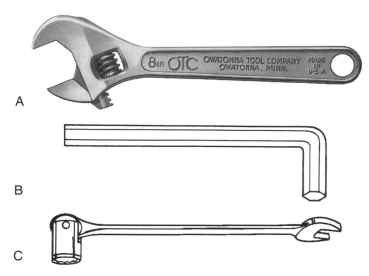

17. Avoid cutting _____ with pliers.   17. _____
    (A) tubing
    (B) wire
    (C) hardened bolts
    (D) None of the above.

18. A(n) _____ screwdriver is shown in the following illustration.   18. _____

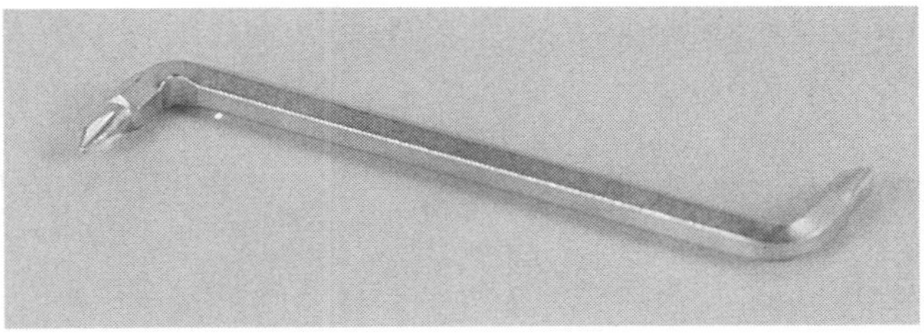

19. Identify the pliers shown in the following illustration.

    (A) _____
    (B) _____
    (C) _____
    (D) _____
    (E) _____
    (F) _____

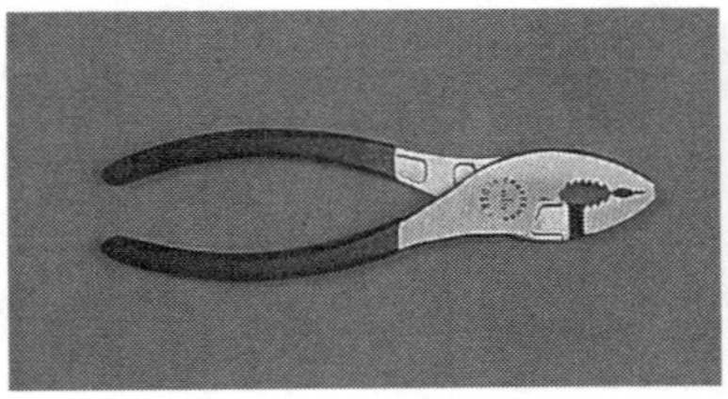

A 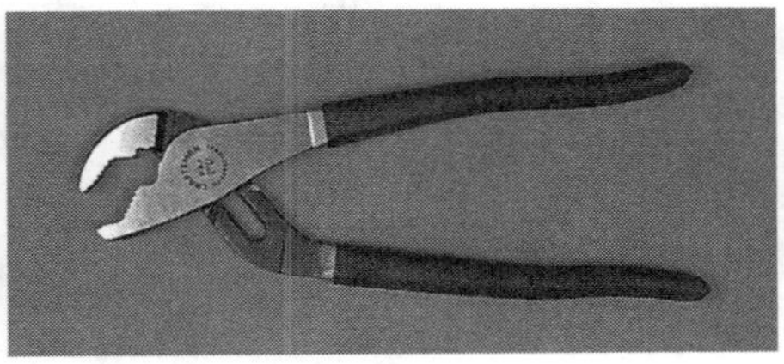   B 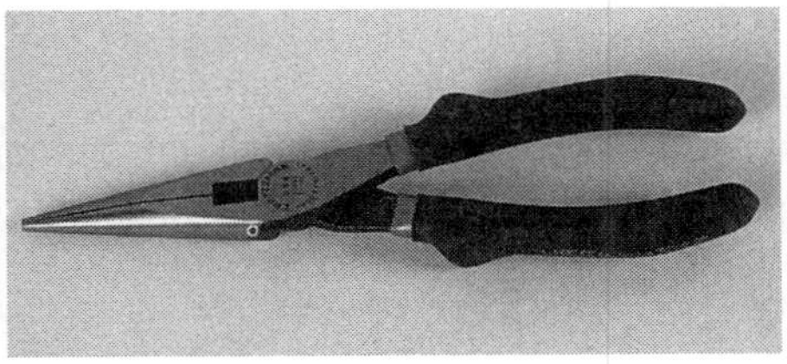   C

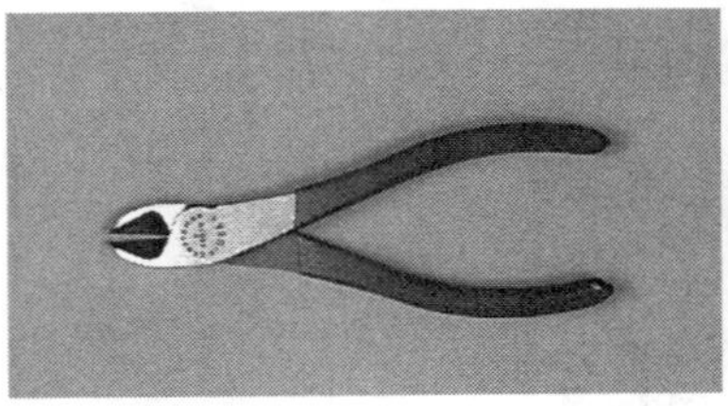

D 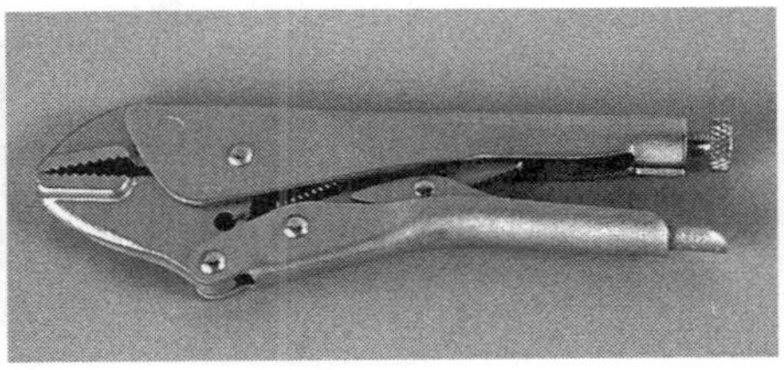   E 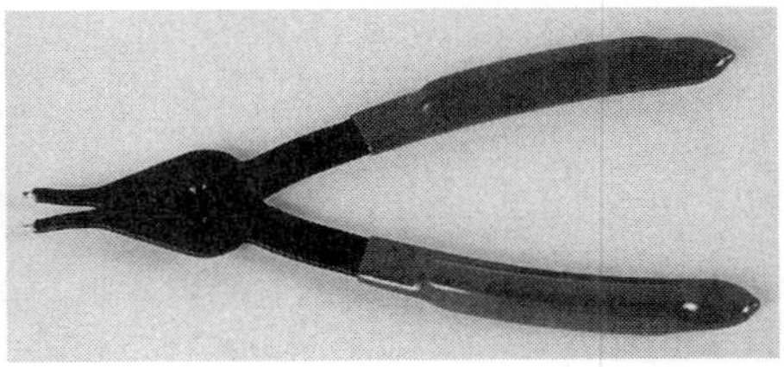   F

20. A few sections of round _____ are useful when driving   20. _____
    parts that may be damaged by steel punches.

21. Explain the purpose of a starting punch.

    _____

    _____

22. A center punch is needed for _____.
    (A) marking parts before drilling
    (B) marking parts for reassembly
    (C) All of the above.
    (D) None of the above.

22. _____

23. The aligning punch is used to _____.

23. _____

24. When sharpening punches and chisels, _____ slowly and often to prevent drawing the temper.

24. _____

25. Identify the damage to the chisel on the left. Identify the repairs made to the chisel on the right.
    (A) _____
    (B) _____
    (C) _____

26. Identify the punches shown in the following illustration.
    (A) _____
    (B) _____
    (C) _____

27. _____ files are commonly used in the auto shop.
    (A) Flat mill
    (B) Round
    (C) Square
    (D) All of the above.

27. _____

28. Be sure the file has a(n) _____ firmly affixed to its tang before use.

28. _____

29. Fine files work well on _____.
    - (A) brass
    - (B) steel
    - (C) aluminum
    - (D) None of the above.

29. _____

30. Identify the file cuts shown in the following illustration.
    - (A) _____
    - (B) _____
    - (C) _____
    - (D) _____

A   B   C   D

(Deere & Co.)

31. Reamers are used to accurately _____.

31. _____

32. _____ are used for cutting internal threads.
    - (A) Taps
    - (B) Dies
    - (C) Heli-coils
    - (D) None of the above.

32. _____

33. To cut external threads on bolts, screws, and pipe, _____ are used.

33. _____

34. Name a special-purpose tap and a special-purpose die that are commonly used in the automotive shop.

    _____

    _____

35. The tool pictured below is a typical _____.

35. _____

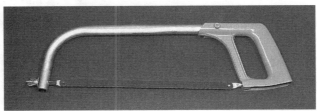

36. The 18-tooth hacksaw blade is used for cutting _____.

36. _____

37. The 32-tooth hacksaw blade is used for cutting _____.

37. _____

38. The following illustration shows a(n) _____-type impact wrench. These wrenches almost always have a(n) _____" drive.

38. _____

_____

39. Quality drill bits are made of _____.
    (A) carbon steel
    (B) high-speed steel
    (C) brass
    (D) All of the above.

39. _____

40. In the following illustration, identify the mechanisms used to tighten the drill bit.
    (A) _____
    (B) _____

41. The tool shown in the following illustration is a(n) _____.

41. _____

42. As tools are obtained, they should be marked in areas that are difficult to _____.

42. _____

43. _____ manuals cover one common system for many vehicle makes and models.

43. _____

44. Troubleshooting charts contain the logical steps needed to determine the _____ of a problem.

44. _____

45. The Internet allows technicians around the world to contact each other with _____.

45. _____

ASE-Type Questions

1. Technician A says that the time it takes to keep your tools clean and orderly is wasted time. Technician B says that open-end wrenches and box-end wrenches should never be stored together. Who is right?
    (A) A only.
    (B) B only.
    (C) Both A and B.
    (D) Neither A nor B.

1. _____

2. The _____ screwdriver is useful in tight quarters where even a "stubby" cannot be used.
    (A) Torx®
    (B) clutch
    (C) offset
    (D) Phillips

2. _____

3. Technician A says that an adjustable wrench tends to loosen and slip. Technician B says that an adjustable wrench should be used in place of an open-end wrench whenever possible. Who is right?
    (A) A only.
    (B) B only.
    (C) Both A and B.
    (D) Neither A nor B.

3. _____

4. Fuel system, vacuum, and brake fittings should be loosened and tightened with a(n) _____ wrench.
    (A) Torx
    (B) offset
    (C) flare-nut
    (D) box end

4. _____

5. Punches are used for all of the following purposes, *except:*
    (A) cutting off rivet heads.
    (B) driving pins from holes.
    (C) aligning holes.
    (D) marking spots for drilling.

5. _____

6. Coarse files are best for filing _____.
   - (A) steel
   - (B) brass
   - (C) hard metals
   - (D) None of the above.

6. _____

7. Cutting oil is helpful when drilling _____.
   - (A) steel
   - (B) cast iron
   - (C) aluminum
   - (D) brass

7. _____

8. A hacksaw is used to cut which of the following?
   - (A) tubing
   - (B) bolts
   - (C) sheet metal
   - (D) All of the above.

8. _____

9. The metal tail of a file, to which the handle is affixed, is called the _____.
   - (A) tongue
   - (B) tail
   - (C) tang
   - (D) point

9. _____

10. Technician A says that a factory service manual contains information on one make and model of vehicle for a given year. Technician B says that some factory service manuals are divided into diagnosis and overhaul manuals. Who is right?
    - (A) A only.
    - (B) B only.
    - (C) Both A and B.
    - (D) Neither A nor B.

10. _____

Name _____

Date _____ Period _____

Instructor _____

Score _____

Text pages 61–77

4

Precision Tools and Test Equipment

Objectives: After studying Chapter 4 in the textbook and completing this section of the workbook, you will be able to:

- Identify common measuring tools.
- Select the appropriate measuring tool for a given job.
- Use precision measuring tools.
- Properly maintain precision measuring tools.

Tech Talk: A competent automotive technician uses precision measuring tools on the job. The ability to use these tools properly distinguishes the good technician from the "parts changer." Knowing when, where, and how to measure is essential to anyone seeking a career as an automotive technician.

Review Questions

Instructions: Study Chapter 4 of the text and then answer the following questions in the spaces provided.

1. Precision measuring tools should be checked for _____ on a regular basis.

1. _____

2. The outside micrometer can be used to check the diameter of all of the following, *except:*
   - (A) piston pins.
   - (B) cylinder bores.
   - (C) crankshaft journals.
   - (D) pistons.

2. _____

3. Individual micrometers are designed to produce readings over a range of _____. It is recommended that a technician get a set of six.

3. _____

4. Micrometers are made so that every full turn of the thimble moves the spindle _____.
    (A) 0.010″
    (B) 0.100″
    (C) 0.001″
    (D) 0.025″

4. _____

5. Each line on the micrometer's thimble edge equals _____ inches.

5. _____

6. List the four steps that should be followed when reading a micrometer.

_____

_____

_____

_____

7. Identify the parts of the micrometer shown in the following illustration.
    (A) _____
    (B) _____
    (C) _____
    (D) _____
    (E) _____
    (F) _____
    (G) _____
    (H) _____

8. The inside micrometer is used for making measurements in _____.

8. _____

9. The micrometer depth gauge is a handy tool for reading the depth of _____.
    (A) slots
    (B) splines
    (C) counterbores
    (D) All of the above.

9. _____

10. When measuring with a micrometer, you should never clamp it _____.

10. _____

11. The _____ micrometer in the following illustration is accurate to one one-thousandth of an inch.

11. _____

(Central Tools)

12. List three things that an inside micrometer is used to measure.

_____

_____

_____

13. The tool in the following illustration is a(n) _____-type micrometer. Label the parts indicated.

(A) _____

(B) _____

(C) _____

(D) _____

(E) _____

(F) _____

(L. S. Starrett)

14. When using a dial indicator, be certain that it is mounted firmly and that the actuating rod is _____ to the plane (direction) of movement to be measured.

14. _____

15. When accuracy is not critical, inside and outside _____ are/is useful tools for quick measurements.

15. _____

16. Describe thickness gauges and explain their use. _____

_____

_____

17. Wire gauges are commonly used to _____.
    (A) measure the gage of a wire
    (B) check spark plug gap
    (C) measure current
    (D) measure bores

17. _____

18. The _____ gauge is the proper tool for checking the number of threads per inch on bolts, screws, and studs.

18. _____

19. The telescoping gauge is an accurate tool for measuring _____.
    (A) connecting rod bores
    (B) main bearing bores
    (C) valve guide diameters
    (D) All of the above.

19. _____

20. A steel straightedge can be used to check a part for _____.

20. _____

21. Identify the measuring tools shown in the following illustrations.
    (A) _____
    (B) _____
    (C) _____
    (D) _____
    (E) _____
    (F) _____

(Snap-on Tools) (L. S. Starrett)

22. The compression gauge measures the pressure developed in the engine _____.

22. _____

23. The gauge shown in the following illustration measures _____ air pressure.

23. _____

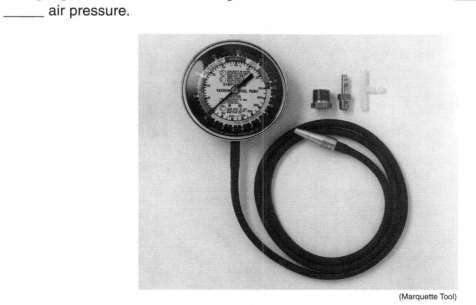

(Marquette Tool)

24. An oil pressure tester can measure the pressure developed by the oil pumps of which three engine systems?

_____

_____

25. When using an ohmmeter, _____ is not important unless a diode is being checked.
    - (A) voltage
    - (B) current
    - (C) polarity
    - (D) None of the above.

25. _____

26. A voltmeter is used to check the electrical potential between two points in a circuit that is _____.

26. _____

27. An ammeter is used to check the electrical property known as _____.

27. _____

28. Multimeters can be used to measure _____.
    - (A) resistance
    - (B) voltage
    - (C) current
    - (D) All of the above.

28. _____

29. In addition to the parts shown in the illustration, a powered test light contains an internal _____.

29. _____

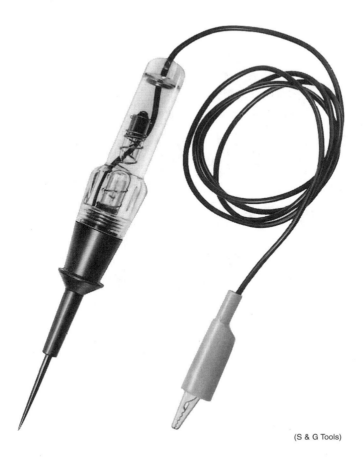

(S & G Tools)

Matching

Match the test to be performed with the test equipment. Some choices will be used more than once.

_____ 30. Measuring the waveform of a fuel injector solenoid.

_____ 31. Retrieving trouble codes from the vehicle computer.

_____ 32. Measuring voltage and resistance.

_____ 33. Reprogramming a vehicle computer.

_____ 34. Measuring the voltage pattern produced by the engine ignition system.

_____ 35. Check wires for continuity.

(A) Multimeter.
(B) Scan tool.
(C) Digital storage oscilloscope.

ASE-Type Questions

1. Technician A says that precision tools should be kept in a protective case. Technician B says that precision tools should be stored in an area that will not be subjected to excessive moisture. Who is right?
    (A) A only.
    (B) B only.
    (C) Both A and B.
    (D) Neither A nor B.

1. _____

2. The thimble of a metric micrometer must be turned _____ full revolution(s) to measure one millimeter.
    (A) 1
    (B) 2
    (C) 4
    (D) 10

2. _____

3. Technician A says that a dial indicator is commonly used to check endplay. Technician B says that a dial indicator is commonly used to measure cylinder diameter. Who is right?
    (A) A only.
    (B) B only.
    (C) Both A and B.
    (D) Neither A nor B.

3. _____

4. An inside micrometer is used to check all of the following, *except:*
    (A) cylinder bores.
    (B) brake drums.
    (C) brake rotors.
    (D) large bushings.

4. _____

5. Technician A says that an inside caliper can be used to accurately measure the diameter of holes. Technician B says that an outside caliper is frequently used to measure the diameter of holes. Who is right?
    (A) A only.
    (B) B only.
    (C) Both A and B.
    (D) Neither A nor B.

5. _____

6. All metals expand and contract in direct proportion to their _____.
    (A) size
    (B) weight
    (C) temperature
    (D) mass

6. _____

7. The tool shown in the following illustration is used to check
   all of the following, *except:*
   - (A) shaft endplay.
   - (B) shaft diameter.
   - (C) gear backlash.
   - (D) cylinder taper.

7. _____

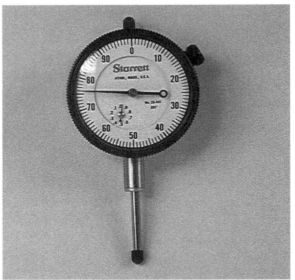

(L. S. Starrett)

8. Technician A says that the fuel pressure tester can be
   used to check for a defective fuel pump or clogged fuel
   filter. Technician B says that the fuel pressure tester is
   used only on carbureted engines. Who is right?
   - (A) A only.
   - (B) B only.
   - (C) Both A and B.
   - (D) Neither A nor B.

8. _____

9. An ohmmeter measures the electrical value known as
   _____, which is the opposition to current flow.
   - (A) resistance
   - (B) impedance
   - (C) capacitance
   - (D) amperage

9. _____

10. All of the following statements about scan tools are true,
    except:
    - (A) Scan tools can be used to obtain trouble codes.
    - (B) No scan tools will work with more than one
      manufacturer's vehicles.
    - (C) Some scan tools can be used to reprogram the
      vehicle computer.
    - (D) The scan tool is attached to the vehicle diag-
      nostic connector.

10. _____

Name _____

Date _____ Period _____

Instructor _____

Score _____

Text pages 79–87

5

Jacks, Lifts, and Holding Fixtures

Objectives: After studying Chapter 5 in the textbook and completing this section of the workbook, you will be able to:

- Identify the most commonly used lifting and holding equipment.
- Select the correct type of lifting or holding equipment for the job.
- Describe safety precautions for jacks, lifts, and holding fixtures.

Tech Talk: If used correctly, jacks, lifts, hoists, and repair stands make potentially cumbersome and tiring jobs easy. On the other hand, if used improperly, this same equipment can injure or kill. Tons of force can be exerted by a jack or lift. This force can crush your hand or foot, or it can cause severe vehicle or equipment damage. Improperly used hoists or repair stands can cause an engine block or transmission to fall, causing personal injury or ruining parts. Learn to use lifting equipment and holding fixtures properly.

Review Questions

Instructions: Study Chapter 5 of the text and then answer the following questions in the spaces provided.

1. List the three ways that lifting equipment can be operated.

   _____

   _____

   _____

2. A floor jack can be used to raise the _____ of a car.   2. _____
     (A) front
     (B) rear
     (C) side
     (D) All of the above.

3. Never work under a vehicle supported only by a(n) _____.   3. _____

33

4. Jack stands should be placed under the vehicle's _____ or another component that can support the _____.

4. _____

_____

5. The end lift can be operated by _____ pressure.
   (A) pneumatic
   (B) hydraulic
   (C) Both A and B.
   (D) Neither A nor B.

5. _____

6. A(n) _____ frame lift raises the car while leaving both the front and rear of the vehicle completely exposed.

6. _____

7. The _____ frame lift eliminates the single central post, leaving the central portion of the car more accessible.

7. _____

8. Remember that many cars weigh _____ or more.

8. _____

9. Identify the parts of the single-post frame lift shown in the following illustration.
   (A) _____
   (B) _____
   (C) _____

10. A(n) _____ is essential to the safe, efficient removal and installation of automatic transmissions.

10. _____

11. Identify the parts of the transmission jack in the following illustration.
    (A) _____
    (B) _____
    (C) _____
    (D) _____
    (E) _____
    (F) _____
    (G) _____

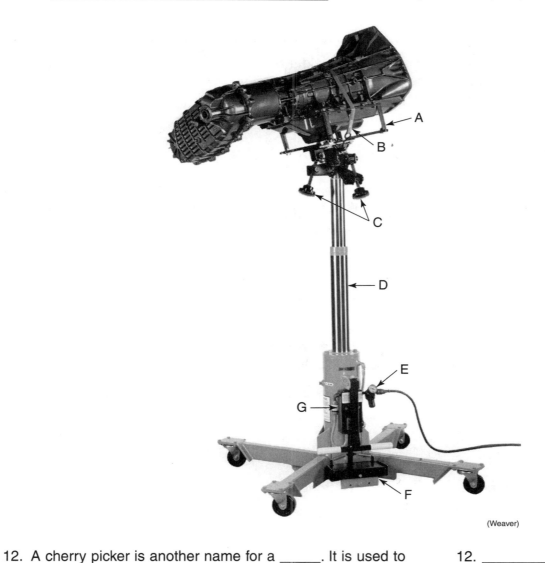

(Weaver)

12. A cherry picker is another name for a _____. It is used to remove _____ and lift other heavy parts.

12. _____

13. A large repair stand can be used to hold _____ for repairs.
    - (A) engine blocks
    - (B) differentials
    - (C) transmissions
    - (D) All of the above.

13. _____

14. What is the one vital rule to remember when using a repair stand?

_____

_____

15. What four safety considerations should be kept in mind when hand-lifting parts?

_____

_____

_____

_____

ASE-Type Questions

1. Technician A says that a hydraulic hand jack should never be used to raise a vehicle. Technician B says that a hydraulic jack can be used to bend parts. Who is right?
   - (A) A only.
   - (B) B only.
   - (C) Both A and B.
   - (D) Neither A nor B.

1. _____

2. It is acceptable to lift a vehicle by positioning a jack under the _____.
   - (A) oil pan
   - (B) frame
   - (C) transmission pan
   - (D) radiator support

2. _____

3. Technician A says that a parallelogram lift eliminates the conventional lift posts. Technician B says that a drive-on lift allows more room to work under the vehicle. Who is correct?
   - (A) A only.
   - (B) B only.
   - (C) Both A and B.
   - (D) Neither A nor B.

3. _____

4. A power train lift can remove the _____ from a front-wheel drive vehicle as an assembly.
   - (A) engine and radiator
   - (B) engine and subframe
   - (C) engine and transaxle
   - (D) transaxle and subframe

4. _____

5. All of the following are correct methods of lifting heavy objects, *except:*
   - (A) lift with your back, not your legs.
   - (B) keep your legs as close together as possible.
   - (C) ask for help if the part is too heavy to lift alone.
   - (D) get a firm grip before lifting.

5. _____

Name _____

Date _____ Period _____

Instructor _____

Score _____

Text pages 89–97

6

Cleaning Equipment and Techniques

Objectives: After studying Chapter 6 in the textbook and completing this section of the workbook, you will be able to:

- List the most common automotive cleaning techniques.
- Compare the advantages and disadvantages of different cleaning methods.
- Select the correct cleaning method for a given job.
- Describe the safety rules that apply to various cleaning techniques.

Tech Talk: A professional auto technician understands the importance of part cleanliness. Although cleaning is time-consuming, it is vital. When working with precision components, the smallest amount of dirt can cause a mechanical failure. A particle of metal the size of a pinhead can damage a cylinder wall or engine bearing, cause an automatic transmission valve or a hydraulic lifter to stick, or clog a carburetor passage or fuel injector nozzle. A quality technician will make sure that all parts are perfectly clean before installing them.

Review Questions

Instructions: Study Chapter 6 of the text and then answer the following questions in the spaces provided.

1. Careless cleaning of parts during teardowns of engines, transmissions, and rear ends may ruin the job and cause _____.

1. _____

2. Automotive body and welding shops occasionally use a(n) _____ to remove rust, paint, and welding scale.

2. _____

3. A bead blast cleaner differs from a conventional sand-blaster in that the blaster unit is placed in a(n) _____.

3. _____

4. Identify the items indicated in the following illustration.

(A) _____

(B) _____

(C) _____

5. Why is it important to use a solvent with a high flash point?_____

_____

_____

6. Why should technicians thoroughly wash their hands and arms when a cleaning job is finished?

_____

_____

7. The steam cleaner is used for cleaning _____.   7. _____

8. List two of the safety rules for steam cleaning.

_____

_____

9. High-pressure spray cleaning equipment uses _____   9. _____
   water.

10. Pressure at the nozzle of a high-pressure spray cleaner   10. _____
    can reach approximately _____ psi.

    (A) 500
    (B) 1000
    (C) 1500
    (D) None of the above.

11. A low-pressure spray-cleaning gun is operated by _____ pressure.

11. _____

12. Although small parts can be cleaned in a can or a bucket, a far more efficient job can be accomplished by using a(n) _____.

12. _____

13. After rinsing components in a parts washer, let the parts _____ and then blow them dry with compressed air.

13. _____

14. Large shops that specialize in rebuilding usually have a hot tank for heavy cleaning of _____.
    (A) engine blocks
    (B) transmission cases
    (C) radiators
    (D) All of the above.

14. _____

15. Identify the parts of the cold solution parts washer pictured in the following illustration.
    (A) _____
    (B) _____
    (C) _____
    (D) _____

(Graymills)

ASE-Type Questions

1. All the following components are subject to accumulations of hard carbon, *except:*
    (A) piston heads.
    (B) valves.
    (C) main bearings.
    (D) combustion chambers.

1. _____

2. Which of the following materials are often plastic coated?
    (A) Cast iron
    (B) Stainless steel
    (C) Aluminum
    (D) Magnesium

2. _____

3. Technician A says that a rotary wire brush can be used to clean deposits from cylinder heads and valves. Technician B says that a rotary wire brush can be used to clean deposits from pistons and bearings. Who is right?
    (A) A only.
    (B) B only.
    (C) Both A and B.
    (D) Neither A nor B.

3. _____

4. Under what circumstances is it acceptable to clean parts in gasoline?
    (A) When the part operates in gasoline.
    (B) When the part is extremely dirty.
    (C) When the cleaning is being done outside.
    (D) Never.

4. _____

5. All of the following statements about steam cleaning are true, *except:*
    (A) steam cleaners can use a flame or electric heat.
    (B) the technician must turn on the heat before turning on the water.
    (C) too much steam will melt the lubricant in chassis components.
    (D) the steam should be "wet" for maximum cleaning.

5. _____

6. Technician A says that goggles or a face shield should be worn when using cleaning solvents. Technician B says that cleaning solutions should only be used in well-ventilated areas. Who is right?
    (A) A only.
    (B) B only.
    (C) Both A and B.
    (D) Neither A nor B.

6. _____

7. A high-pressure parts washer functions similarly to a home _____.
    (A) clothes washer
    (B) clothes dryer
    (C) sink
    (D) dishwasher

7. _____

8. Technician A says that the solution in a vapor cleaner is heated to clean parts. Technician B says that a cold soak solution requires that the parts be suspended in the cleaning solution. Who is right?
    (A) A only.
    (B) B only.
    (C) Both A and B.
    (D) Neither A nor B.

8. _____

9. Cold soak solutions usually come in a special bucket that   9. _____
   contains a _____.
   - (A) brush
   - (B) basket
   - (C) drain
   - (D) vent

10. A brake washer collects _____, which must be disposed of   10. _____
    as hazardous waste.
    - (A) oil
    - (B) grease
    - (C) asbestos
    - (D) carbon monoxide

Name _____

Date _____ Period _____

Instructor _____

Score _____

Text pages 99–112

7

Welding Equipment and Techniques

Objectives: After studying Chapter 7 in the textbook and completing this section of the workbook, you will be able to:

- Describe the equipment needed for welding metal.
- Describe the equipment needed for brazing metal.
- Describe the procedures for brazing metal.

Tech Talk: The technician is often called on to perform welding jobs. Although most welding is performed by auto body technicians, skill in welding is often necessary to repair cracked brackets or aluminum castings. Welding skills are also needed to install trailer hitches. Brazing is often used to repair sheet metal parts, such as oil pans or timing covers. The skilled technician will make the effort to learn the basics of welding and brazing.

Review Questions

Instructions: Study Chapter 7 of the text and then answer the following questions in the spaces provided.

Arc Welding

1. Define *arc welding*. _____

_____

_____

2. The MIG filler wire passes through the center of the _____
as welding takes place.

2. _____

3. Define *plasma cutting*. _____

_____

_____

4. Is plasma cutting cheaper or more expensive than oxyacetylene welding? Why?

   _____

   _____

5. Welding rods are usually coated to provide a(n) _____ shield around the arc.

   5. _____

6. Define *striking an arc.* _____

   _____

   _____

7. When using an arc welder, always wear a(n) _____ to protect your eyes and face.

   7. _____

8. Your eyes can suffer damage from the _____ produced during arc welding.

   8. _____

9. A good arc has a sound like _____.

   9. _____

10. Excessive arc length will cause a high humming noise and a lot of _____.

    10. _____

11. Define *slag.* _____

    _____

12. Some metals may produce deadly _____ when welded.

    12. _____

13. The following illustration shows the tip of an electric arc welder during use. Identify the parts indicated.

    (A) _____

    (B) _____

    (C) _____

    (D) _____

    (E) _____

14. Identify the parts of the arc welding setup shown in the following illustration.

(A) _____

(B) _____

(C) _____

(D) _____

(E) _____

(F)_____

(G) _____

(H) _____

(I) _____

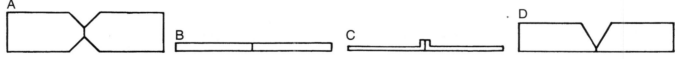

(Lincoln Electric Co.)

Oxyfuel Gas Welding

15. Complete the following statements about welding joints based on the following illustration.

(A) A(n) _____ joint is used when metal thickness is greater than 3/8″.

15A. _____

(B) A(n) _____ joint may be used on parts not exceeding 1/8″ in thickness.

15B. _____

(C) A(n) _____ joint is often used on metal of 1/32″ thickness or less.

15C. _____

(D) A(n) _____ joint is used when thickness ranges from 1/8″–3/8″.

15D. _____

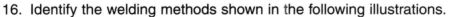

16. Identify the welding methods shown in the following illustrations.

(A) _____

(B) _____

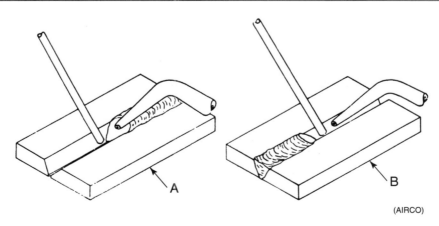

(AIRCO)

17. When backhand welding, the torch tip is _____ the direction of travel.

17. _____

18. The weld should _____ through the joint.

18. _____

19. The regulators reduce _____ to a usable level.

19. _____

20. The acetylene hose is normally _____ in color.

20. _____

21. The oxygen hose is generally _____ in color.

21. _____

22. If the flame is anything other than neutral, the resulting weld will be _____.

22. _____

23. The technician should always wear protective equipment when welding. Examples of safety equipment are _____ and _____.

23. _____

_____

24. Weld only in areas where there is good _____.

24. _____

Brazing

25. Brazing consists of heating the work to a temperature high enough to melt the _____ without melting the _____.

25. _____

_____

26. Define *capillary action.* _____

_____

27. Parts should be held together during brazing and cooling to avoid internal _____.

27. _____

28. Brazing flux should be compatible with the type of _____ being used.

28. _____

29. An uncoated bronze rod is heated and dipped in _____.

29. _____

30. What type equipment will produce sufficient heat for brazing?

_____

_____

ASE-Type Questions

1. TIG and MIG are two types of welding in which the welding joint is protected by a shield of _____.
    (A) inert gas
    (B) corrosive gas
    (C) oxygen
    (D) carbon

1. _____

2. All of the following statements about plasma cutting are true, *except:*
   - (A) plasma cutting reduces warping.
   - (B) a plasma cutter resembles a MIG or TIG welder.
   - (C) the plasma cutter arc causes the air to become ionized.
   - (D) plasma cutting is not as clean as oxyacetylene cutting.

2. _____

3. Technician A says that before welding, the arc-welding machine must be properly grounded. Technician B says that arc-welding equipment should never be used when standing in water. Who is right?
   - (A) A only.
   - (B) B only.
   - (C) Both A and B.
   - (D) Neither A nor B.

3. _____

4. Technician A says that a short arc will make a humming noise. Technician B says that a short arc will tend to cause the rod to stick to the work. Who is right?
   - (A) A only.
   - (B) B only.
   - (C) Both A and B.
   - (D) Neither A nor B.

4. _____

5. Technician A says that all oxyacetylene hose fittings should be greased before they are attached. Technician B says that the lines should be purged before lighting the torch. Who is right?
   - (A) A only.
   - (B) B only.
   - (C) Both A and B.
   - (D) Neither A nor B.

5. _____

6. Technician A says that acetylene and oxygen cylinder valves should be opened as rapidly as possible. Technician B says that it is important to know what type of material you will be welding before starting the welding process. Who is right?
   - (A) A only.
   - (B) B only.
   - (C) Both A and B.
   - (D) Neither A nor B.

6. _____

7. Technician A says a neutral flame is recommended for gas welding. Technician B says an oxidizing flame is recommended for gas welding. Who is right?
   (A) A only.
   (B) B only.
   (C) Both A and B.
   (D) Neither A nor B.

7. _____

8. A carburizing flame has too much _____ in the mixture.
   (A) oxygen
   (B) acetylene
   (C) moisture
   (D) vapor

8. _____

9. Technician A says that a leaking oxygen hose should be replaced. Technician B says that a leaking oxygen hose can be repaired. Who is right?
   (A) A only.
   (B) B only.
   (C) Both A and B.
   (D) Neither A nor B.

9. _____

10. A regular bronze or manganese bronze rod can be used to braze all of the following, *except:*
    (A) steel.
    (B) cast iron.
    (C) malleable iron.
    (D) stainless steel.

10. _____

Name _____

Date _____ Period _____

Instructor _____

Score _____

Text pages 113–142

8

Fasteners, Gaskets, and Sealants

Objectives: After studying Chapter 8 in the textbook and completing this section of the workbook, you will be able to:

- Identify automotive fasteners.
- Properly select fasteners.
- Torque fasteners to specifications as necessary.
- Remove damaged or broken fasteners.
- Describe gasket construction, materials, and application.
- Describe the construction and installation of seals.
- Describe the types and selection of sealants and adhesives.

Tech Talk: Fasteners, such as bolts, nuts, screws, and clips, hold the parts of a vehicle together. Thousands of fasteners are used in the construction of the average vehicle. A technician must constantly remove, install, and repair fasteners when performing any type of automotive work. To prevent an improper repair job, threaded fasteners must be carefully selected according to diameter, thread pitch, and strength. Many fasteners must be carefully tightened to the correct torque values. Non-threaded fasteners may also be used for the intended fastening job.

Modern vehicles rely on gaskets, sealants, seals, and adhesives to hold parts together and prevent leaks of engine oil, brake fluid, gear oil, greases, transmission fluid, power steering fluid, and coolant. Defective gaskets can cause loss of engine compression or intake manifold vacuum.

Anyone seeking a career as an automotive technician must have a thorough knowledge of automotive fasteners, gaskets, and sealers. This includes both their description and proper installation.

Review Questions

Instructions: Study Chapter 8 of the text and then answer the following questions in the spaces provided.

Types of Fasteners

1. Machine screws are used without _____.

1. _____

2. When using a sheet metal screw, why should you punch, rather than drill, the hole?

   _____

   _____

3. Sketch the missing screw heads in the following illustration.

   HEX   FLAT   OVAL   FILLISTER   ROUND

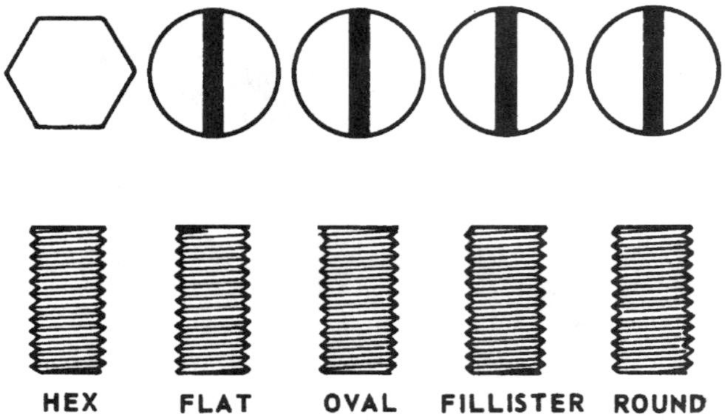

4. Identify the fastener drive types shown in the following illustration.

   (A) _____

   (B) _____

   (C) _____

   (D) _____

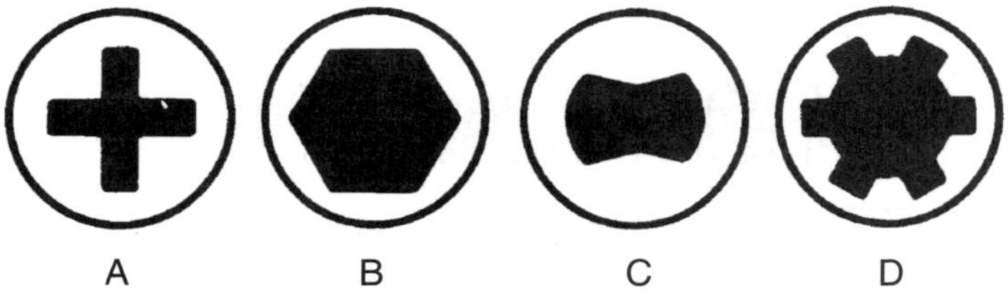

   A   B   C   D

5. A stud is a metal rod that is _____.   5. _____

6. The following illustration shows several methods of removing broken studs. Describe each operation shown.

   (A) _____

   (B) _____

   (C) _____

   (D) _____

   (E) _____

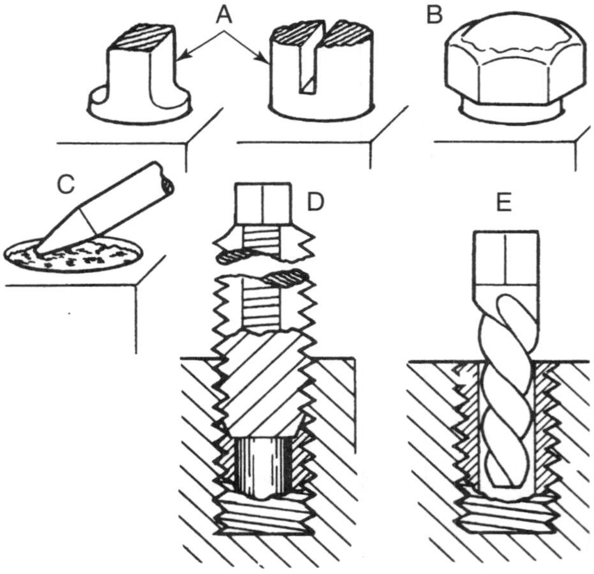

7. Label the fasteners in the following illustration.

   (A) _____

   (B) _____

   (C) _____

   (D) _____

   (E) _____

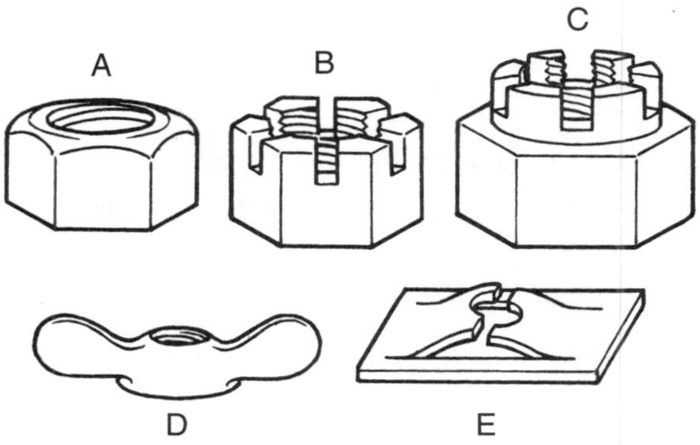

8. Nuts for automotive use are generally _____ in shape.

   8. _____

9. Summarize the three procedures that can be used when threads in holes are damaged beyond repair.

   (A) _____

   (B) _____

   (C) _____

10. Partially stripped threads can be cleaned up and repaired through the use of a thread _____, a thread chaser, or a tap.

   10. _____

11. The following illustration shows a common method for repairing stripped threads. Identify the parts indicated.

   (A) _____

   (B) _____

   (C) _____

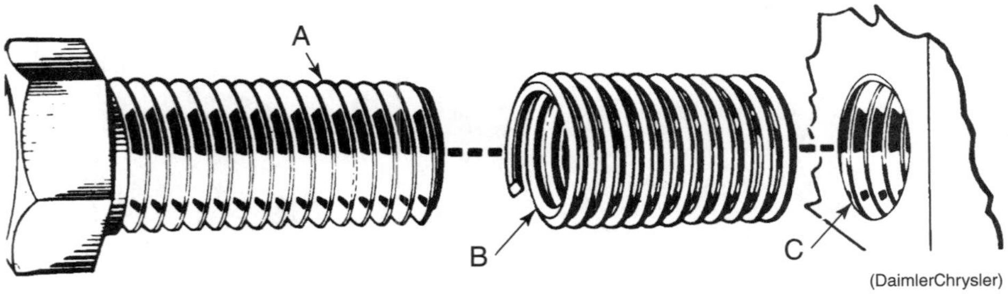

(DaimlerChrysler)

12. Label the fastener parts in the following illustration.

   (A) _____

   (B) _____

   (C) _____

   (D) _____

   (E) _____

   (F) _____

   (G) _____

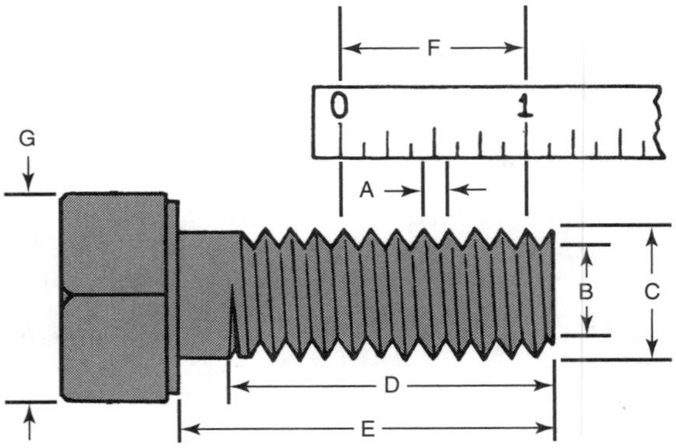

13. Summarize the purpose of bolt head markings. _____

_____

_____

14. All self-locking nuts use the same basic principle, the creation of _____ between two fasteners.

14. _____

15. Define *palnut.* _____

_____

_____

16. _____ are used to attach gears and pulleys to shafts.
    (A) Keys
    (B) Splines
    (C) Pins
    (D) All of the above.

16. _____

17. When a setscrew is used, the shaft will usually have a(n) _____ spot to accept the screw tip.

17. _____

18. Label the fasteners and fastener parts in the following illustration.

    (A) _____
    (B) _____
    (C) _____
    (D) _____
    (E) _____
    (F) _____
    (G) _____
    (H) _____
    (I) _____

Torquing Fasteners

19. List seven problems that can occur when fasteners are torqued improperly.

    (A) _____
    (B) _____
    (C) _____
    (D) _____
    (E) _____
    (F) _____
    (G) _____

20. A torque wrench measures the _____ force (torque) that is applied to a fastener.

20. _____

21. When installing fasteners, always check for the correct _____, thread type, and length.

21. _____

22. Where a number of fasteners are used to secure a part, the proper tightening _____ should be followed.

22. _____

23. If no sequence or tightening chart is available, it is usually advisable to start tightening in the _____ and work _____.

23. _____

_____

24. Tightening fasteners in a _____ pattern draws parts together in such a way that their mating surfaces contact evenly.

24. _____

25. Draw the crisscross tightening pattern on the following illustration using lines and arrows. Also, show the tightening sequence numbers.

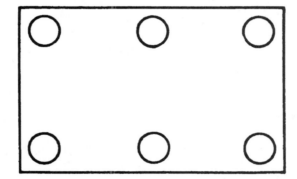

26. Draw the crisscross tightening pattern on the following illustration using lines and arrows. Also, show the tightening sequence numbers.

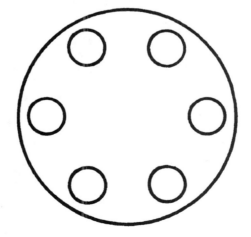

Gaskets and Sealants

27. Define *gasket*. _____
_____
_____

28. Some of the materials used in the construction of gaskets include _____.
    (A) rubber
    (B) copper
    (C) paper
    (D) All of the above.

28. _____

29. If a gasket is reused, it will fail to compress and _____ properly.

29. _____

30. After thorough cleaning, always inspect both part _____ surfaces to detect any nicks, dents, etc.

30. _____

31. On more complicated gaskets, such as cylinder head gaskets, make certain the gasket is _____.

31. _____

32. Many head gaskets have the word _____ and occasionally the word _____ stamped on them.

32. _____
_____

33. Carefully inspect gaskets for _____.

33. _____

34. Explain how a simple paper or combination cork and rubber gasket can be made.
_____
_____
_____
_____

35. Screw holes can be made in gaskets by _____.

35. _____

36. The following illustration shows a technician making a simple gasket. Identify the parts indicated.
    (A) _____
    (B) _____

37. Rubber gaskets are highly _____.

37. _____

38. When a relatively thin, stamped part is bent along its parting edge, it must be _____ before installation.

38. _____

39. To straighten stamped parts, place the edge on a smooth, _____ surface and gently _____ to straighten bent areas.

39. _____

_____

40. Name five defects that should be checked for on gasket mating surfaces.

_____

_____

41. Before installing a new gasket, check it for _____.

41. _____

42. When installing a gasket, fasteners should always be torqued in the proper _____.

42. _____

43. Label the parts of the multilayered head gasket in the following illustration.

    (A) _____

    (B) _____

    (C) _____

    (D) _____

    (E) _____

(Fel-Pro)

44. Explain why a creased head gasket should never be used, even if it has been straightened.

_____

_____

Oil Seals

45. An oil seal is secured to one part while the sealing lip allows the other part to _____ or reciprocate (move) without leakage.

45. _____

46. Identify the parts of the oil seal shown in the following illustration.

    (A) _____

    (B) _____

    (C) _____

    (D) _____

    (E) _____

A   B   C   D   E

47. The following illustration shows various methods of seal removal. Identify the parts indicated.

(A) _____

(B) _____

(C) _____

(D) _____

(E) _____

48. After preparing the seal counterbore with sealer, place the seal squarely against the opening with the seal lip facing _____ .

48. _____

49. If a seal driving set is not available, a section of _____ of the correct diameter can be used to install (drive in) a seal.

49. _____

50. If a hammer and punch are used to install a seal, be careful to strike at different spots near the _____ edge of the seal.

50. _____

51. Why is it a bad idea to use too much sealer?

_____

_____

52. Identify the conditions in the following oil seal illustration.

(A) _____

(B) _____

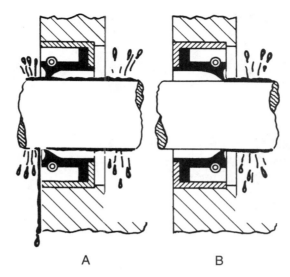

A   B

53. A nonmoving part is referred to as a(n) _____ part.   53. _____

54. An O-ring must be slightly _____ to properly seal.   54. _____

55. Sealants that can be used in place of a precut gasket are   55. _____
    called _____ gaskets.

ASE-Type Questions

1. Technician A says that sheet metal screws are used to hold   1. _____
   thin metal parts together. Technician B says that sheet
   metal screws are used with standard nuts. Who is right?
   - (A) A only.
   - (B) B only.
   - (C) Both A and B.
   - (D) Neither A nor B.

2. Which of the following tools is *not* needed to remove a   2. _____
   broken bolt with a screw extractor?
   - (A) drill
   - (B) drill bit
   - (C) flat chisel
   - (D) center punch.

3. Technician A says that if a portion of a broken bolt pro-   3. _____
   trudes from its hole, the bolt may be removed by gripping
   the protruding portion with locking pliers and turning it out.
   Technician B says that if a portion of a broken bolt does
   not protrude from its hole, the bolt can be removed by
   driving it clockwise with a punch. Who is right?
   - (A) A only.
   - (B) B only.
   - (C) Both A and B.
   - (D) Neither A nor B.

4. Bolts and screws may be identified by all of the following, *except:*
    (A) length.
    (B) tensile strength.
    (C) malleability.
    (D) thread pitch.

4. _____

5. Technician A says that the tool shown in the following illustration is used to measure the major diameter of a bolt. Technician B says that the tool is used to measure the number of threads per inch. Who is right?
    (A) A only.
    (B) B only.
    (C) Both A and B.
    (D) Neither A nor B.

5. _____

6. Snap rings are usually made of _____.
    (A) aluminum
    (B) copper
    (C) brass
    (D) steel

6. _____

7. Setscrews are usually used to hold a pulley or hub to a _____.
    (A) second hub or pulley.
    (B) stationary casting.
    (C) keyway
    (D) shaft.

7. _____

8. If the center of pull on the handle of a torque wrench is exactly one foot from the center of the head, a one pound pull on the handle would produce _____ of torque applied to the fastener.
    (A) 1 ft. lb.
    (B) 1 ft./lb.^{2}
    (C) 1/2 ft. lb.
    (D) 1 N•m

8. _____

9. Technician A says that gaskets and seals are used throughout the engine to confine fuel and oil. Technician B says that gaskets and seals can be used to confine air and vacuum. Who is right?
    (A) A only.
    (B) B only.
    (C) Both A and B.
    (D) Neither A nor B.

9. _____

10. All of the following statements about use of sealant are true, *except:*
    (A) sealant should be applied to both sides of a gasket before installation.
    (B) sealant should be used only when specified by the manufacturer.
    (C) sealant helps hold a gasket in place during installation.
    (D) sealant helps seal small cracks and surface imperfections.

10. _____

11. Technician A says that seals are used to confine fluids. Technician B says that seals prevent the entry of foreign materials. Who is right?
    (A) A only.
    (B) B only.
    (C) Both A and B.
    (D) Neither A nor B.

11. _____

12. All the following components are part of a seal, *except:*
    (A) the garter spring.
    (B) the metal case.
    (C) the sealing element.
    (D) the retainer.

12. _____

13. If a seal begins to tip during installation, strike the _____.
    (A) low side
    (B) sealing tip
    (C) high side
    (D) counterbore

13. _____

14. When installing a seal over a keyway, use _____ to prevent damage.
    (A) large amounts of grease.
    (B) large amounts of the same lubricant that the seal will operate in.
    (C) an oversize driver.
    (D) a mounting sleeve.

14. _____

15. Technician A says that too much sealant is better than no sealant at all. Technician B says that sealer should never be used on transmission valve bodies. Who is right?
    (A) A only.
    (B) B only.
    (C) Both A and B.
    (D) Neither A nor B.

15. _____

Name _____

Date _____ Period _____

Instructor _____

Score _____

Text pages 143–159

9

Tubing and Hose

Objectives: After studying Chapter 9 in the textbook and completing this section of the workbook, you will be able to:

- Identify different types of tubing, hose, and fittings.
- Select the correct type of tubing, hose, or fitting for the job.
- Properly install new tubing and hose.

Tech Talk: A vehicle's metal and plastic tubing and rubber hoses carry critically needed fluids to various parts of the car, just as human arteries and veins carry blood and oxygen to the various parts of the body. A failure of an "artery" (fuel line, brake line, water hose, power steering hose, transmission cooler line) may cause serious mechanical problems. The car could stop "dead" in the road. A complete understanding of hoses and tubing, as well as their related connectors and fittings, is essential if you plan to be a good technician.

Review Questions

Instructions: Study Chapter 9 of the text and then answer the following questions in the spaces provided.

1. Most automotive tubing is made of _____ or _____.

1. _____

   _____

2. Copper tubing is _____, easy to _____, and forms good joints.

2. _____

   _____

3. Although a hacksaw can be used to cut tubing, a faster and better method is to use a(n) _____.

3. _____

4. Label the cut tubing shown in the following illustration.

    (A) _____

    (B) _____

    (C) _____

    (D) _____

5. In a flare connection, the tube _____ securely grasps both sides of the flare, producing a leakproof seal.

    5. _____

6. The O-ring fitting uses _____ threads and depends on the O-ring to prevent leaks.

    6. _____

7. Define *unions.* _____

    _____

    _____

8. Explain the use of elbows. _____

    _____

    _____

9. When several branch lines are served by a single feeder line, a(n) _____ block can be used.

    9. _____

10. The _____ is used to stop flow through a line.

    10. _____

11. A(n) _____ is used to draw off the liquid contents of a tank.

    11. _____

12. Label the parts of the I.S.O. fitting shown in the following illustration.

    (A) _____

    (B) _____

    (C) _____

    (D) _____

(Chevrolet)

13. The following illustration shows a flare being formed. Label the parts indicated.

    (A) _____

    (B) _____

    (C) _____

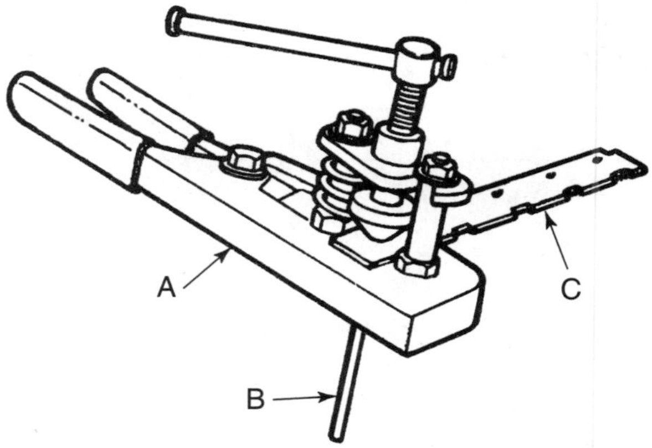

(Dodge)

14. The following illustration shows a(n) _____ fitting. Label the parts indicated.

    (A) _____

    (B) _____

    (C) _____

    (D) _____

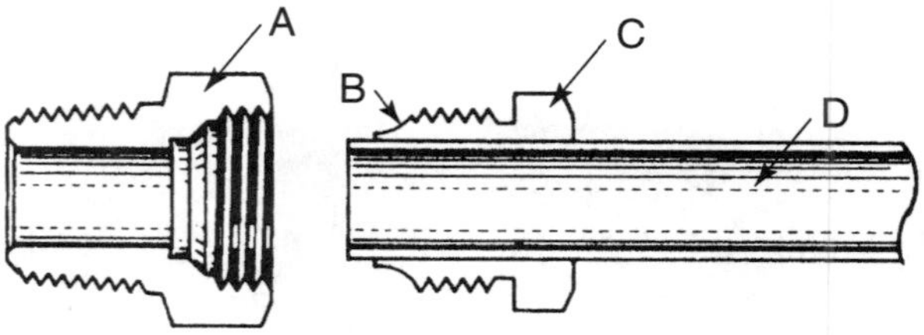

15. The following figure shows a type of _____ fitting.

    15. _____

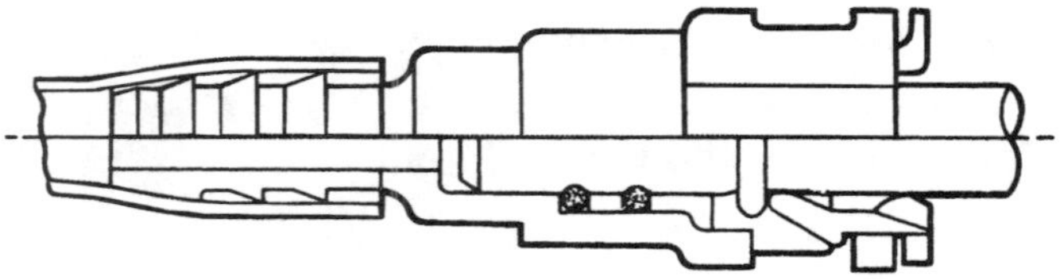

16. The fitting shown in the following illustration is usually called a(n) _____ fitting.

    16. _____

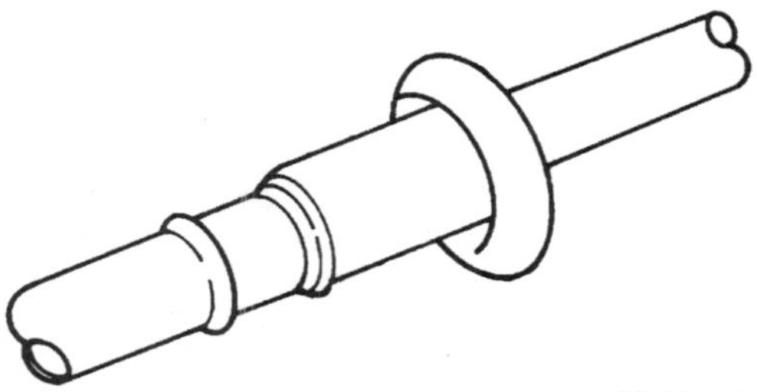

(Kia Motors)

17. Using the following illustration, label each type of bend and draw its cross-section at the end of the arrow.

    (A) _____

    (B) _____

    (C) _____

        A   B   C

18. Flexible hose is used in the following systems of a car:

    _____

    _____

19. Any hose that shows signs of _____, _____, or _____ should be replaced.

    19. _____

    _____

    _____

20. Hoses often deteriorate on the _____, causing portions of hose material to break loose and producing partial or even complete blockage.

    20. _____

ASE-Type Questions

1. Technician A says that copper tubing is commonly used for brake lines. Technician B says that copper tubing should never be used for fuel lines. Who is right?

    (A) A only.
    (B) B only.
    (C) Both A and B.
    (D) Neither A nor B.

    1. _____

2. All of the following are types of basic tubing connections, *except:*

    (A) flare.
    (B) compression.
    (C) push-on
    (D) clamp

    2. _____

3. Technician A says that flared fittings should be used on high-pressure automotive applications. Technician B says that compression fittings can be used on high-pressure automotive applications. Who is right?

    (A) A only.
    (B) B only.
    (C) Both A and B.
    (D) Neither A nor B.

    3. _____

4. Push-on fittings are being discussed. Technician A says that push-on fittings are often used on fuel lines. Technician B says that push-on fittings are often used on air conditioning lines. Who is right?
    (A) A only.
    (B) B only.
    (C) Both A and B.
    (D) Neither A nor B.

4. _____

5. Power steering and brake hose must withstand pressures exceeding _____.
    (A) 500 psi
    (B) 750 psi
    (C) 1000 psi
    (D) 2000 psi

5. _____

Name _____

Date _____ Period _____

Instructor _____

Score _____

Text pages 161–181

10

Friction and Antifriction Bearings

Objectives: After studying Chapter 10 in the textbook and completing this section of the workbook, you will be able to:

- Compare the differences between friction and antifriction bearings.
- Explain the application of different bearing designs.
- Properly install friction bearings.
- Diagnose common reasons for bearing failures.
- List the different kinds of antifriction bearings.
- Explain the advantages of each type of antifriction bearing.
- Describe service procedures for antifriction bearings.

Tech Talk: Friction and antifriction bearings are used throughout the vehicle. Friction bearings are used as engine connecting-rod bearings, main bearings, camshaft bearings, starter bushings, transmission bushings, etc. Worn friction bearings can cause a wide range of problems, such as oil leakage and part knocking.

Antifriction bearings are also used throughout the modern vehicle. These bearings are used in transmissions, differentials, wheel bearings, alternators, and other assemblies. If serviced improperly, antifriction bearings can become noisy. If noisy bearings are ignored, they can lock up, causing serious mechanical damage.

It is important to understand the design, operating principles, and service of both friction and antifriction bearings.

Review Questions

Instructions: Study Chapter 10 of the text and then answer the following questions in the spaces provided.

Major Classes of Bearings

1. Define *journal.* _____

_____

2. Define *antifriction bearing.* _____

_____

3. Identify the bearing types shown in the following illustration.

(A)_____-type bearing.   13A. _____

(B)_____-type bearing.   13B. _____

A   B

4. Identify the types of forces applied to the bearings in the following illustration.

(A) _____

(B) _____

(C) _____

A   B   C

Friction Bearings

5. Define *thrust flange.* _____

_____

6. Bearing spread causes the insert diameter across the   6. _____
parting edges to be slightly larger than the _____.

7. Define *bearing crush.* _____

_____

8. Insert halves come in _____. It is important that they not   8. _____
be _____.

_____

9. One of the most widely used methods of measuring bear-   9. _____
ing clearances involves the use of a(n) _____.
    (A) outside micrometer
    (B) steel rule
    (C) Plastigage
    (D) None of the above.

10. The bearing shown in the following illustration is a(n) _____-type bearing. Label the parts indicated.

    (A) _____

    (B) _____

    (C) _____

    (D) _____

    (E) _____

(AE Clevite)

Antifriction Bearings

11. The antifriction bearing uses _____ elements to reduce friction.

11. _____

12. By using two or more bearings facing in opposite directions, _____ in either direction can be handled.

12. _____

13. The spherical roller will handle _____ loads.

13. _____

14. Identify the parts of the ball bearing assembly shown in the following illustration.

    (A) _____

    (B) _____

    (C) _____

    (D) _____

    (E) _____

    (F) _____

    (G) _____

    (H) _____

    (I) _____

    (J) _____

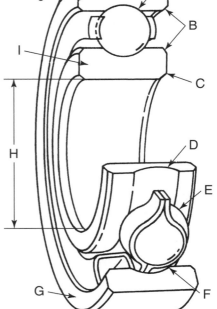

(Deere & Co.)

15. Identify the parts of the straight roller bearing shown in the following illustration.

    (A) _____

    (B) _____

    (C) _____

    (D) _____

    (E) _____

    (F) _____

    (G) _____

    (H) _____

    (I) _____

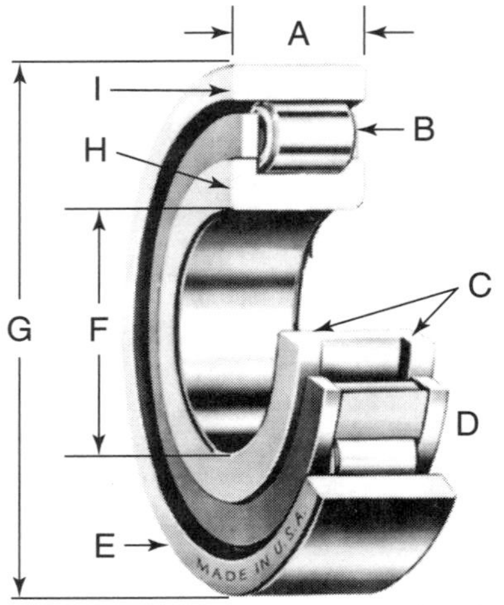

(AFBMA)

16. Define *bearing identification.*

    _____

    _____

17. Identify the correct and incorrect way of pressing the bearing.

    (A) _____

    (B) _____

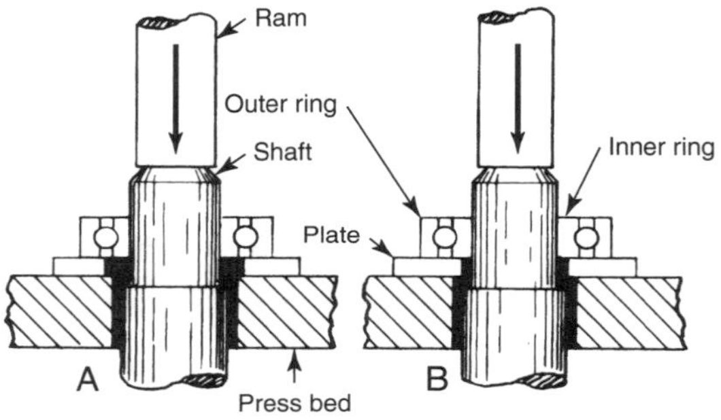

18. The following illustration shows the procedure for removing a bearing retaining race. Label the parts indicated.

    (A) _____

    (B) _____

    (C) _____

    (D) _____

    (E) _____

    (F) _____

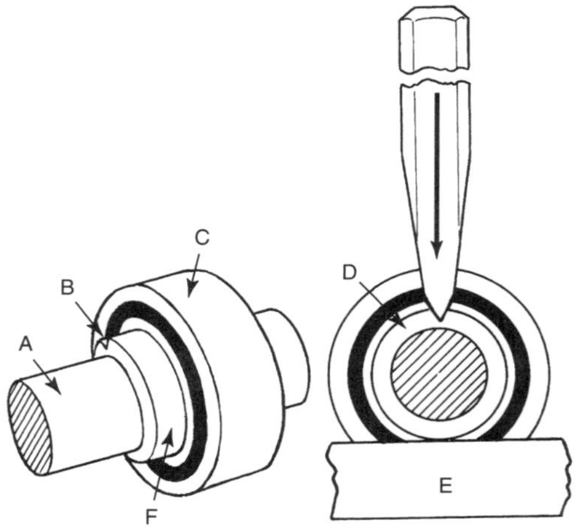

19. Explain how to inspect a nonseparable bearing.

    _____

    _____

20. For separable bearings, carefully inspect the _____ and   20. _____
    each _____ element.

    _____

21. Fine dirt will produce a(n) _____ effect.   21. _____

22. Define *brinelling.* _____

    _____

23. Overheating will break down the _____ of bearings and   23. _____
    cause rapid failure.

24. If a bearing is to be stored for an extended period of time,   24. _____
    coat it with light grease and wrap it in _____ paper.

25. To facilitate installation, the outer race of large bearings   25. _____
    can be placed in _____.

ASE-Type Questions

1. Technician A says that friction bearings slide against a   1. _____
   portion of a shaft. Technician B says that antifriction bearings eliminate friction. Who is right?
       (A) A only.
       (B) B only.
       (C) Both A and B.
       (D) Neither A nor B.

2. Friction-type bearings are used on all of the following,   2. _____
   except:
       (A) crankshaft.
       (B) camshaft.
       (C) connecting rods.
       (D) wheels.

3. Technician A says that radial loads are parallel to the axis   3. _____
   of the bearing. Technician B says the thrust loads occur at
   right angles to the axis of the bearing. Who is right?
       (A) A only.
       (B) B only.
       (C) Both A and B.
       (D) Neither A nor B.

4. Technician A says that excessive clearance in a bearing   4. _____
   insert will cause increased oil pressure. Technician B says
   that excessive bearing clearance can cause increased oil
   consumption. Who is right?
       (A) A only.
       (B) B only.
       (C) Both A and B.
       (D) Neither A nor B.

5. All of the following can cause bearing failure, *except:*
    (A) insufficient clearance.
    (B) dirt.
    (C) a bowed crankcase.
    (D) high oil pressure.

5. _____

6. All of the following are common antifriction bearings, *except:*
    (A) needle bearings.
    (B) self-aligning bearings.
    (C) precision insert bearings.
    (D) roller bearings.

6. _____

7. Technician A says that the deep-groove ball bearing will handle heavy thrust loads and medium radial loads. Technician B says that straight roller bearings are designed to handle heavy radial loads and heavy thrust loads. Who is right?
    (A) A only.
    (B) B only.
    (C) Both A and B.
    (D) Neither A nor B.

7. _____

8. All of the following statements about tapered roller bearings are true, *except:*
    (A) Tapered rollers are the most common roller bearings.
    (B) Tapered roller bearings will carry radial loads.
    (C) Tapered roller bearings will carry thrust loads.
    (D) Tapered roller bearings are non-adjustable.

8. _____

9. Torrington bearings are a type of _____ bearing.
    (A) needle
    (B) ball
    (C) roller
    (D) None of the above.

9. _____

10. Technician A says that spalling can be caused by foreign particles, overloading, or normal wear. Technician B says that once spalling starts, iron oxide powder is formed. Who is right?
    (A) A only.
    (B) B only.
    (C) Both A and B.
    (D) Neither A nor B.

10. _____

Name _____

Date _____ Period _____

Instructor _____

Score _____

Text pages 183–207

11

Preventive Maintenance

Objectives: After studying Chapter 11 in the textbook and completing this section of the workbook, you will be able to:

- Describe the purpose of preventive maintenance.
- Describe oil classifications.
- Explain how to change engine oil and filter.
- Explain how to lubricate suspension fittings.
- Explain how to check air filters.
- Identify body parts that require lubrication
- Explain how to check vehicle fluid levels.
- Explain how to inspect tires.
- Describe tire rotation procedures.
- Explain how to inspect the brake system for defects.
- Explain how to inspect steering and suspension parts.
- Describe engine coolant replacement procedures.
- Explain the procedure for changing transmission/transaxle fluid.
- Identify battery maintenance procedures.

Tech Talk: As the name implies, preventive maintenance involves preventing problems by properly maintaining a vehicle. Thanks to improved designs and materials, modern automobiles and trucks require fewer preventive maintenance services than the vehicles of 20 or 30 years ago. However, the preventive maintenance services that must still be performed are vital. Study this chapter to learn how to perform preventive maintenance on modern vehicles.

Review Questions

Instructions: Study Chapter 11 of the text and then answer the following questions in the spaces provided.

Protecting the Engine Lubrication System

1. Vehicle manufacturers publish two kinds of oil change schedules, one for _____ operation and one for _____, or heavy duty, service.

1. _____

_____

2. The better the job that a(n) _____ does, the sooner it will need replacement.

2. _____

3. Motor oil must seal the area between the _____ and _____.

3. _____

_____

4. Motor oil must carry away _____ and keep impurities _____ until they can be trapped by the oil filter.

4. _____

_____

5. Viscosity is oil's resistance to _____.
   (A) heat
   (B) dirt
   (C) flow
   (D) sludging

5. _____

6. The W in the term 10W-30 motor oil stands for _____.

6. _____

7. Light (lower viscosity) motor oils should be used in engines that are _____.
   (A) worn
   (B) in good condition
   (C) smoking
   (D) All of the above.

7. _____

8. When motor oil is capable of holding impurities in suspension, it is called a(n) _____ oil.

8. _____

9. A reasonable oil change interval would be _____.
   (A) 500 miles (805 km).
   (B) 7500 miles (12,000 km).
   (C) Somewhere between A and B.
   (D) None of the above.

9. _____

10. If the engine operating conditions are not ideal, oil change intervals should be _____.

10. _____

11. What is the difference between the API certification starburst and the API service symbol? _____

_____

_____

12. The best time to check the oil level is when the engine has been turned off _____ and the vehicle is _____.

12. _____

_____

Changing the Oil and Filter

13. Change the oil when the engine is _____.

13. _____

14. Always check the drain plug _____ for damage.

14. _____

15. After changing the oil and filter, run the engine for a few minutes and check for leaks around the _____ and _____.

15. _____

_____

Other Maintenance Services

16. The PCV system should be checked at every _____.

16. _____

17. _____ and _____ filters should be changed if they show any sign of restriction.

17. _____

_____

18. Modern vehicles use _____ air cleaner elements.

18. _____

19. Some air cleaners can be cleaned with _____.

19. _____

20. Front suspension and steering parts often contain _____, which should be lubricated on a regular basis.

20. _____

21. Modern greases are called _____ greases.

21. _____

22. If a joint has a sealed boot, apply the grease until the boot _____ slightly.

22. _____

23. If grease squirts from around the grease gun and fitting, the fitting is _____.

23. _____

24. While greasing the front suspension, check for grease fittings on the driveshaft _____. If the vehicle is a front-wheel drive model, check the _____ boots for damage.

24. _____

_____

25. Why does the parking brake cable sometimes stick?

_____

_____

Look at the chart on page 194 of your textbook and match the vehicles below with the type of automatic transmission fluid that they should use. Some choices may be used more than once.

_____ 26. Lincoln

_____ 27. Jeep

_____ 28. Accura

_____ 29. Mazda

_____ 30. Isuzu

(A) Honda Gen ATF
(B) Mercon V
(C) ATF+3
(D) ATF+4

31. Some power steering and brake reservoirs are made of _____ so the level can be checked without removing the cap.

31. _____

32. What is a hydrometer used to check?

_____

_____

33. To drain coolant from the engine block, the engine must be operated until the _____ opens. To further drain the engine, some engine blocks have _____. Some engines have a(n) _____ valve to remove air from the engine.

33. _____

_____

_____

34. Why should the radiator cap be loosened before the engine is warmed up to drain the coolant?

_____

_____

35. Coolant test strips are used to check the _____ and _____ of the coolant.

35. _____

_____

36. Some manufacturers recommend draining the _____ when the transmission is drained.

36. _____

37. Some transmission/transaxle pan gaskets can be _____.

37. _____

38. It is common for the oil in the transmission, transfer case, and rear axle to become contaminated with _____.

38. _____

39. Some rear axles and transfer cases require special oil because they have internal friction _____.

39. _____

40. To change brake fluid, have an assistant press on the brake pedal as you loosen the _____ screws.

40. _____

41. When changing brake fluid, add new fluid to the _____.

41. _____

42. A common tire _____ interval is 5000 miles (8000km).

42. _____

43. Brakes should be inspected every _____ miles or _____ km.

43. _____

_____

Matching

Match the vehicle component with a defect that it should be checked for.

_____ 44. V-belts

_____ 45. Coolant hoses

_____ 46. Tires

_____ 47. Brakes

_____ 48. Front suspension

(A) Tread separation.
(B) Scored rotor.
(C) Oil leaks.
(D) Electrochemical degradation.
(E) Tension.
(F) Excessive movement.

49. Battery components can be cleaned with _____ and water.   49. _____

50. Before disconnecting the battery terminals, install a(n) _____ in the cigarette lighter socket.   50. _____

ASE-Type Questions

1. Maintenance schedules are being discussed. Technician A says that maintenance should be performed after a certain mileage has been reached, regardless of the time that has passed. Technician B says that maintenance should be performed after a certain time has passed, regardless of mileage. Who is right?
    (A) A only.
    (B) B only.
    (C) Both A and B.
    (D) Neither A nor B

    1. _____

2. Single weight oils should be used only when the temperature is above _____.
    (A) 0° F (–17° C).
    (B) 32° F (0° C).
    (C) 50° F (10° C).
    (D) 70° F (20° C).

    2. _____

3. Non-detergent motor oils may be specified for all of the following uses, *except:*
    (A) late-model engines.
    (B) manual gearboxes.
    (C) 2-cycle engines.
    (D) very old engines.

    3. _____

4. Technician A says that the S in motor oil service grades indicates that it is designed for use in a diesel engine. Technician B says that the C in motor oil service grades indicates that it is designed to be in an engine with compression ignition. Who is right?
    (A) A only.
    (B) B only.
    (C) Both A and B.
    (D) Neither A nor B

    4. _____

5. All of the following statements are true, *except:*
    (A) some filters must be filled with oil before installation.
    (B) oil should be changed when the engine is cold.
    (C) before installing a new spin-on filter, a thin film of oil should be rubbed on the sealing ring.
    (D) over-tightening a spin-on engine oil filter may split the gasket.

    5. _____

6. Technician A says that a loose suspension part should be overgreased to compensate for wear. Technician B says that a loose suspension part should be replaced. Who is right?
    (A) A only.
    (B) B only.
    (C) Both A and B.
    (D) Neither A nor B.

6. _____

7. The components in the following illustration should be lubricated with _____.
    (A) manual gearbox oil
    (B) 10W-30 motor oil
    (C) chassis grease
    (D) Dextron® transmission fluid

7. _____

(DaimlerChrysler)

8. A low level in which of the following reservoirs does *not* indicate a leak?
    (A) Brake master cylinder.
    (B) Power steering.
    (C) Windshield washer.
    (D) Coolant recovery.

8. _____

9. To check the fluid level on a rear-wheel drive automatic transmission, the shift lever should be placed in _____ or _____ with the engine warm and running.
    (A) Neutral, Drive
    (B) Neutral, Park
    (C) Park, Reverse
    (D) Reverse, Drive

9. _____

10. Technician A says that oil should be used to lubricate lock cylinders. Technician B says that dry graphite should be used to lubricate lock cylinders. Who is right?
    (A) A only
    (B) B only.
    (C) Both A and B.
    (D) Neither A nor B.

10. _____

Name _____

Date _____ Period _____

Instructor _____

Score _____

Text pages 209–223

12

Wire and Wiring

Objectives: After studying Chapter 12 in the textbook and completing this section of the workbook, you will be able to:

• Identify different types of automotive wiring.

• Select the correct type of wiring for the job.

• Make basic wiring repairs.

• Read wiring diagrams.

• Perform basic circuit tests.

Tech Talk: Early automobiles had only a few feet of wire connecting the headlights, taillights, starter, generator, and ignition system. Modern automobiles and trucks have hundreds of feet of wire connecting hundreds of different electrical and electronic components. It is almost impossible to work on any part of a vehicle without dealing with an electrical component or connection. For this reason, the automotive technician must learn the basics of automotive wire and wiring.

Review Questions

Instructions: Study Chapter 12 of the text and then answer the following questions in the spaces provided.

Primary Wire

1. Every wire is assigned a number. The larger the number, the _____ the wire's conductor.

2. As wire size increases, electrical resistance _____.

3. As wire length increases, electrical resistance _____.

4. A wiring harness has several wires either pulled through a(n) _____ or _____ together.

1. _____

2. _____

3. _____

4. _____

_____

5. Explain why wire color coding is used.

_____

_____

6. Define *wiring diagram.* _____

_____

_____

7. Primary terminals include _____ terminals.
   (A) blade
   (B) lug
   (C) bullet
   (D) All of the above.

7. _____

8. Plug-in connectors are designed with _____ lugs to ensure that they are not improperly connected.

8. _____

9. Plug-in connectors used on many computer-controlled vehicles have rubber _____ with as many as three sealing rings to protect the connection from moisture.

9. _____

10. Identify the components in the following illustration.
    (A) _____
    (B) _____
    (C) _____
    (D) _____
    (E) _____

11. Define *soldering.* _____

_____

_____

12. It is the job of flux to remove _____ and prevent it from reforming.

12. _____

13. Before soldering, all traces of _____, corrosion, and _____ must be removed from the pieces to be joined.

13. _____

    _____

14. To allow for good heat transfer from the soldering iron tip to the work, the soldering iron tip should be _____ and _____.

14. _____

    _____

15. Pieces to be joined by soldering should be _____ so the solder is melted by the _____ in the metals to be soldered together.

15. _____

_____

16. The following illustration shows the two types of wire used in an automobile. Complete the following statements and identify the parts indicated.

    (A) This is _____ wire.

    (B) This is _____ wire.

    (C) This is the wire's _____.

    (D) This is the wire's _____.

16A. _____

16B. _____

16C. _____

16D. _____

17. Identify the electrical symbols in the following illustration.

    (A) _____

    (B) _____

    (C) _____

    (D) _____

    (E) _____

    (F) _____

    (G) _____

    (H) _____

    (I) _____

    (J) _____

    (K) _____

    (L) _____

(DaimlerChrysler)

18. Identify the soldering tip conditions indicated in the following drawings.

    (A) _____

    (B) _____

19. Complete the following statements based on the following illustration.

    (A) This soldering tip position is wrong because

    _____

    _____

    (B) This position is right because the tip is in

    _____

    _____

A

B

Secondary Wire

20. What causes crossfiring in secondary wires? _____

    _____

    _____

21. Resistance is built into modern secondary wires to reduce   21. _____

    _____.

22. To remove a secondary wire, grasp it at the _____.   22. _____

Printed Circuits

23. Define *printed circuit.* _____

    _____

    _____

24. The printed circuit replaces the maze of individual   24. _____
    _____ that are otherwise needed to connect numerous
    components.

25. Printed circuits allow a great number of _____ in a very   25. _____
    small area.

Troubleshooting Wiring

26. Why should the technician never pierce or cut a wire during testing?

    _____

    _____

27. List three wiring problems that commonly occur in automobiles.

    _____

    _____

    _____

28. List three types of test equipment that can be used to test wiring.

_____

_____

_____

29. When a wire is electrically continuous, it is said to have _____.

29. _____

30. A small, battery-operated test light is useful in tracing wires when the wires are not _____.

30. _____

ASE–Type Questions

1. All the following are factors in determining correct wire size, *except:*
    (A) conductor location.
    (B) line voltage.
    (C) electrical load.
    (D) wire length.

1. _____

2. Technician A says that it is almost impossible to improperly connect plug-in connectors. Technician B says that because of the low voltages involved, the connectors in computer-related systems are not adversely affected by moisture. Who is right?
    (A) A only.
    (B) B only.
    (C) Both A and B.
    (D) Neither A nor B.

2. _____

3. A fuse block may contain all of the following, *except:*
    (A) minifuses.
    (B) maxifuses.
    (C) fusible links.
    (D) relays.

3. _____

4. Technician A says that primary wire handles battery voltage. Technician B says that secondary wire is used in the high-tension circuit. Who is right?
    (A) A only.
    (B) B only.
    (C) Both A and B.
    (D) Neither A nor B.

4. _____

5. A small test light can be used to check wires for _____.
    (A) impedance
    (B) resistance
    (C) continuity
    (D) proper gage

5. _____

Name _____

Date _____ Period _____

Instructor _____

Score _____

Text pages 225–250

13

Chassis Electrical Service

Objectives: After studying Chapter 13 in the textbook and completing this section of the workbook, you will be able to:

- Identify and define chassis wiring and related components.
- Explain the differences between chassis wiring and engine wiring.
- Identify lighting systems and components.
- Troubleshoot and replace system components.
- Identify solenoids and relays.
- Troubleshoot and replace solenoids and relays.
- Identify motor-operated components.
- Troubleshoot and replace motor-operated components.
- Identify vehicle security systems and components.
- Troubleshoot and replace vehicle security system components.
- Work safely on vehicles equipped with air bag systems.
- Troubleshoot and service air bag systems.

Tech Talk: Chassis electrical equipment includes many vehicle systems that are not related to the ignition, starting, or charging systems. However, these systems are important to the overall operation of the vehicle and the safety of the passengers. In addition to lighting, windshield wipers, and horns, chassis electrical equipment includes the radios and CD players, electric windows and seats, door locks, rear window defrosters, cruise control systems, and air bag systems. Studying this chapter will prepare you to service chassis electrical components and systems.

Review Questions

Instructions: Study Chapter 13 of the text and then answer the following questions in the spaces provided.

Chassis Wiring

1. What does chassis wiring consist of? _____

    _____

    _____

2. Define *printed circuit.* _____

    _____

    _____

3. Chassis wires are _____ to make identification easier.

    3. _____

4. Almost all factory installed fuses are installed in a(n) _____.

    4. _____

5. What is the difference between a fusible link and a fuse?

    _____

    _____

6. What is the advantage of a circuit breaker over a fuse?

    _____

    _____

7. Common causes of blown fuses include _____.
    (A) shorted wires
    (B) shorts in electrical devices
    (C) motor overload
    (D) All of the above.

    7. _____

8. Wires may short inside the _____ if they are pinched between body parts.

    8. _____

Vehicle Lights and Switches

9. The two main classes of lights are _____ and _____.

    9. _____

    _____

10. The headlight switch contains a(n) _____ for adjusting the brightness of the dashboard lights.

    10. _____

11. Most halogen headlight bulbs can be replaced from the _____ of the headlight assembly.

    11. _____

12. Daytime running lamps are controlled by the _____ switch

    12. _____

13. Backup lightbulbs contain _____ filament(s).

    13. _____

14. The turn signal wiring is always routed through a(n) _____ unit.

    14. _____

15. The two classes of dashboard lights are _____ and
    _____.

15. _____
    _____

16. A fiber optic harness carries _____ instead of electricity.

16. _____

17. What is the most common vehicle light system problem?

    _____

    _____

18. Some brake switches must be adjusted when they are
    _____.

18. _____

19. If a gauge on a digital instrument panel fails to light up, it
    may be necessary to replace the _____.
    (A) wiring harness
    (B) instrument panel fuse or fuses
    (C) entire instrument ground wire(s)
    (D) entire instrument panel

19. _____

Chassis-Mounted Motors

20. The modern vehicle may have as many as _____ dc
    motors.

20. _____

21. Modern windshield washer pumps are located in the
    _____.

21. _____

22. Some motors may require _____ when they are replaced
    in the vehicle.

22. _____

23. Power window and tailgate motors usually have built-in
    overload switches or circuit breakers to prevent _____.

23. _____

24. Most defective electric motors are exchanged for _____
    units.

24. _____

25. _____ straps must be in place when a motor is reinstalled.

25. _____

Chassis-Mounted Solenoids and Relays

26. Relays are a type of _____ that close electrical contacts.

26. _____

27. A solenoid can be tested by bypassing the solenoid's
    _____ with a jumper wire.

27. _____

28. Many relays used on late-model vehicles are _____.

28. _____

29. A common Pass-Key problem is loss of tension at the igni-
    tion-switch _____ contacts.

29. _____

30. Pass-Lock and later Pass-Key systems can be checked
    with a(n) _____.

30. _____

31. The remote keyless entry system relies on a(n) _____ transmitter to unlock the vehicle door.

31. _____

32. The transmitter used in remote keyless entry systems is _____ to the vehicle.

32. _____

33. If a keyless entry system does not operate from the hand-held locking device, what should the technician check first? _____

_____

_____

34. Static in the sound system that only occurs when the engine is running means that the problem is in the _____ or _____.

34. _____

_____

35. Most horn problems are caused by defects in the _____ contacts.

35. _____

36. Most rear window defroster service consists of finding and correcting breaks in the _____.

36. _____

Body Computer Systems

37. The _____ sensor keeps the air bag system from being activated by a faulty impact sensor.

37. _____

38. The air bag is inflated by _____-generating material.

38. _____

39. Many air bag components are not _____.

39. _____

40. When an air bag is deployed, _____ powder is released into the interior.

40. _____

41. The powder released during air bag deployment can cause _____ of the skin.

41. _____

42. Identify the components of the air bag system shown in the following illustration.

(A) _____

(B) _____

(C) _____

(D) _____

(E) _____

(F) _____

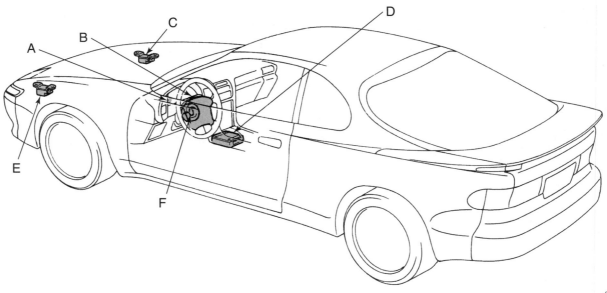

(Toyota)

43. In the spaces provided, list the most common cruise control components.

    (A) _____

    (B) _____

    (C) _____

    (D) _____

44. The majority of cruise control problems involve the _____   44. _____
    linkage or the _____ system.   _____

45. On new vehicles, throttle actuators are _____ that are   45. _____
    directly operated by the control module.

ASE-Type Questions

1. Technician A says that a wiring harness is a group of wires   1. _____
   wrapped together for ease of installation. Technician B says
   that wiring harnesses have molded connectors that plug into
   other harnesses or electrical components. Who is right?
     (A) A only.
     (B) B only.
     (C) Both A and B.
     (D) Neither A nor B.

2. All of the following statements about halogen headlights   2. _____
   are true, *except:*
     (A) Halogen is a gas.
     (B) Some halogen headlights are part of a sealed
         beam unit.
     (C) The halogen bulb can be replaced without
         removing the entire headlight assembly.
     (D) Halogen lights last longer, but are not as bright
         as sealed beams.

3. Technician A says that the emergency lights and turn signal flashers use the same relay. Technician B says that modern vehicles usually have separate bulbs for the tail and brake lights. Who is right?
    (A) A only.
    (B) B only.
    (C) Both A and B.
    (D) Neither A nor B.

3. _____

4. All of the following statements about instrument panel gauges are true, *except:*
    (A) When a gauge has a problem, always check the gauge fuse first.
    (B) Check the voltage limiter when digital gauges are used.
    (C) Never ground an electronic sensor connection.
    (D) Unplugging the related sensor may help to isolate a defective gauge.

4. _____

5. If the burned out element of a dual-filament bulb contacts the working filament, what can happen?
    (A) A fuse can blow.
    (B) Electrical feedback may occur.
    (C) A wire can melt.
    (D) The alternator may be overloaded.

5. _____

6. Technician A says that an electric motor can be tested by bypassing the control switch. Technician B says that when a small electric motor is defective, it is usually repaired. Who is right?
    (A) A only.
    (B) B only.
    (C) Both A and B.
    (D) Neither A nor B.

6. _____

7. To replace some power window motors, which of the following procedures must be performed?
    (A) Remove the door from the vehicle.
    (B) Remove the outer door skin.
    (C) Remove one hinge and tilt the door inward.
    (D) Drill holes in the door frame.

7. _____

8. Sound system static is present when the engine is running, and when the key is on with the engine not running. Which of the following is the *least* likely cause of the problem?
    (A) Sound system internal parts
    (B) Ignition system
    (C) Speakers
    (D) Ground straps

8. _____

9. Technician A says that body computers control functions that do not affect the operation of the engine. Technician B says that the sensors, control units, and actuators used in body computer systems are similar to those found in computerized engine-control systems. Who is right?

    (A) A only.
    (B) B only.
    (C) Both A and B.
    (D) Neither A nor B.

9. _____

10. All of the following rules should be followed when servicing air bag systems, *except:*

    (A) if any air bag system component is accidentally dropped, the component should be replaced.
    (B) when placing a live inflator module on a workbench, always face the bag and trim cover down.
    (C) always disable the air bag system before attempting to service any component on or near the system.
    (D) never subject the inflator module to temperatures greater than 175°F (79.4°C).

10. _____

Name _____

Date _____ Period _____

Instructor _____

Score _____

Text pages 251–278

14

Charging and Starting System Service

Objectives: After studying Chapter 14 in the textbook and completing this section of the workbook, you will be able to:

- Install, test, and service a battery.
- Use jumper cables correctly.
- Test, service, and repair a charging system.
- List the major internal parts of an alternator.
- Test, service, and repair a starting system.
- List the major internal parts of a starter.
- List the major internal parts of a starter solenoid.

Tech Talk: Batteries, starters, and alternators have not changed as much as other automotive electrical equipment. Except for the use of electronic voltage regulators, the starting and charging systems are largely unchanged over the last 30 years. However, their effect on the operation of the vehicle is as important as ever. The alternator and battery must handle the increased electrical load caused by optional equipment and electronic control systems. Modern starters must be lighter and more heat resistant than older models, but they still must produce enough torque to start the engine. Studying this chapter will prepare you for starting and charging system service.

Review Questions

Instructions: Study Chapter 14 of the text and then answer the following questions in the spaces provided.

Battery Service

1. A visual inspection of the battery will help determine service needs, which might include _____.
    (A) cleaning the cable terminals
    (B) replacing the battery
    (C) tightening the hold-down
    (D) All of the above.

1. _____

2. Battery electrolyte contains _____ acid, which can cause serious burns.

2. _____

3. When the temperature drops, battery capacity _____.

3. _____

4. When a battery has an open circuit voltage of 12.0 volts, it is about _____% charged.

4. _____

5. A charging battery gives off _____ and _____ gases, which are flammable.

5. _____

_____

6. Specific gravity testing is a handy method of determining _____.

6. _____

7. If water is added to a battery, it cannot be immediately tested with a(n) _____.

7. _____

8. A difference of more than _____ points (specific gravity) between cells indicates that the battery is defective.

8. _____

9. The preferred method of battery charging is the _____ charge.

9. _____

10. The standard battery rating measurement for modern batteries is _____.

10. _____

11. Reserve capacity is a measure of how long the battery can operate the vehicle's electrical system if the _____ system fails.

11. _____

12. If battery's polarity is not observed, what could happen?

_____

_____

13. Label the battery parts in the following illustration.

(A) _____

(B) _____

(C) _____

(D) _____

(E) _____

(F) _____

(G) _____

(H) _____

(I) _____

(J) _____

(K) _____

(General Motors)

Charging System Service

14. Overcharging may be the cause of short _____ life.

14. _____

15. A light _____ noise when the alternator is charging is normal.

15. _____

16. The two electrical properties that must be checked to determine alternator condition are output _____ and _____.

16. _____

_____

17. If the alternator does not charge when the _____ is bypassed, the alternator is defective.

17. _____

18. Some regulators are installed _____.
    (A) on the engine firewall
    (B) on the alternator
    (C) in the alternator
    (D) All of the above.

18. _____

19. Most _____ charging systems have self-diagnostic capabilities.

19. _____

20. Brushes should be replaced if they show any _____.

20. _____

21. An ohmmeter is used to test the rotor for _____, _____, and _____.

21. _____

_____

_____

22. A(n) _____ diode will cause the test lamp to light in both directions.

22. _____

23. Alternator brushes are often held in place with a pin until the _____ is installed.

23. _____

24. The following illustration shows an alternator regulator being _____.

24. _____

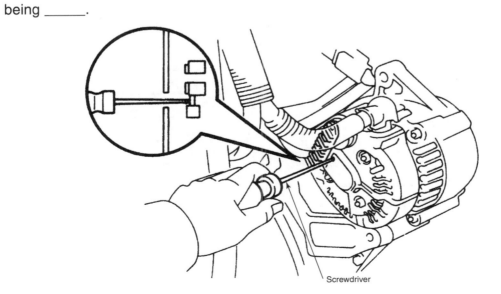

Screwdriver

(Toyota)

Starting System Service

25. Many complaints of poor starter performance are traced to the _____.

25. _____

26. If the starter solenoid clicks but does not operate the starter, the problem may be in the internal solenoid _____.

26. _____

27. If the starter spins but does not engage the flywheel, the solenoid may not be operating the _____ mechanism.

27. _____

28. Define *starter load test*.

_____

_____

29. To test a starter armature for short circuits, place the armature in a(n) _____.

29. _____

30. If a short circuit is encountered while testing the armature, the thin strip of metal will _____.

30. _____

31. Prolonged cranking of the starter will often overheat the starter and cause _____.

31. _____

32. If the starter is _____, replace the brushes regardless of their length.

32. _____

33. The starter mounting pad should be cleaned to ensure a good _____ and accurate _____.

33. _____

_____

34. If the flywheel-to-pinion clearance is incorrect, it is often possible to adjust it with _____.

34. _____

35. Identify the parts of the starter in the following illustration.

(A)_____   (I) _____

(B)_____   (J) _____

(C) _____   (K)_____

(D) _____   (L) _____

(E)_____   (M) _____

(F)_____   (N) _____

(G) _____   (O) _____

(H) _____

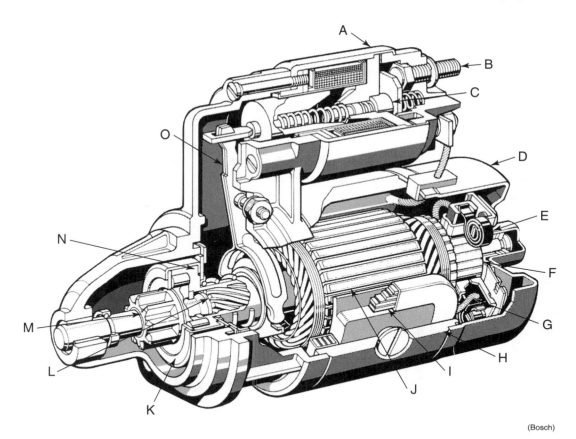

(Bosch)

ASE-Type Questions

1. Technician A says that skin should be flushed with cold water if it comes in contact with electrolyte. Technician B says that acid can be neutralized with baking soda and water. Who is right?
   - (A) A only.
   - (B) B only.
   - (C) Both A and B.
   - (D) Neither A nor B.

1. _____

2. A vehicle battery loses its charge after two weeks without use. Which of the following is the *least* likely cause?
   - (A) Parasitic draw by the vehicle clock.
   - (B) Parasitic draw by the vehicle ECM.
   - (C) Corrosion on the battery terminals.
   - (D) An internal battery short.

2. _____

3. Technician A says that the condition of a battery is determined by its state of charge. Technician B says that cold weather increases the efficiency of a battery. Who is right?
   - (A) A only.
   - (B) B only.
   - (C) Both A and B.
   - (D) Neither A nor B.

3. _____

4. Battery groups are based on all of the following, *except:*
    (A) size.
    (B) weight.
    (C) terminal type.
    (D) voltage.

4. _____

5. Which of the following is most likely to cause high resistance in the charging circuit?
    (A) A corroded connection at the battery's negative cable.
    (B) Frayed or missing battery cable insulation.
    (C) Excessive charging voltage.
    (D) Low charging voltage.

5. _____

6. Technician A says that dirty slip rings can be cleaned with 400 grit emery cloth. Technician B says that diodes can be tested with an ohmmeter. Who is right?
    (A) A only.
    (B) B only.
    (C) Both A and B.
    (D) Neither A nor B.

6. _____

7. All of the following statements about alternator stator service are true, *except:*
    (A) the stator windings can be tested for opens with a test light.
    (B) if an ohmmeter check of the stator shows zero resistance, the stator is shorted.
    (C) the technician should visually inspect the stator for signs of overheating.
    (D) if all other parts test good, suspect the stator.

7. _____

8. Which of the following is *not* part of the starting system?
    (A) The neutral safety switch.
    (B) The battery.
    (C) The ignition switch.
    (D) The regulator.

8. _____

9. A neutral safety switch has failed open. The switch will allow the starter to crank in which of the following gears?
    (A) Neutral and Park.
    (B) Drive and Neutral.
    (C) Park and Reverse.
    (D) The starter will not crank in any gear.

9. _____

10. All of the following statements about starter repair are true, *except:*
    (A) It is often cheaper to install a new starter than fix the old one.
    (B) Brushes should always be changed when the starter is taken apart.
    (C) Overrunning clutch drives should be soaked in solvent and dried with air pressure.
    (D) An armature that is oily should be wiped off with a rag soaked in solvent.

10. _____

Name _____

Date _____ Period _____

Instructor _____

Score _____

Text pages 279–311

15

Computer System Diagnosis and Repair

Objectives: After studying Chapter 15 in the textbook and completing this section of the workbook, you will be able to:

- Retrieve trouble codes.
- Identify trouble code formats.
- Interpret trouble codes.
- Use a scan tool to check components and systems.
- Use a multimeter to check computer system components.
- Identify computer system component waveforms.
- Explain how to adjust throttle and crankshaft position sensors.
- Explain how to replace computer control system input sensors.
- Explain how to replace a computer control system ECM.
- Explain how to replace an ECM memory chip.
- Explain how to program an ECM with an erasable PROM.
- Explain how to replace computer control system actuators.

Tech Talk: The untrained technician cannot possibly diagnose and repair today's electronics-heavy vehicles. The modern technician must understand the electronics of the modern vehicle if he or she intends to make a career of automotive service.

Review Questions

Instructions: Study Chapter 15 of the text and then answer the following questions in the spaces provided.

1. What is the purpose of the MIL?

   _____

   _____

2. List four computer related complaints that could cause the vehicle to be brought in for service. __

_____

_____

3. Computer input sensors operate on very _____ voltages.

3. _____

4. Before performing involved diagnostic routines on computer-controlled vehicles, always check for and retrieve trouble codes from the _____.

4. _____

5. OBD II data link connectors always have _____ pins.

5. _____

6. How do OBD I systems use the MIL to display trouble codes?

_____

_____

7. OBD II trouble codes have _____ letter(s) and _____ number(s).

7. _____

_____

8. What is meant by a scan tool snapshot? _____

_____

_____

Matching

For questions 9–16, match the type of sensor with the value that it should be tested for. Some values are used more than once.

_____ 9. Zirconia oxygen sensor.

_____ 10. Titania oxygen sensor.

_____ 11. Oxygen sensor heater.

_____ 12. Analog MAF sensor.

_____ 13. MAP sensor.

_____ 14. Temperature sensor.

_____ 15. Pressure sensor.

_____ 16. Throttle position sensor.

(A) Voltage.
(B) Amperage.
(C) Resistance.

17. If the engine operates erratically when a connector is flexed or lightly pulled on, the connector is _____.

17. _____

18. The graph in the following illustration is produced by a properly operating _____ sensor.

18. _____

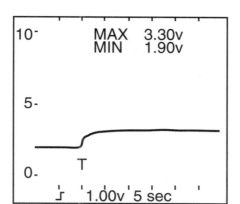

```
10⁻        MAX   3.30v
           MIN   1.90v


 5-


              T
 0-
       ⅃  '1.00v' 5 sec'   '
```
Ferret Instruments

19. Which of the following cannot be adjusted?
    - (A) Throttle position sensor.
    - (B) MAF sensor.
    - (C) Crankshaft position sensor.
    - (D) All of the above.

19. _____

20. The most delicate of all input sensors is the _____ sensor.

20. _____

21. Before replacing any computer system part, the technician should make sure that the ignition switch is turned to the _____ position.

21. _____

22. Which of the following temperature sensors is under pressure when the engine is off?
    - (A) Air.
    - (B) Coolant.
    - (C) Exhaust.
    - (D) All of the above.

22. _____ _____

23. How do you replace a PROM?

_____

_____

24. The type of reprogramming shown in the following figure is called _____ programming.

24. _____

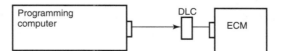

```
┌─────────────┐              DLC   ┌──────┐
│ Programming │            ▭       │ ECM  │
│ computer    │──▭──────▶ ▯ ▯──    │      │
└─────────────┘                    └──────┘
```

25. Solenoids can sometimes be made to work by removing and _____ them.

25. _____

26. Where are vehicle relays usually installed?

_____

_____

27. After replacing any sensor or actuator, erase _____ from the ECM memory.

27. _____

28. If the repair involved correcting a(n) _____ condition, verify that the exhaust emissions output is within specifications.

28. _____

29. After service, the computer system may require a(n) _____ procedure.

29. _____

30. An OBD II drive cycle test takes about _____ minutes, starting with a(n) _____ engine.

30. _____

_____

ASE-Type Questions

1. All of the following defects could affect internal ECM operation, *except:*
    (A) bad ground.
    (B) low charging voltage.
    (C) high charging voltage.
    (D) high idle speed.

1. _____

2. Technician A says that the first step in diagnosing computer control system problems is to check for obvious system defects. Technician B says that the ECM's fuse should be pulled before retrieving trouble codes. Who is right?
    (A) A only.
    (B) B only.
    (C) Both A and B.
    (D) Neither A nor B.

2. _____

3. Technician A says that in many cases, a computer system problem can be detected by a visual inspection. Technician B says that since input sensors operate on very low voltages, even slight wiring problems can create inaccurate readings. Who is right?
    (A) A only.
    (B) B only.
    (C) Both A and B.
    (D) Neither A nor B.

3. _____

4. All of the following statements about retrieving OBD II trouble codes are true, *except:*
   (A) the trouble code can be displayed as a series of flashes of the MIL.
   (B) DLC terminals should never be grounded to retrieve trouble codes.
   (C) a scan tool must be used to retrieve the trouble codes.
   (D) OBD II systems use a standardized data link connector.

4. _____

5. Technician A says that after erasing trouble codes, a hard code will usually reset immediately after the engine is started. Technician B says that an intermittent code will not generally reset immediately. Who is right?
   (A) A only.
   (B) B only.
   (C) Both A and B.
   (D) Neither A nor B.

5. _____

6. Which of the following sensors does *not* produce a waveform?
   (A) Titania oxygen sensor.
   (B) Analog MAF sensor.
   (C) Digital MAF sensor.
   (D) Transmission pressure sensor.

6. _____

7. The waveform shown in the following illustration indicates which of the following?
   (A) A defective rear oxygen sensor.
   (B) A defective front oxygen sensor.
   (C) A defective catalytic converter.
   (D) An exhaust leak at the front oxygen sensor.

7. _____

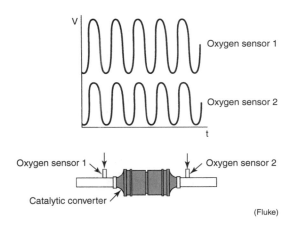

(Fluke)

8. What is the most likely problem when the scan tool reads a trouble code that does not exist?
    (A) Defective ECM.
    (B) Defective scan tool.
    (C) Ungrounded data link connector.
    (D) Blown ECM fuse.

8. _____

9. All of the following statements about ECM diagnosis and replacement are true, *except:*
    (A) sometimes the only way to diagnose an ECM is to substitute a known good ECM.
    (B) replace the ECM before checking other computer system parts.
    (C) a defective output device could damage a new ECM.
    (D) an ECM that produces a trouble code for a known good part is probably defective.

9. _____

10. A distributorless (DIS) engine will not start. A scan tool shows no crankshaft position sensor signal when the engine is cranked. Which is the *least* likely cause of this reading?
    (A) Defective crankshaft position sensor.
    (B) Defective ECM.
    (C) Loose crankshaft position sensor connector.
    (D) Shorted fuel injector.

10. _____

Name _____

Date _____ Period _____

Instructor _____

Score _____

Text pages 313–340

16

Ignition System Service

Objectives: After studying Chapter 16 in the textbook and completing this section of the workbook, you will be able to:

- Inspect, test, and repair ignition systems.
- Describe the purpose of firing order information.
- Adjust ignition timing.
- Remove, test, and replace distributorless ignition components.
- Explain the use of an ignition oscilloscope.
- Remove, test, and replace a distributor assembly.
- Clean, inspect, test, and replace spark plugs.

Tech Talk: Servicing modern electronic and computer-controlled ignition systems is simpler than servicing older systems. However, troubleshooting these systems can be more difficult since there is no way to detect problems in the electronic components by looking at them. Since electronic systems produce higher voltages at the spark plugs, worn or defective plugs may be overlooked until they fail completely. Studying this chapter will enable you to take a logical approach to diagnosing and repairing ignition problems.

Review Questions

Instructions: Study Chapter 16 of the text and then answer the following questions in the spaces provided.

Ignition System Problem Diagnosis

1. A DIS system does not have _____.
    (A) an ignition coil
    (B) a distributor
    (C) spark plugs
    (D) a secondary system

1. _____

2. List three types of damage for which primary wiring should be checked. _____

_____

3. Problems in the _____ circuit can affect the voltage output   3. _____

   of the _____ circuit.

   _____
                                                                        _____

4. On engines that have a distributor, for what three types of damage should the cap and coil be

   checked? _____

5. The following illustration shows a(n) _____ test being   5. _____

   set up.

6. Following a visual inspection, what should be done first if an engine cranks but will not start?

   _____

7. Electronic and computerized ignition systems produce   7. _____

   between _____ volts.
      (A) 3000 and 5000
      (B) 5000 and 15,000
      (C) 15,000 and 20,000
      (D) 50,000 and 100,000

8. Many modern ignition systems operate on battery voltage   8. _____

   and do not use a(n) _____ in the primary circuit.

9. Why should the technician *not* remove a plug wire when the engine is running?

   _____

   _____

10. In the following figure, illustration A shows the coil being checked for _____ resistance. Illustration B shows the coil being checked for _____ resistance.

10. _____

_____

Positive and negative terminals

Coil

A

Ohmmeter leads

High tension terminal (tower)

B

(Toyota)

11. A 4-cylinder engine with a DIS system is missing on the cylinders that are supplied spark by one of the coils. Swapping coils causes the miss to move to the other two cylinders. What ignition part is defective?
    (A) Coil.
    (B) Ignition module.
    (C) Two plugs.
    (D) Cannot tell from the information provided.

11. _____

12. Secondary plug wires can be checked with a(n) _____ to determine their condition.

12. _____

13. Remember that resistance wire is easily damaged. Grasp the _____, not the wire.

13. _____

14. If the engine is positioned to fire the number one_____, the rotor (chalk mark) will be aligned with the number one _____ in the cap.

14. _____

_____

15. During a distributor cap inspection, check inside the cap for signs of _____.

15. _____

16. The tips of some replacement rotors should be coated with _____ before they are installed.

16. _____

17. Most Hall-effect and magnetic _____ can be checked with an ohmmeter.

17. _____

18. Explain how distributor air gap is adjusted. _____

_____

_____

_____

19. If the engine dies or begins running roughly when the ignition module is _____, the module is defective.

19. _____

20. On an oscilloscope, the amount of voltage required to fire the spark plug is indicated by the _____ of the firing line.

20. _____

21. An oscilloscope produces a(n) _____ on a screen

21. _____

22. What do oscilloscope pattern irregularities indicate?

_____

23. Identify the pattern irregularities in the following illustration.

   (A) _____

   (B) _____

   (C) _____

   (D) _____

   (E) _____

24. Special adapters may be required to allow a(n) _____ ignition system to be checked with an oscilloscope.

24. _____

25. On older point-type systems, the breaker points should be _____ when the ignition system is serviced, especially if the points are burned or pitted.

25. _____

26. Note the location and positioning of the primary lead wire and _____ terminals before removing the old points.

26. _____

27. If new contact points cannot be obtained, the old points can be reused after they are _____.

27. _____

28. Before setting ignition timing, check for preliminary steps listed on the _____ located under the hood.

28. _____

29. If the distributor has a vacuum advance unit, the _____ to the unit is usually disconnected and plugged before checking initial timing.

29. _____

30. On many vehicles with engine computers, the _____ must be disconnected from the computer before the timing is set.
    (A) coil
    (B) ignition module
    (C) distributor
    (D) None of the above.

30. _____

31. The distributor body must be _____ to set timing.

31. _____

32. If a vehicle has provisions for a magnetic probe, a(n) _____ timing light can be used.

32. _____

33. The timing scale and pointer should be _____ before setting timing.

33. _____

34. To quickly test the centrifugal advance, slowly increase engine speed from idle to about 2500 rpm. The timing mark should advance smoothly _____ the direction of engine rotation.

34. _____

35. As a quick test of the distributor vacuum advance, run the engine at a steady 1200 rpm. Note the position of the timing mark with a timing light. When the vacuum line is connected to the distributor diaphragm, the timing mark should _____ immediately.

35. _____

36. Computer-controlled ignitions make use of which of the following to adjust timing?
    (A) Vacuum advance.
    (B) Centrifugal advance.
    (C) Input sensors.
    (D) All of the above.

36. _____

37. A(n) _____ pickup on timing light can be connected at any point along a spark plug wire without piercing the wire.

37. _____

Overhauling Distributors

38. To simplify reinstallation, mark the positions of the distributor housing and the _____ relative to the engine before removing the distributor.

38. _____

39. If a distributor will not bottom during installation, the distributor shaft is probably not aligned with the _____.

39. _____

40. Identify the parts of the electronic ignition distributor shown in the following illustration.

(A) _____

(B) _____

(C) _____

(D) _____

(E) _____

(F) _____

(G) _____

(H) _____

(I) _____

(J) _____

(K) _____

(DaimlerChrysler)

Servicing Spark Plugs

41. If the cylinder head is made of _____, allow it to cool thoroughly before removing the spark plugs.

41. _____

42. A careful study of used spark plugs is helpful in determining

_____.

    (A) engine condition
    (B) plug heat range selection
    (C) Both A and B.
    (D) Neither A nor B.

42. _____

43. Oil fouling may be caused by _____.
    (A) a ruptured vacuum pump diaphragm
    (B) excessive choking
    (C) advanced ignition timing
    (D) excessive fuel additives

43. _____

44. When gapping plugs, bend only the _____ electrode.

44. _____

45. Plug gap should be checked using a(n) _____.

45. _____

46. If torque specs are not available, tighten a gasket-equipped plug finger tight. Then, give it another _____ turn(s) more.

46. _____

47. The heat range of a plug is controlled by the _____ of the insulator.

47. _____

48. Spark plug size is determined by the _____ of the threaded section.

48. _____

49. Spark plug reach is determined by the _____ of the threaded section.

49. _____

50. Why is it important to tighten spark plugs to the proper specifications?

_____

_____

ASE-Type Questions

1. Technician A says that the first step in ignition system problem diagnosis is to perform a spark test. Technician B says that the first step in ignition system diagnosis is to visually inspect the ignition components. Who is right?
    (A) A only.
    (B) B only.
    (C) Both A and B.
    (D) Neither A nor B.

1. _____

2. Technician A says that the carbon track in the following illustration was caused by flashover. Technician B says that the coil can be saved by carefully filing away the carbon track along its entire length. Who is right?
    (A) A only.
    (B) B only.
    (C) Both A and B.
    (D) Neither A nor B.

2. _____

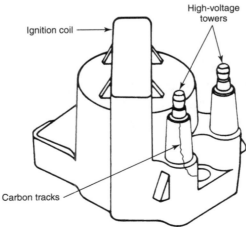

High-voltage towers

Ignition coil

Carbon tracks

3. A DIS coil is being tested with an ohmmeter. Readings are 1.5 ohms between the primary terminals, and 0 ohms between a primary terminal and a secondary terminal. What should the technician do next?
    (A) Check the resistance between both secondary terminals.
    (B) Check the resistance between a secondary terminal and the coil body.
    (C) Check the resistance between a primary terminal and the coil body.
    (D) Replace the coil.

3. _____

4. All of the following should be observed when arranging spark plug wires in their holders, *except:*
    (A) always run wires parallel to each other when possible.
    (B) separate wires going to adjacent cylinders that fire in succession.
    (C) route wires to prevent pinching and fraying by other engine parts.
    (D) keep wires away from heat and oil.

4. _____

5. All of the following can be checked with an ohmmeter, *except:*
    (A) pickup coils
    (B) ignition coils
    (C) secondary wires
    (D) spark plugs

5. _____

6. Technician A says that the engine should be run at idle to set timing because higher speeds would cause the automatic advance mechanisms to operate. Technician B says that the engine should be run at idle to set timing because the timing mark is difficult to see at higher speeds. Who is right?
    (A) A only
    (B) B only.
    (C) Both A and B.
    (D) Neither A nor B.

6. _____

7. Technician A says that distributors must be overhauled when bushing, shaft, gear, or cam wear becomes excessive. Technician B says that some new distributors are not repairable and must be replaced when defective. Who is right?
    (A) A only.
    (B) B only.
    (C) Both A and B.
    (D) Neither A nor B.

7. _____

8. In some late-model engines, the maximum useful life span for spark plugs can be up to _____ miles.
    (A) 25,000
    (B) 50,000
    (C) 75,000
    (D) 100,000

8. _____

9. Several spark plugs are gas fouled. All of the following could be the cause, *except:*
    (A) clogged air cleaner.
    (B) sticking valves.
    (C) plugs that are too hot for the engine.
    (D) plugs that are too cold for the engine.

9. _____

10. Technician A says that a tapered seat spark plug should be turned hand tight and then tightened an additional 1/16 turn. Technician B says that a spark plug using a gasket should be hand tightened only. Who is right?
    (A) A only.
    (B) B only.
    (C) Both A and B.
    (D) Neither A nor B.

10. _____

Name _____

Date _____ Period _____

Instructor _____

Score _____

Text pages 341–355

17

Fuel Delivery

Objectives: After studying Chapter 17 in the textbook and completing this section of the workbook, you will be able to:

- Describe the cleaning, removal, repair, and replacement of fuel tanks.
- Clean, repair, and install fuel lines.
- Test, remove, repair, and replace mechanical fuel pumps.
- Test, remove, repair, and replace electric fuel pumps.
- Service fuel filters.
- Explain vapor lock.
- List the safety rules involved in fuel delivery system service.

Tech Talk: A fuel delivery system must provide a sufficient quantity of clean, filtered fuel to the engine to supply the carburetor, fuel injectors, or diesel injection pump. The fuel delivery system consists of the fuel tank, mechanical or electric pump, filter, and lines. The fuel delivery system affects the performance and dependability of the engine. Defects in the fuel supply system can leave the vehicle stranded. Study this system carefully so that you will be prepared to work on automotive fuel delivery systems.

Review Questions

Instructions: Study Chapter 17 of the text and then answer the following questions in the spaces provided.

Fuel Tanks and Lines

1. Even an empty fuel tank will contain _____ that, if ignited, will produce an explosion.

2. A leaking metal fuel tank may be repaired by _____.

3. A dent in a fuel tank can be removed by the use of air pressure with the tank filled with _____.

1. _____

2. _____

3. _____

4. To clean a fuel line, direct an air blast _____ the direction of fuel flow.

4. _____

5. The tool shown in the following illustration is being used to remove the _____.

5. _____

(Volvo)

Mechanical Fuel Pumps

6. Before testing a mechanical fuel pump, inspect lines for damage and clean or replace all _____.

6. _____

7. When testing a mechanical fuel pump, it should be checked for _____ pressure and _____ pressure.

7. _____

_____

8. Average mechanical fuel pump pressure will be from _____ and will be constant.

8. _____

9. When volume or pressure does not meet specifications, the pump _____ should be determined before condemning the pump.

9. _____

10. When installing a mechanical fuel pump, make certain the rocker arm bears against the _____.

10. _____

11. The following illustration shows a way to check fuel pump pressure and volume. Identify the items indicated.

(A) _____

(B) _____

(C) _____

(D) _____

(E) _____

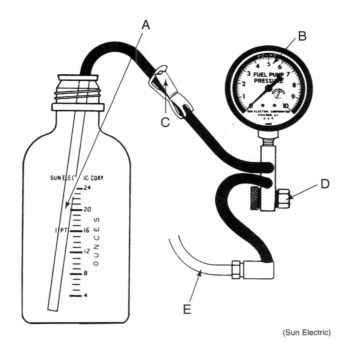

(Sun Electric)

Electric Fuel Pumps

12. Electric fuel pump pressures are _____ than those of mechanical pumps.

12. _____

13. Tapping a fuel pump that is _____ may get it working again.

13. _____

14. Most electric fuel pumps cannot be repaired and are _____ as a unit.

14. _____

15. Before replacing a(n) _____ fuel pump, bleed off the fuel pressure.

15. _____

16. What is the purpose of the inertia switch?_____

_____

_____

17. Identify the parts of the inertia switch shown in the following illustration.

   (A) _____

   (B) _____

   (C) _____

   (D) _____

   (E) _____

(Ford)

Fuel Filter Service

18. Modern fuel filters are _____ instead of being cleaned and reinstalled.

18. _____

19. Some diesel engine filters contain a(n) _____ trap in addition to the filtering element.

19. _____

20. Define *vapor lock*. _____

_____

_____

ASE-Type Questions

1. All of the following are ways to drain a fuel tank, *except:*
    - (A) using a siphoning hose.
    - (B) blowing fuel out with compressed air.
    - (C) removing a drain plug in the tank.
    - (D) using the fuel pump to drain the tank.

1. _____

2. Technician A says that the negative battery terminal should be removed before servicing a fuel tank. Technician B says that residual fuel system pressure should be released before servicing a fuel tank. Who is right?
    - (A) A only.
    - (B) B only.
    - (C) Both A and B.
    - (D) Neither A nor B.

2. _____

3. A kinked plastic fuel line should be _____.
    - (A) replaced with a section of neoprene hose
    - (B) softened with a heat gun to remove the kink
    - (C) discarded
    - (D) repaired with a brass union

3. _____

4. Technician A says that most mechanical fuel pumps are rebuilt when problems are encountered. Technician B says that when rebuilding fuel pumps, the metal parts should be soaked in carburetor cleaner for at least 30 minutes. Who is right?
    - (A) A only.
    - (B) B only.
    - (C) Both A and B.
    - (D) Neither A nor B.

4. _____

5. In the following illustration, a technician is checking the fuel pump for _____.

5. _____

    (A) leakage
    (B) pressure
    (C) flow
    (D) vacuum

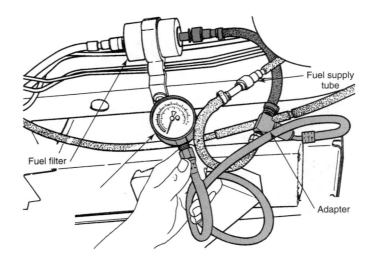

Fuel supply tube

Fuel filter

Adapter

Name _____

Date _____ Period _____

Instructor _____

Score _____

Text pages 357–385

18

Fuel Injection System Service

Objectives: After studying Chapter 18 in the textbook and completing this section of the workbook, you will be able to:

- Identify types of fuel injection systems.
- Identify fuel injection system components.
- Describe the operation of mechanical gasoline fuel injection systems.
- Identify safety rules to be followed when servicing fuel system components.
- Explain how to check fuel injection system pressure.
- Explain how to check fuel injector solenoid and fuel pump operation.
- Explain how to use waveform meters to check fuel injection component condition.
- Explain how to remove and replace fuel system components.
- List the major steps to installing an aftermarket fuel injection system.
- Identify diesel fuel injection system components.
- Explain how to check diesel fuel injection system operation.
- Explain how to remove and replace diesel fuel injection components.

Tech Talk: Fuel injection systems closely control the amount of fuel entering each engine cylinder. Diesel engines use high-pressure injection systems to inject fuel directly into the combustion chamber. Both gasoline and diesel injection systems are controlled by the ECM, based on sensor inputs. To be prepared to service today's cars and trucks, you must know how to work on fuel injection systems. Study this chapter carefully!

Review Questions

Instructions: Study Chapter 18 of the text and then answer the following questions in the spaces provided.

1. List the three basic types of electronic fuel injection systems.

_____

2. Changes in engine load affect intake manifold _____, which assists a spring inside the pressure regulator to control fuel pressure.

2. _____

3. Fuel system safety precautions should always be taken because fuel is both _____ and _____.

3. _____

_____

4. Which of the following steps should the technician take before measuring fuel pressure?
   (A) Obtain trouble codes.
   (B) Check the exhaust system for kinks or clogs.
   (C) Check vacuum lines for poor connections, improper arrangements, kinks, and leaks.
   (D) Check the air intake and air cleaner for clogs and obstructions.
   (E) Check the level, type, and octane rating of the fuel being used.
   (F) All of the above

4. _____

5. If a fuel pressure gauge shows no pressure when attached to a running engine, what is the most likely problem? _____

_____

_____

6. An electric fuel pump is not working. Which of the following is *not* a likely cause?
   (A) Oil pressure switch disconnected.
   (B) Fuel pressure regulator vacuum line disconnected.
   (C) Fuel pump relay defective.
   (D) Fuel pump fuse blown.

6. _____

7. Explain how to use a noid light, and what can be determined from its use.

_____

_____

8. During an injector balance test, the #5 injector shows a drop that is much greater than the average of all fuel injectors. In your own words, what does this mean?_____

_____

_____

9. Identify the types of injector circuits that create each of the following waveforms.
   (A) _____
   (B) _____
   (C) _____
   (D) _____

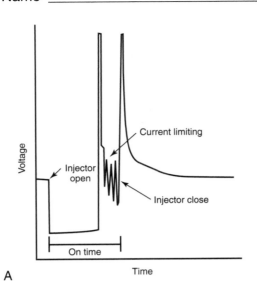

A

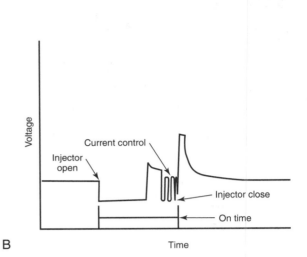

B

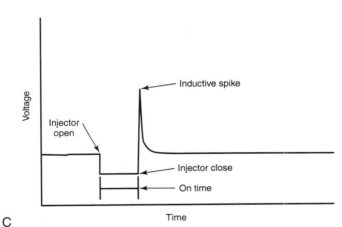

C

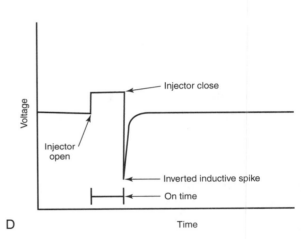

D

10. Buildup of what substance can cause problems in the throttle body and idle air control (IAC) valve?

_____

_____

11. Some fuel injection systems have a(n) _____ valve for removing fuel pressure.

11. _____

12. An on-vehicle injector cleaner attaches to the fuel system and cleans the injectors as the engine _____.

12. _____

13. Fuel injectors should never be soaked in _____.

13. _____

14. Some multiport injectors must be removed as an assembly with the fuel _____.

14. _____

15. Lubricate the injector _____ before reinstalling them.

15. _____

16. After fuel injectors are serviced, it is important that the engine be started and checked carefully for _____.

16. _____

17. Most fuel pressure regulators are _____ and must be replaced if they will not properly control fuel pressure.

17. _____

Matching

For questions 18–22, match the type of throttle body and the service that can be performed on it. There may be more than one answer to some of the following questions.

_____ 18. Adjust or replace the throttle position sensor.

_____ 19. Replace the fuel injector(s).

_____ 20. Replace the pressure regulator.

_____ 21. Split throttle body halves.

_____ 22. Replace entire unit, cannot be serviced.

(A) Throttle body injection.
(B) Multiport injection.
(C) Central point injection.

23. If a mechanical fuel injection engine runs poorly only when it is cold, what is the most likely problem? _____

_____

24. Some mechanical fuel injection systems have two idle speed adjustments—one on the _____ and another on the _____.

24. _____

_____

25. You are installing an aftermarket fuel injection system. Be certain that the power supply wire is energized _____.
    (A) at all times
    (B) only when the ignition switch is on
    (C) only when the engine is running
    (D) only when the engine is cold

25. _____

26. List four problems caused by dirty, damaged, or sticking diesel injectors.

_____

_____

27. Before removing a fuel injection pump, mark the position of the pump housing on the _____.

27. _____

28. Explain how to purge the fuel injectors of a diesel engine.

_____

_____

29. A defective glow plug or glow plug control system can cause hard starting when the diesel engine is _____.

29. _____

30. Glow plug power wires can cause _____ if the technician is not careful around them.

30. _____

ASE-Type Questions

1. A fuel pressure tester is attached to a front-wheel drive
   vehicle's fuel injection system. Fuel pressure is 40 PSI
   with the key on and the engine off and 37 PSI with the
   engine running. All of the following statements about this
   fuel injection system are true, *except:*
   - (A) the fuel system has a vacuum-assisted pressure
     regulator.
   - (B) the fuel pump is operating properly.
   - (C) the pressure regulator vacuum line is disconnected.
   - (D) the pressure gauge is properly attached.

1. _____

2. Technician A says that a defective injector makes a click-
   ing noise as the engine operates. Technician B says that if
   a miss goes away when the alternator is disconnected,
   one of the injectors is probably defective. Who is right?
   - (A) A only.
   - (B) B only.
   - (C) Both A and B.
   - (D) Neither A nor B.

2. _____

3. All of the following statements about fuel injection system
   service are true, *except:*
   - (A) some manufacturers recommend removing fuel
     pressure by disconnecting a fuel line and operat-
     ing the engine until it stops.
   - (B) diagnosis of a returnless injections system is simi-
     lar to that for a system with a pressure regulator.
   - (C) it is sometimes necessary to remove the intake
     manifold or plenum to gain access to the fuel
     injectors.
   - (D) fuel injectors can be cleaned on the engine.

3. _____

4. The spray pattern of a central fuel injection system can be
   made more visible by using a _____ light to illuminate the
   spray.
   - (A) drop
   - (B) fluorescent
   - (C) halogen
   - (D) timing

4. _____

5. Two fuel injectors on a V6 engine are not operating. When the injectors are checked with an ohmmeter, one of the operating injectors is found to be shorted and the two non-functioning injectors check OK. All of the following statements about this fuel injection system are true, *except:*
    (A) one ECM driver operates three injectors.
    (B) a shorted injector will pull current from the two good injectors.
    (C) all three injectors must be replaced.
    (D) the ECM is OK.

5. _____

6. Which of the following can be used to clean a throttle plate and throttle body?
    (A) An on-vehicle injector cleaner.
    (B) Compressed air.
    (C) Engine cleaner and a toothbrush.
    (D) Carburetor cleaner and a toothbrush.

6. _____

7. When a calculated charge PROM is installed, the ECM ignores input from the _____.
    (A) MAF sensor
    (B) MAP sensor
    (C) fuel pressure regulator
    (D) rpm sensor

7. _____

8. Which of the following components would *not* be a part of an aftermarket fuel injection system?
    (A) Fuel tank.
    (B) Surge tank.
    (C) Fuel pressure regulator.
    (D) Engine speed sensor.

8. _____

9. The diesel injector can be tested for all of the following, *except:*
    (A) chatter.
    (B) current draw.
    (C) spray pressure.
    (D) opening pressure.

9. _____

10. Technician A says that the glow plug wires can become very hot. Technician B says that current can flow to the glow plugs for up to 10 minutes after the control light goes out. Who is right?
    (A) A only.
    (B) B only.
    (C) Both A and B.
    (D) Neither A nor B.

10. _____

Name _____

Date _____ Period _____

Instructor _____

Score _____

Text pages 387–401

19

Exhaust System Service

Objectives: After studying Chapter 19 in the textbook and completing this section of the workbook, you will be able to:

- Describe exhaust manifold removal and installation.
- Explain heat control valve service, and repair.
- Identify and use special exhaust system service tools.
- Remove and install a muffler.
- Remove and install an exhaust pipe and tailpipe.
- Remove and install a catalytic converter.
- Remove and install a turbocharger waste gate diaphragm.
- Remove and install a turbocharger.
- Remove and install a supercharger.

Tech Talk: Although the job of the exhaust system is relatively simple, to remove engine exhaust gases, improper service of the exhaust system can cause engine damage, objectionable noise, or even death from carbon monoxide poisoning. Some parts of the exhaust system also help with engine warm-up and air pollution reduction. This chapter covers the proper techniques for servicing exhaust components. By using these techniques, the technician can prevent poor engine performance, part damage, and personal injury. Some vehicles have turbochargers and superchargers. The technician should be familiar with turbocharger and supercharger operation in order to service these vehicles.

Review Questions

Instructions: Study Chapter 19 of the text and then answer the following questions in the spaces provided.

Exhaust System Service

1. Exhaust manifold fasteners should be retightened after the engine has been operated, unless the manifold gasket is made of _____.

1. _____

2. Before working on a heat control valve, allow the manifold to _____.

2. _____

3. Never lubricate a heat control valve with engine oil, since the oil will form _____.

3. _____

4. Define *exhaust back pressure.*_____

_____

_____

5. A(n) _____ is used to restore the round shape to a pipe end that has been crimped or dented.
   (A) chain wrench
   (B) flange
   (C) straightening cone
   (D) end reformer

5. _____

6. If one exhaust pipe nipple (pipe) enters its mating nipple too deeply, the _____ may be too high.

6. _____

7. If one exhaust pipe nipple does not enter its mating nipple deeply enough, the joint may _____ or _____.

7. _____

_____

8. Exhaust alignment is important. Make a careful check of the entire system to make certain that all exhaust system parts have sufficient operating _____.

8. _____

9. As a final step in exhaust system testing, always operate the engine and check each _____ for any signs of _____.

9. _____

_____

10. If there is an exhaust leak under the vehicle, _____ can build up in the passenger compartment.
    (A) carbon monoxide
    B) catalyst
    (C) carbon deposits
    (D) cyanide

10. _____

Superchargers and Turbochargers

11. Turbocharger seal failure will cause large amounts of _____ to enter the exhaust.
    (A) intake air
    (B) engine oil
    (C) fuel
    (D) coolant

11. _____

12. Which of the following do *not* have to be removed to replace turbochargers?

    (A) Turbocharger oil lines.
    (B) Turbocharger exhaust connections.
    (C) Waste gate vacuum lines.
    (D) Intake manifolds.

12. _____

13. A clogged intercooler may result in engine _____.

13. _____

14. When a supercharged engine is running, there will be a(n) _____ between the throttle valve and the supercharger intake.

14. _____

15. Too much supercharger pressure indicates a stuck _____ valve.

15. _____

ASE-Type Questions

1. Exhaust manifolds are being discussed. Technician A says that once the exhaust manifold is installed on the cylinder head, periodic service is not necessary. Technician B says that, on all modern vehicles, a heat control valve is installed in the exhaust manifold. Who is right?

    (A) A only.
    (B) B only.
    (C) Both A and B.
    (D) Neither A nor B.

1. _____

2. All of these statements about the exhaust system connection in the following illustration are true, *except:*

    (A) a gasket must be used with this kind of connection.
    (B) this connection uses two flat surfaces to seal.
    (C) this connection can be loosened with a pipe cutter.
    (D) proper torque is important when reassembling this connection.

2. _____

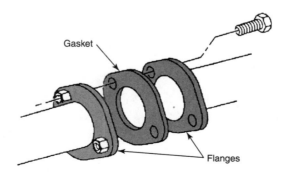

Gasket

Flanges

3. Technician A says that the exhaust-pipe-to-manifold gasket can be reused if it is not damaged. Technician B says that when the exhaust pipe is part of the muffler, it should never be cut off during service. Who is right?
    (A) A only.
    (B) B only.
    (C) Both A and B.
    (D) Neither A nor B.

3. _____

4. When a catalytic converter is replaced, what should be done with the old converter?
    (A) It should be wrapped in plastic and discarded with the regular trash.
    (B) It should be wrapped in plastic and discarded with other toxic waste.
    (C) It should be saved for recycling.
    (D) It should be cleaned in non-petroleum solvent and reused.

4. _____

5. The *last* check to be performed during exhaust system service is to _____.
    (A) Start the engine and check for leaks.
    (B) Start the engine and check part alignment.
    (C) Recheck pipe joint tightness.
    (D) Recheck exhaust manifold flange tightness.

5. _____

Name _____

Date _____ Period _____

Instructor _____

Score _____

Text pages 403–415

20

Driveability Diagnosis

Objectives: After studying Chapter 20 in the textbook and completing this section of the workbook, you will be able to:

- Define driveability diagnosis.
- Explain the differences between driveability diagnosis and tune-ups.
- Explain the principles of drivability diagnosis.
- List the seven steps in drivability diagnosis.
- Explain the modern use of the term tune-up.
- List the basic steps for a maintenance tune-up.
- Explain the need for road and dynamometer tests.

Tech Talk: One of the most important areas of modern automotive service is driveability diagnosis. The technician must learn what causes driveability problems and how to cure them. The technician must also discard the old notion of a tune-up as a cure for driveability problems. Tune-ups today are maintenance procedures, similar to an oil change. Study this chapter to become proficient at driveability diagnosis.

Review Questions

Instructions: Study Chapter 20 of the text and then answer the following questions in the spaces provided.

Matching

For questions 1–7, match the step of the seven-step diagnostic process with the action that is performed as part of that test.

_____ 1. Verify the complaint.

_____ 2. Check for obvious defects.

_____ 3. Determine which system is the most probable cause of the problem.

_____ 4. Perform pinpoint tests.

_____ 5. Check for related problems.

_____ 6. Correct the defect.

_____ 7. Recheck system operation.

(A) Check fuel trim.
(B) Make repairs as needed.
(C) Look over the engine for loose hoses or wires.
(D) Take a break and consider all possible problem causes.
(E) Use a scan tool to confirm that the repairs have been successful.
(F) Perform a road test with the owner.
(G) Check spark plug condition.

8. What is the difference between a guess and an educated guess?

_____

_____

9. When fuel trim is measured by counts, the usual range is from 0 counts to _____ counts, with the ideal count being _____.

9. _____

_____

10. If the fuel trim counts are above or below the ideal setting, the ECM could be compensating for a _____.
(A) rich mixture
(B) lean mixture
(C) All of the above.
(D) None of the above.

10. _____

11. The OBD II misfire monitor information can be obtained with what kind of test equipment? _____

_____

_____

12. When should the technician assume that the vehicle's owner would want repairs made without his or her prior approval?_____

_____

_____

13. The maintenance tune-up is sometimes called a(n) _____ tune-up.

13. _____

14. Parts that are replaced during a maintenance tune-up include _____ and _____.

14. _____

_____

15. Before beginning a tune-up, the technician should obtain the proper engine _____.

15. _____

16. Before beginning a tune-up, always check the engine _____.

16. _____

17. If the distributor cap is replaced, the _____ should also be replaced.

17. _____

18. Follow the procedure on the underhood _____ to set the timing.

18. _____

19. Before finishing the tune-up, check the _____ system voltage.

19. _____

20. Always conduct a(n) _____ test before returning the vehicle to the customer.

20. _____

ASE-Type Questions

1. Driveability diagnosis is being discussed. Technician A says that driveability diagnosis is a matter of changing defective parts. Technician B says that when diagnosing a problem, educated guesses should always be avoided. Who is right?
     (A) A only.
     (B) B only.
     (C) Both A and B.
     (D) Neither A nor B.

1. _____

2. All of the following problems can be detected by following the procedures in Step 2, *except:*
     (A) rich fuel trim.
     (B) low coolant level.
     (C) restricted air filter.
     (D) blown fuse or fusible link.

2. _____

3. Technician A says that knowing the exact problem is not as important as knowing what usually goes bad on a particular engine. Technician B says that it is just as important to correctly interpret test equipment readings as to have the right equipment. Who is right?
     (A) A only.
     (B) B only.
     (C) Both A and B.
     (D) Neither A nor B.

3. _____

4. Misfire monitoring is being discussed. Technician A says that a scan tool can display the number of misfires that have occurred on a particular cylinder in the recent past. Technician B says that a type B misfire will not set a trouble code. Who is right?
     (A) A only.
     (B) B only.
     (C) Both A and B.
     (D) Niether A nor B.

4. _____

5. Technician A says that after a drivability problem has been found and corrected, the vehicle should be road tested to check operation. Technician B says that a repaired vehicle should be road tested to determine whether trouble codes reset. Who is right?
   (A) A only.
   (B) B only.
   (C) Both A and B.
   (D) Neither A nor B.

5. _____

6. All of the following statements about follow-up after correcting a drivability problem are true, *except:*
   (A) follow-up is an important part of a maintenance tune-up.
   (B) a hidden defect can cause an obvious problem.
   (C) hidden defects are common.
   (D) do not return the vehicle to its owner until you are reasonably sure that there is no hidden problem.

6. _____

7. A specification chart would include which of the following?
   (A) Idle speed.
   (B) Electrical schematics.
   (C) Troubleshooting charts.
   (D) Common problem areas.

7. _____

8. Which of the following would *not* be done as part of a maintenance tune-up?
   (A) Check compression.
   (B) Replace spark plugs.
   (C) Replace the ignition coil(s).
   (D) Replace air and fuel filters.

8. _____

9. All of the following statements about service records are true, *except:*
   (A) the life span of many parts can be gauged by how many miles they have been in service.
   (B) the life span of many parts can be gauged by how long they have been in service.
   (C) warranty claims will be based on previous service records.
   (D) minor repairs and maintenance items generally do not need to be listed on service records.

9. _____

10. When using a chassis dynamometer, make sure that _____.
    (A) all road rules are followed
    (B) the area is well ventilated
    (C) the vehicle is not restrained in any way
    (D) the owner is present

10. _____

Name _____

Date _____ Period _____

Instructor _____

Score _____

Text pages 417–438

21

Emission System Testing and Service

Objectives: After studying Chapter 21 in the textbook and completing this section of the workbook, you will be able to:

- Identify the two major types of emissions tests.
- Explain general procedures for conducting emissions test.
- Prepare a vehicle for emissions testing.
- Use an exhaust gas analyzer to check vehicle emissions.
- Explain how OBD II systems monitor the operation of various emission control devices.
- Diagnose emission control problems.
- Service and repair the emission control devices used on late-model vehicles.

Tech Talk: Automotive emission control systems are designed to eliminate the pollutants that enter the atmosphere from a motor vehicle. They do this in two ways: by reducing the amount of pollutants produced by the internal combustion process and by changing engine pollutants into harmless substances. This helps keep the air we breathe clean. By maintaining emission control systems, the automotive technician protects an extremely vital resource—the air that supports life on earth.

Review Questions

Instructions: Study Chapter 21 of the text and then answer the following questions in the spaces provided.

1. In the United States, emissions tests are conducted by _____ to comply with _____ emissions laws.

1. _____

_____

2. A(n) _____ emissions test is conducted with the engine idling.

2. _____

3. The vehicle must be operated on a dynamometer to conduct a(n) _____ emissions test.

3. _____

4. The most failure prone part of the enhanced emissions system is the _____.

4. _____

5. A vehicle cannot be emissions tested if the _____.
   (A) exhaust smoke is visible
   (B) brake warning light is on
   (C) engine is visibly smoking
   (D) OBD II drive cycle was not performed

5. _____

6. Name five vehicle defects that could cause severe damage if the vehicle is operated on a dynamometer. _____

_____

_____

7. Once the vehicle is on the dynamometer, install the _____ devices.

7. _____

8. There are two methods of testing evaporative emissions systems. The _____ test is made with the engine running. The _____ is made using nitrogen.

8. _____

_____

9. More advanced exhaust gas analyzers can measure five gases. List them.

_____

_____

10. When a vacuum hose is removed during an exhaust gas test, the exhaust readings should show a momentary _____ mixture, followed by a(n) _____ mixture.

10. _____

_____

11. In the following illustration, a vacuum pump is being used to test the operation of a(n) _____.

11. _____

Vacuum pump

(DaimlerChrysler)

12. The early fuel evaporation (EFE) system transfers heat from the _____ system to the intake manifold.

12. _____

13. A throttle body coolant passage may fail to warm the throttle body if the _____ sticks open.

13. _____

14. During idle and wide-open throttle, the EGR valve remains _____.

14. _____

15. To test a vacuum-operated EGR valve, depress the EGR valve diaphragm with your finger as the engine idles. The engine should lose about _____ rpm and idle _____ when this is done.

15. _____

_____

16. Should a positive backpressure EGR valve open when vacuum is applied with the engine off? Should a negative backpressure EGR valve open when vacuum is applied with the engine off?

_____

_____

17. The integrated electronic EGR valve contains an internal _____ sensor.

17. _____

18. Which of the following cleaning methods is recommended for removing carbon from an EGR valve?
    (A) Soaking in non-petroleum solvent.
    (B) Soaking in petroleum-based solvent.
    (C) Brushing.
    (D) Sandblasting.

18. _____

19. The air injection system diverter valve prevents backfiring by momentarily diverting the _____.

19. _____

20. Normal air injection system pump pressure will be about _____.

20. _____

21. The catalytic converter can be ruined by a(n) _____ air-fuel mixture.

21. _____

22. A damaged catalytic converter should be _____.

22. _____

23. The PCV system should be checked at every _____.

23. _____

24. An older evaporation control system may have a filter in the _____.

24. _____

25. Enhanced evaporation systems are in vehicles with the _____ computer control system.

25. _____

ASE-Type Questions

1. To make an effective emissions test, the engine should be _____.
   (A) cold
   (B) warming up
   (C) at its normal operating temperature
   (D) close to overheating

   1. _____

2. Technician A says that maximum simulated speeds for emissions tests vary between states. Technician B says that the test may require several sequences of acceleration and braking. Who is right?
   (A) A only.
   (B) B only.
   (C) Both A and B.
   (D) Neither A nor B.

   2. _____

3. All of the following are basic classes of emission controls, *except:*
   (A) internal engine modifications.
   (B) internal cleaning systems.
   (C) fuel vapor controls.
   (D) external cleaning systems.

   3. _____

4. A temperature override switch allows vacuum to pass to the EGR valve when the engine is warmed up. Technician A says that the temperature override switch should be replaced. Technician B says that the EGR valve should be replaced. Who is right?
   (A) A only.
   (B) B only.
   (C) Both A and B.
   (D) Neither A nor B.

   4. _____

5. All of the following statements about exhaust gas analyzers are true, *except:*
   (A) the simplest exhaust gas analyzer will measure carbon monoxide only.
   (B) the engine should idle for a few minutes before using the analyzer.
   (C) when air flow is restricted, the analyzer should show a momentary lean mixture.
   (D) mixture readings should be taken at different engine speeds.

   5. _____

6. Technician A says that the ECM controls all types of digital EGR valves. Technician B says that some digital EGR
valves are operated by engine vacuum. Who is right?
    (A) A only.
    (B) B only.
    (C) Both A and B.
    (D) Neither A nor B.

6. _____

7. Catalytic converters can be damaged by all of the following, *except:*
    (A) using unleaded gas.
    (B) shorting out plugs during engine tests.
    (C) using carburetor cleaner with the engine running.
    (D) backfiring in the exhaust system.

7. _____

8. Which of the following is the *least* likely cause of a rough
idle?
    (A) Defective PCV system.
    (B) Leaking EGR valve.
    (C) Disconnected thermostatic air cleaner vacuum
        line.
    (D) Stuck diverter valve.

8. _____

9. What is the *most* important reason for the technician to
use nitrogen to pressurize an evaporation control system?
    (A) Nitrogen is cheap.
    (B) Nitrogen will not form a combustible mixture with
        the fuel.
    (C) Nitrogen will not react with system metals.
    (D) Nitrogen is EPA approved.

9. _____

10. In the following illustration, what emissions system is
being checked for proper operation?
    (A) EGR.
    (B) Air injection.
    (C) PCV.
    (D) Evaporation control.

10. _____

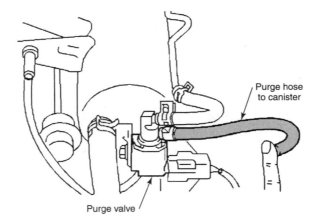

Purge hose
to canister

Purge valve

Name _____

Date _____ Period _____

Instructor _____

Score _____

Text pages 439–452

22

Engine Mechanical Troubleshooting

Objectives: After studying Chapter 22 in the textbook and completing this section of the workbook, you will be able to:

- Summarize preliminary test steps.
- Perform a compression test.
- Use a cylinder leakage detector.
- Perform a vacuum test.
- Check engine oil pressure.
- Diagnose engine mechanical problems.

Tech Talk: The importance of working logically when diagnosing an engine problem cannot be overemphasized. Jumping to conclusions will only prolong the job, wasting both time and money. When diagnosing an apparent engine problem, consider all possible sources of the problem and check the basic engine components and systems before testing the more complex systems.

Review Questions

Instructions: Study Chapter 22 of the text and then answer the following questions in the spaces provided.

Steps to Diagnosing Engine Problems

1. The technician should begin engine diagnosis by asking the _____ to describe the problem.

1. _____

2. Explain how to make a compression test. _____

_____

_____

3. If an engine has low compression, be sure to inform the customer that the engine needs _____ repairs.

3. _____

4. If the compression readings are too high, the cylinder head may have an excessive buildup of _____.

4. _____

5. A cylinder leakage detector can be used to pinpoint the cause of low _____.

5. _____

6. At idle speed, a vacuum gauge should read _____ inches of mercury.

6. _____

Identify the engine problems associated with the vacuum gauge readings shown in the following illustrations.

7. _____

8. _____

9. _____

(Nissan)

Engine and Engine System Problem Diagnosis Charts

10. If an engine has hydrostatic lock, there is _____ in the combustion chamber.

10. _____

11. Noise in a power steering pump or alternator can be isolated by disconnecting the related _____.

11. _____

12. Exhaust leaks are often mistaken for _____ noise.

12. _____

13. A technician's _____ is helpful in pinpointing the source of an engine noise.

13. _____

14. For diagnosis, bearing knocks can be quieted and therefore isolated by _____ spark plugs.

14. _____

15. The two classes of combustion knocks are _____ and _____.

15. _____

_____

Matching

For questions 16–20, match the engine defect with the sound it makes. (Refer to the engine noise diagnosis section at the end of Chapter 22.)

_____ 16. Light clicking or tapping noise.

_____ 17. Heavy thud as the engine is accelerated.

_____ 18. Light metallic rap as the engine is floating.

_____ 19. Variety of noises, may quiet down when engine warms up.

_____ 20. Sharp metallic pinging sound.

(A) Tight timing chain.
(B) Loose connecting rod bearing.
(C) Loose crankshaft main bearing.
(D) Piston slap.
(E) Loose valve adjustment.
(F) Detonation.

ASE-Type Questions

1. All of the following should be checked before checking engine mechanical condition, *except:*
    (A) oil and coolant levels.
    (B) compression.
    (C) air and fuel filters.
    (D) computer trouble codes.

1. _____

2. Compression testing is being discussed. Technician A says that on most engines, the lowest cylinder reading should not be more than 25% lower than the highest cylinder reading. Technician B says that checking compression in one or two cylinders is generally sufficient. Who is right?
    (A) A only.
    (B) B only.
    (C) Both A and B.
    (D) Neither A nor B.

2. _____

3. If compression on a cylinder goes up after oil is added to the cylinder, the _____ are probably worn.
    (A) piston rings
    (B) valve guides
    (C) valve seats
    (D) piston skirts

3. _____

4. Technician A says that oil level should be checked before testing oil pressure. Technician B says that idle speed should be checked before testing oil pressure. Who is right?
    (A) A only.
    (B) B only
    (C) Both A and B.
    (D) Neither A nor B.

4. _____

5. To test oil pressure, the oil pressure gauge can be installed in place of the _____.
   (A) oil pressure sender
   (B) oil pump
   (C) pressure regulator
   (D) lifter gallery plug

5. _____

6. All of the following are possible causes of low oil pressure, *except:*
   (A) internal leaks in the lubrication system.
   (B) clogged oil pickup.
   (C) high idle speed.
   (D) pressure regulator stuck open.

6. _____

7. Technician A says that excessively high oil pressure can be caused by an oil that is too heavy. Technician B says that excessively high oil pressure can be caused by a stuck pressure regulator. Who is right?
   (A) A only.
   (B) B only.
   (C) Both A and B.
   (D) Neither A nor B.

7. _____

8. Which of the following is the *least* likely result of using engine oil that is too light or becomes diluted?
   (A) Preignition.
   (B) Excessive oil consumption.
   (C) Low oil pressure.
   (D) Engine noise.

8. _____

9. Which of the following engine mechanical problems can cause backfiring in the intake manifold?
   (A) Worn piston rings.
   (B) Broken piston rings.
   (C) Loose intake valve adjustment.
   (D) Burned intake valve.

9. _____

10. Technician A says that loose piston pins will cause a sharp double knock. Technician B says that the noise created by loose piston pins is most noticeable during periods of heavy acceleration. Who is right?
    (A) A only.
    (B) B only.
    (C) Both A and B.
    (D) Neither A nor B.

10. _____

Name _____

Date _____ Period _____

Instructor _____

Score _____

Text pages 453–477

23

Cooling System Service

Objectives: After studying Chapter 23 in the textbook and completing this section of the workbook, you will be able to:

*   Explain the role of antifreeze in an engine cooling system.
*   Properly clean a cooling system.
*   Detect leaks in a cooling system.
*   Test a radiator pressure cap.
*   List the safety rules dealing with cooling systems.
*   Inspect and replace cooling system hoses.
*   Inspect, replace, and adjust drive belts.
*   Test and replace a thermostat.
*   Inspect, repair, and replace a coolant pump.

Tech Talk: Without a properly functioning cooling system, a car's engine may overheat in a matter of minutes. Overheating may cause serious damage to the engine. Engine parts can get hot enough to score, crack, or warp. Thousands of people have made the mistake of trying to "make it home" with an overheated engine. Many ended up paying hundreds of dollars to repair blown head gaskets, burned valves, warped cylinder heads, cracked blocks, and other major problems. Study this chapter to learn how to service a cooling system and prevent engine damage.

Review Questions

Instructions: Study Chapter 23 of the text and then answer the following questions in the spaces provided.

Cooling System Service

1. About _____ of the heat produced by combustion is absorbed by the metal parts of an engine and must be disposed of by the cooling system.

1. _____

2. Label the cooling system parts in the following illustration.

(A) _____

(B) _____

(C) _____

(D) _____

(E) _____

(F) _____

(G) _____

(H) _____

(I) _____

(J) _____

(K) _____

(L) _____

(M) _____

(N) _____

(O) _____

(P) _____

(Q) _____

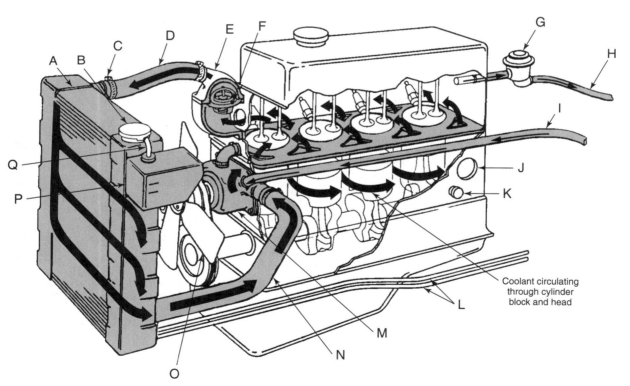

Coolant circulating
through cylinder
block and head

(Gates Rubber)

For questions 3–5, match the three most common types of antifreeze with their properties.

_____ 3. Ethylene glycol.

_____ 4. Propylene glycol.

_____ 5. Organic acid technology (OAT).

(A) Environmentally safe.
(B) Can be used for 100,000 miles (161,000 km).
(C) 50-50 mixture has a freezing point of −35°F (−37°C).

6. Common coolant ratios are around _____ water and _____ ethylene glycol.

6. _____

_____

7. List the three devices that are used to test the amount of freeze protection (freezing point) of a coolant solution.

_____

_____

8. Which of the coolant test devices listed in the previous question can also test pH level?

_____

9. If the radiator filler neck on a cooling system is below the engine and a bleed valve is not used, it may be necessary to _____ the front of the vehicle to allow trapped air to escape.

9. _____

10. Any time the cooling system slowly loses water, the system should be _____.

10. _____

11. When pressure testing, check the pressure marking on the _____ or check the service manual for system limits.

11. _____

12. If the pressure tester shows an increase in pressure when the engine is running, a(n) _____ is indicated.

12. _____

13. When testing for combustion (internal) leaks with test chemicals, the test fluid will change _____ if there are combustion leaks in the system.

13. _____

14. If you suspect a combustion leak, always watch for _____.

14. _____

Radiator Pressure Caps

15. The operation of a pressure cap can be tested with a(n) _____.

15. _____

16. The cap should retain a pressure within _____ of its rating.

16. _____

17. Check the condition of the _____ inside of the filler neck.

17. _____

Hose Inspection

18. Name five things that can be spotted by looking at the outside of a coolant hose.

_____

_____

19. Define *electrochemical degradation (ECD).*_____

_____

_____

20. When using sealer at a hose and fitting, always apply the sealer to the _____.

20. _____

Drive Belt Inspection

21. When inspecting belts for signs of failure, look for _____.
    - (A) cracking
    - (B) oil soaking
    - (C) glazing
    - (D) All of the above.

21. _____

22. Some belts are automatically tightened by an engine-mounted _____.

22. _____

23. Identify the belts in the following illustrations.
    - (A) _____
    - (B) _____
    - (C) _____

A

B

C

(Gates)

Fan Service

24. Never attempt to repair a fan by straightening the _____.

24. _____

25. To test a fluid clutch fan, run an engine to normal operating temperature and then turn it off. If the fan can be hand-turned more than _____ revolution(s) with only minor resistance, it is defective.

25. _____

26. If an electric fan motor runs when its control system is bypassed with jumper wires (motor operated directly from the battery), what is defective?
    (A) Fan motor.
    (B) Fan blade assembly.
    (C) Control system.
    (D) Fan clutch.

26. _____

Thermostat Service

27. A thermostat can cause overheating by _____.

27. _____

28. Always install a thermostat so the thermostatic element will be in contact with the _____ in the block.

28. _____

29. Identify the parts of the thermostat setup shown in the following illustration.
    (A) _____
    (B) _____
    (C) _____
    (D) _____
    (E) _____
    (F) _____

(Honda)

Coolant Pumps

30. Inspect the coolant pump for signs of leakage at the seal drain hole and the _____ area.

30. _____

31. When mounting the water pump, use a new _____.

31. _____

32. When removing core hole plugs (freeze plugs), drive a sharp-nosed _____ through the plug and pry to remove the plug.

32. _____

33. Some replacement freeze plugs are made of _____ for easy installation.

33. _____

34. When servicing air-cooled engines, always check the condition and tension of the blower-fan _____.

34. _____

35. All of the following can cause overheating, *except:*
    (A) loose belt.
    (B) late ignition timing.
    (C) low-temperature thermostat.
    (D) thermostat stuck closed.

35. _____

ASE-Type Questions

1. When should antifreeze be used in a vehicle cooling system?
   (A) when the temperature could drop below 32°F (0°C).
   (B) when the temperature could drop below 0°F (−9°C).
   (C) when the temperature could drop below −9°F (−14°C).
   (D) At all times.

1. _____

2. Technician A says that many electric radiator cooling fans are turned on and off by a thermoswitch. Technician B says that on late-model vehicles, the fan is often controlled by the computer. Who is right?
   (A) A only.
   (B) B only.
   (C) Both A and B.
   (D) Neither A nor B.

2. _____

3. A thermostat that is stuck open or removed can cause all of the following, *except:*
   (A) overcooling.
   (B) oil dilution.
   (C) increased fuel vaporization.
   (D) crankcase sludging.

3. _____

4. Which of the following is *not* a coolant pump problem?
   (A) Worn bearings.
   (B) Leaking fan clutch.
   (C) Worn impeller blades.
   (D) Leaking seals.

4. _____

5. All of the following statements about cooling system service are true, *except:*
   (A) corrosion buildup can be removed with a chemical cooling-system cleaner.
   (B) any hose that collapses as the engine cools off should be replaced.
   (C) a fan blade showing any damage should be replaced.
   (D) a loose belt will place a heavy strain on the bearings of the driven unit.

5. _____

Name _____

Date _____ Period _____

Instructor _____

Score _____

Text pages 479–496

24

Engine Removal, Disassembly, and Inspection

Objectives: After studying Chapter 24 in the textbook and completing this section of the workbook, you will be able to:

- Describe general procedures for removing an engine from a car.
- Explain the use of engine removal equipment.
- List safety rules that apply to engine removal.
- Describe general procedures for disassembling an engine.
- Explain how to make visual checks of major engine parts.

Tech Talk: Removing an engine from a late-model vehicle requires skill and patience. Engine removal on modern vehicles involves many variables, including engine size, type of engine mounting (front or transverse), engine mount locations, transmission and transaxle variations, emission control placement, and accessory equipment location. The vast range of available engines further complicates engine disassembly.

All of the variable mentioned must be considered when preparing to remove and disassemble an engine. If you are unprepared, a considerable amount of time and effort can be wasted. However, after studying the material in this chapter, you will find the job of engine removal and disassembly much easier.

Review Questions

Instructions: Study Chapter 24 of the text and then answer the following questions in the spaces provided.

General Engine Removal

1. If just the engine is to be removed, be certain to support the _____ or _____.

   1. _____

   _____

2. Once a wire, control rod, or other part has been removed, it is good practice to put the _____ back into place.

   2. _____

3. The wire in the following illustration has been _____ to facilitate reinstallation.

3. _____

Engine Removal from Top of Vehicle

4. Attach the puller so that the weight of the engine, or engine and transmission, will be _____ at the desired angle.

4. _____

5. After removing the engine from the car, immediately _____ it until it is the _____ floor.

5. _____

_____

6. Identify the parts of the lifting device shown in the following illustration.

(A) _____

(B) _____

Engine Removal from under the Vehicle

7. When an engine is removed from beneath the vehicle, the engine and _____ are removed as a unit.

7. _____

8. When an engine is removed from beneath the vehicle, keep the engine from falling off of the jack by installing

_____.

8. _____

9. The engine removal jack is used to _____ the engine assembly slightly so the tension can be removed from the

_____.

9. _____

_____

Engine Disassembly and Inspection

10. Describe how to remove a shaft-type rocker arm assembly.

_____

_____

11. To remove the rocker arms on engines using ball stud–type rockers, loosen each ball nut until the rocker arm can be _____ to clear the pushrod.

11. _____

12. To remove a cylinder head, _____ the order of the recommended tightening sequence and loosen each cylinder head cap screw.

12. _____

13. If the cylinder head is stuck, carefully use _____ to free it from the engine.
   - (A) pry bars
   - (B) screwdrivers
   - (C) Either A or B.
   - (D) Neither A nor B.

13. _____

14. Do not force or hammer a tapered object between the _____ and _____ mating surfaces, as the slightest nick or dent can cause serious damage.

14. _____

_____

15. After removal, the cylinder head should be placed in a(n) _____.

15. _____

16. In the following illustration, the hammer and socket are being used to _____.
   - (A) remove the valve guide
   - (B) remove carbon buildup
   - (C) loosen the valve spring
   - (D) loosen the valve spring keeper

16. _____

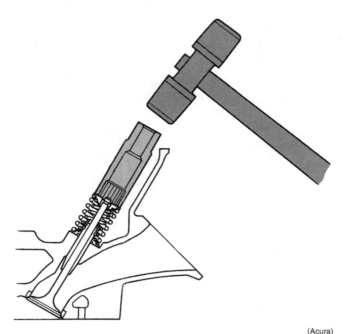

(Acura)

17. Never use a hot tank to clean a(n) _____ cylinder head.   17. _____
    (A) overhead cam
    (B) overhead valve
    (C) aluminum
    (D) cast iron

18. Identify the parts in the following illustration.
    (A) _____
    (B) _____
    (C) _____

19. To remove the camshaft timing mechanism from the   19. _____
    engine, the _____ must usually be removed first, followed
    by the timing cover.

20. When any timing gear part shows wear or damage, it must   20. _____
    be _____.

21. When it is necessary to replace any timing mechanism part, what else should be replaced? ____
    _____

    _____

22. When removing a crankshaft, lift it straight up, being care-   22. _____
    ful to avoid any damage to the _____ surfaces.

23. Circle the illustration below that shows a properly removed ring ridge.

24. Heavy scoring that cannot be corrected by reboring will require the installation of a(n) _____ or replacement of the block.

24. _____

25. A block that is _____ must usually be discarded.

25. _____

ASE-Type Questions

1. To make reinstallation of the hood easier after the engine has been reinstalled, the technician should _____.
   (A) scribe around the hood latch
   (B) scribe around the hood hinges
   (C) replace the hood bolts with studs
   (D) loosen the hood latch at the front of the vehicle

1. _____

2. When lifting any heavy assembly, make certain that the fastener is threaded into the hole for a distance of at least _____ times its diameter.
   (A) 1/2
   (B) 1
   (C) 1 1/2
   (D) 2

2. _____

3. To remove an engine from beneath the vehicle, all of the following steps must be performed, *except:*
   (A) remove alternator wiring.
   (B) remove CV axles.
   (C) secure transmission to body.
   (D) use jack to lower engine assembly.

3. _____

4. Technician A says that waiting for an engine to cool before removing a cylinder head is a waste of time. Technician B says that waiting for an engine to cool before removing the head will prevent warpage. Who is right?
   (A) A only.
   (B) B only.
   (C) Both A and B.
   (D) Neither A nor B.

4. _____

5. A technician is removing a dual overhead camshaft cylinder head from a block. The technician must remove and mark which of the following components?
   (A) Pushrods.
   (B) Valve springs.
   (C) Valve seals.
   (D) Camshaft bearing caps.

5. _____

6. Technician A says that some camshaft timing mechanisms contain only gears. Technician B says that all timing mechanisms use a chain or a belt. Who is right?
    (A) A only.
    (B) B only.
    (C) Both A and B.
    (D) Neither A nor B.

6. _____

7. After removing the timing mechanism but before removing the camshaft, you may have to remove the following additional engine parts, *except:*
    (A) oil pump.
    (B) fuel pump.
    (C) cylinder head(s).
    (D) distributor.

7. _____

8. Technician A says that before removing the main bearing caps, each cap and the corresponding crankshaft web should be marked with a punch or a number stamp. Technician B says that marking main bearing caps is not necessary because the caps are interchangeable. Who is right?
    (A) A only.
    (B) B only.
    (C) Both A and B.
    (D) Neither A nor B.

8. _____

9. A cylinder that is worn or scored past the maximum boring dimension can be saved by _____.
    (A) careful welding
    (B) installing a sleeve
    (C) Either A or B.
    (D) Neither A nor B.

9. _____

10. In the following illustration, the technician is checking the cylinder head for _____.
    (A) cracks
    (B) overheating between cylinders
    (C) combustion chamber size
    (D) warping

10. _____

(Fel-Pro)

Name _____

Date _____ Period _____

Instructor _____

Score _____

Text pages 497–515

25

Cylinder Head and Valve Service

Objectives: After studying Chapter 25 in the textbook and completing this section of the workbook, you will be able to:

- Inspect the cylinder head and parts for defects.
- Properly grind valve seats and valves.
- Test valve springs.
- Service valve guides.
- Service valve seats.
- Reassemble and install a cylinder head.
- Install a cylinder head.

Tech Talk: The valves are among the hardest working parts in the engine. They must be able to open and close in a fraction of a second while sealing off tremendously hot, high-pressure combustion gases. Therefore, the valves must be very carefully cleaned, inspected, refaced, and reinstalled. Valves that cannot be successfully serviced must be replaced. Carelessness can cause a quick comeback.

Review Questions

Instructions: Study Chapter 25 of the text and then answer the following questions in the spaces provided.

Cylinder Head and Valve Service

1. As the valves are removed from the head, place the valves in a(n) _____ so they may be replaced in their original guides.

1. _____

2. If the cylinder head coolant passages are badly clogged, give the head an initial cleaning in a(n) _____.

2. _____

Valve Reconditioning

3. All _____ valves should be discarded.
   - (A) burned
   - (B) warped
   - (C) cracked
   - (D) All of the above.

3. _____

4. Identify the parts of the grinding machine in the following illustration.
   - (A) _____
   - (B) _____
   - (C) _____
   - (D) _____
   - (E) _____
   - (F) _____

(Sunnen)

5. The stones on a valve grinder must be _____ to the correct angle.

5. _____

6. A(n) _____ angle provides a narrow contact area between the valve and the seat.

6. _____

7. The margin on a valve should be approximately _____.
   - (A) 1/32″
   - (B) 1/16″
   - (C) 3/32″
   - (D) None of the above.

7. _____

8. The valve stem should always be trued by _____.

8. _____

9. Never remove more than _____ from the end of a valve stem.

9. _____

10. The technician should never attempt to grind a(n) _____-filled valve.

10. _____

11. Label the following illustration showing a valve face angle.

   (A) _____

   (B) _____

   (C) _____

   (D) _____

   (E) _____

   (F) _____

   (G) _____

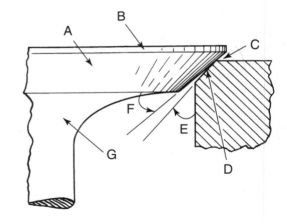

12. Label the following valve grinding illustrations.

   (A) _____

   (B) _____

   (C) _____

   (D) _____

   (E) _____

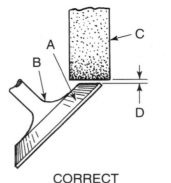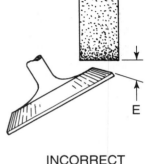

CORRECT   INCORRECT

Cylinder Head Reconditioning

13. Valve stem clearance is generally considered excessive when it is more than _____.

13. _____

14. The valve seat must be the correct _____ and engage the _____ of the valve near its central portion.

14. _____

   _____

15. Seat width varies but will average around _____ for both the intake and exhaust valves.

15. _____

16. When grinding seats, remember that the finished seat will be only as accurate as the _____.

16. _____

17. The following illustration shows correct and incorrect valve seating. Label the parts indicated.

(A) _____

(B) _____

(C) _____

(D) _____

(E) _____

(F) _____

(G) _____

(H) _____

(I) _____

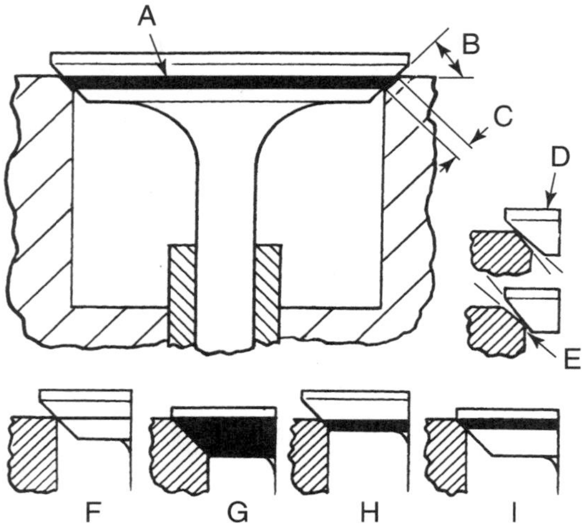

18. The following figure shows pencil marks being used to check valve seat _____.

18. _____

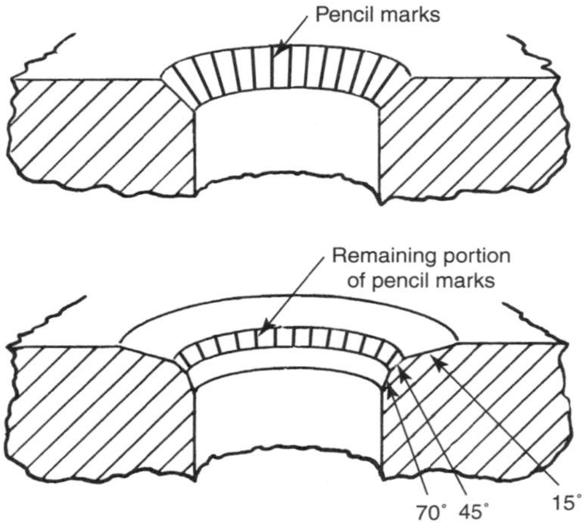

19. The device in the following illustration is called a valve seat _____. It is used in place of valve seat _____.

19. _____

_____

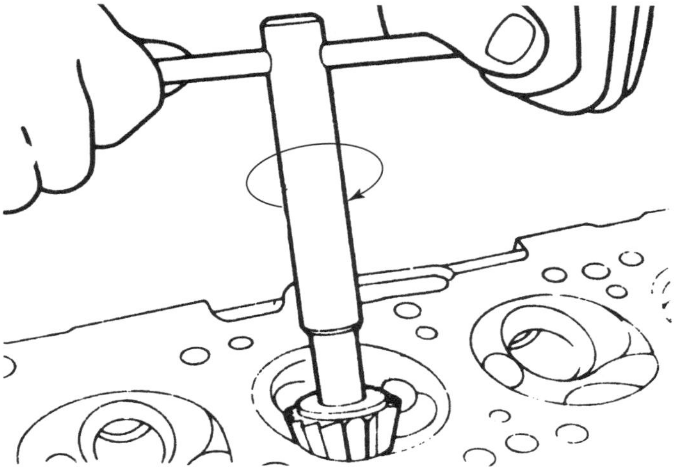

(Hyundai)

20. Valve lapping requires the use of lapping _____ between   20. _____
the valve face and the seat.

21. Use a thin film of _____ rubbed on the valve face or _____   21. _____
marks about 1/4″ (6.35 mm) apart to check the concen-
tricity of the seat and valve face.

_____

22. Most rocker arm studs are _____ into the head. Some are   22. _____
_____ into the head.

_____

23. After extended service, valve springs tend to lose _____.   23. _____

24. A weak valve spring may cause valve_____.   24. _____

25. Valve stem height must be checked on engines with no   25. _____
provision for valve _____.

26. The following illustration shows valve spring tension being tested. Identify the parts indicated.

(A) _____

(B) _____

(C) _____

(DaimlerChrysler)

Cylinder Head Assembly

27. What will happen if the valve spring keepers are not installed properly?

_____

_____

28. Excessive valve spring installed height can be corrected   28. _____
by the use of _____ installed between the spring and
head.

29. Some engine blocks have _____ guide pins, which do not   29. _____
need to be removed after cylinder head installation.

30. Torque-plus-angle tightening is used on engines with   30. _____
_____ blocks and _____ heads.

_____

ASE-Type Questions

1. Technician A says that heavy carbon deposits under the head of an intake valve may indicate worn valve guides. Technician B says that heavy carbon deposits under the head of an intake valve may be a sign of clogged oil drain holes in the cylinder head. Who is right?
   (A) A only.
   (B) B only.
   (C) Both A and B.
   (D) Neither A nor B.

1. _____

2. All of the following statements about the valve face angle are true, *except:*
   (A) on some engines, the valve and seat are ground to the same angle.
   (B) on some engines, the valve is ground to 5° less than the seat
   (C) an interference angle is designed to provide quick seating.
   (D) use of an interference angle varies between manufacturers.

2. _____

3. Technician A says that excessive valve guide clearance will often promote excess oil consumption. Technician B says that valve guide clearance is greatest in the center of the guide. Who is right?
   (A) A only.
   (B) B only.
   (C) Both A and B.
   (D) Neither A nor B.

3. _____

4. Which of the following is *not* true of removable valve guide service?
   (A) The old valve guide can be driven out.
   (B) Chilling the new valve guides makes them easier to install.
   (C) A replacement valve guide has the same shape on both ends.
   (D) Some removable valve guides must be reamed after installation.

4. _____

5. To narrow a valve seat, you must remove metal from the _____ of the seat.
   (A) lower portion
   (B) middle portion
   (C) upper portion
   (D) entire surface

5. _____

6. Technician A says that if valve spring squareness does not meet specifications, the spring can be bent back into shape. Technician B says that if valve spring's compressed pressure is not within specifications, the spring must be replaced. Who is right?
    (A) A only.
    (B) B only.
    (C) Both A and B.
    (D) Neither A nor B.

6. _____

7. An umbrella seal is installed in what part of the valve assembly?
    (A) Over the valve stem between the spring retainer and valve guide.
    (B) Inside of the valve spring retainer, on top of the keepers.
    (C) Inside of the valve guide on the valve head side.
    (D) Inside of the valve guide on the valve stem side.

7. _____

8. A valve spring compressor must be used to install all of the following parts on the head, *except:*
    (A) the valve springs.
    (B) the valve spring retainers.
    (C) the keepers.
    (D) the valve seat.

8. _____

9. The engine block should be checked for _____ before installing the head.
    (A) ring ridge
    (B) warping
    (C) timing mark position
    (D) compression ratio

9. _____

10. Which of the following sequences should be used to tighten cylinder head bolts?
    (A) Start at the outward head bolts and work inward.
    (B) Start at the inside head bolts and work outward.
    (C) Follow the manufacturer's sequence exactly.
    (D) Start at any point and tighten in at least three stages.

10. _____

Name _____

Date _____ Period _____

Instructor _____

Score _____

Text pages 517–549

26

Engine Block, Crankshaft, and Lubrication System Service

Objectives: After studying Chapter 26 in the textbook and completing this section of the workbook, you will be able to:

- Measure cylinder wear.
- Hone a cylinder wall.
- Describe cylinder reboring.
- Check crankshaft and main bearing bores for problems.
- Install new main bearing inserts and rear seal.
- Measure main bearing clearance and crankshaft endplay.
- Service connecting rods.
- Service automotive pistons.
- Properly install piston rings.
- Correctly install a piston and rod assembly in its cylinder.
- Measure bearing clearance with Plastigage.
- Service engine oil pumps.

Tech Talk: It is amazing that the average engine is as durable as it is. Each piston accelerates from zero to 60 mph and then decelerates back to zero on every stroke. This process occurs in less than four inches of piston travel and is repeated thousands of times every minute at highway speeds. It occurs millions of times during the life of the engine. At high engine speeds, each piston and rod assembly can exert over a ton of force on the crankshaft. For this reason, the block, pistons, crankshaft, rods, and bearings must be serviced properly and reassembled carefully. If the technician is careless, an overhauled engine can literally "blow up" (parts break and fly apart) in a matter of minutes.

Review Questions

Instructions: Study Chapter 26 of the text and then answer the following questions in the spaces provided.

Cylinder Service

1. A study of the _____ will help indicate crankshaft bore alignment problems.

1. _____

2. If a 0.0015″ feeler gauge can be slid between the _____ and _____ bore, bore alignment must be corrected.

2. _____

_____

3. After extended use, engine cylinders may become _____ and _____ to the extent that machining is required.

3. _____

_____

4. To order the correct new rings, it is essential that the cylinder _____ be determined.

4. _____

5. Identify the parts A and B. Describe what is happening to the piston rings at locations C and D in the following illustration.

   (A) _____
   (B) _____
   (C) _____
   (D) _____

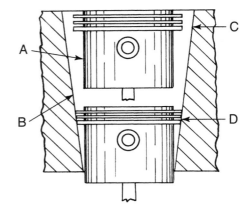

6. Define *piston slap.* _____

_____

_____

7. Label the cylinder wear patterns shown in the following illustration.

   (A) _____
   (B) _____
   (C) _____
   (D) _____
   (E) _____

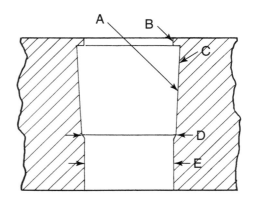

8. If the cylinder is not scored, cracked, or scuffed, taper up   8. _____
   to a maximum of _____ is permissible.

9. What is being performed in the following illustration?

   _____

   _____

Inside micrometer

10. Cylinder out-of-roundness should not exceed _____.   10. _____
    (A) 0.005″
    (B) 0.025″
    (C) 0.05″
    (D) None of the above.

11. Cylinders with minor taper and out-of-roundness can be   11. _____
    deglazed with a(n) _____.

12. Never pull a spinning hone out of a cylinder. Parts can   12. _____
    _____.

13. Heavy scoring of a cylinder (scoring beyond maximum   13. _____
    reboring) will require the installation of a(n) _____.

14. After installing an in-block camshaft bearing, check the   14. _____
    alignment of the bearing and block _____.

15. Journal out-of-roundness must not exceed _____.   15. _____

16. If the crankshaft is only slightly out of alignment, it may be   16. _____
    _____.

17. A crankshaft journal must be polished to a finish of _____   17. _____
    or smoother.

18. When installing bearing inserts, the bearing _____ and   18. _____
    the insert backs must be clean and dry.

19. Check each insert to be sure that the oil holes _____ with   19. _____
    the oil passageways.

20. When installing a crankshaft, wipe a heavy _____ on all bearing surfaces.

20. _____

21. To check bearing clearance, place a length of Plastigage on the _____, about 0.25″ (6.35 mm) from top center.

21. _____

22. This illustration shows a crankshaft being checked for _____.

22. _____

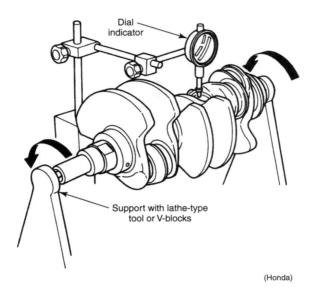

(Honda)

23. The illustration below shows a technician checking bearing clearance. Identify the parts indicated.

(A) _____

(B) _____

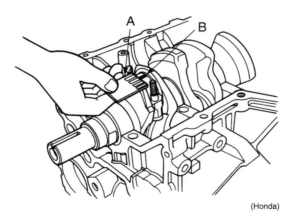

(Honda)

Rear Main Bearing Oil Seals

24. It is good practice to soak a wick seal in _____ prior to installation.

24. _____

25. When installing a synthetic crankshaft rear main seal, make sure that the seal lip faces _____.

25. _____

26. In the following illustration, the technician is installing a(n) _____. Label the parts indicated.

(A) _____

(B) _____

(C) _____

(DaimlerChrysler)

27. Average crankshaft endplay ranges from _____ to _____.

27. _____

_____

28. To remove bearing inserts in an emergency, a plug can be made from a(n) _____.

28. _____

29. Make certain the insert locating lug engages the lug _____ in the bore.

29. _____

30. It is possible to install a new rear main oil seal without removing the _____.
   (A) rear main cap
   (B) oil pan
   (C) old seal
   (D) crankshaft

30. _____

Connecting Rod Service

31. Before removing any rods, check to see that all rods are _____ and that both upper and lower bearing halves are marked with the same _____.

31. _____

_____

32. Connecting rods should always be checked for _____ and _____ distortion.

32. _____

_____

33. Connecting rod bore out-of-roundness should not exceed _____.
   (A) 0.0005″
   (B) 0.001″
   (C) 0.015″
   (D) None of the above.

33. _____

Piston Service

34. When pistons show signs of corrosion, look carefully for possible _____ leaks.

34. _____

35. Explain how ring groove width should be checked.

_____

_____

36. A piston pin must have ample _____ for oil; yet it must not have looseness that will result in pin knock and ultimate failure.

36. _____

37. Since pin-to-piston clearances are extremely small, proper _____ is an exacting job.

37. _____

38. If the pin is a press fit in the rod or held to the rod with a clamp setup, the pin must be carefully _____ so that rod side movement will not cause the pin to strike the cylinder wall.

38. _____

39. On full-floating pin installation, make certain the end locks are not _____ and check that they are _____ in their grooves.

39. _____

_____

40. It is advisable to allow at least _____ of ring gap for each inch (25.4 mm) of cylinder diameter.

40. _____

41. Explain the procedure for checking a ring's side clearance.

_____

_____

42. Many rings are marked with the word "_____."

42. _____

43. If rubber hoses or other protective devices are not slipped over the connecting rod bolts before piston installation, the _____ can be damaged.

43. _____

44. The following illustration shows a piston being installed in a cylinder. Identify the items indicated.

(A) _____

(B) _____

(C) _____

(D) _____

(E) _____

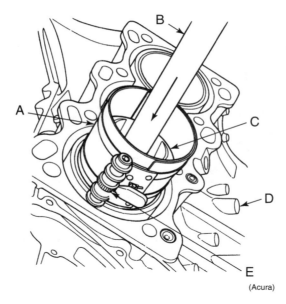

(Acura)

45. The setup in the following illustration is used to check the connecting rod's _____.

45. _____

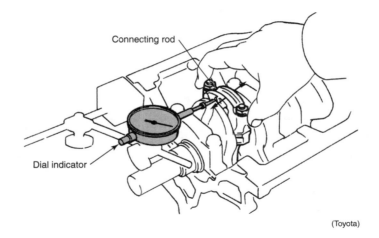

(Toyota)

Oil Pumps

46. When servicing rotor-type oil pumps, replace rotors in _____.

46. _____

47. The following illustration shows rotor end clearance being checked. Identify the parts indicated.

(A) _____

(B) _____

(C) _____

(D) _____

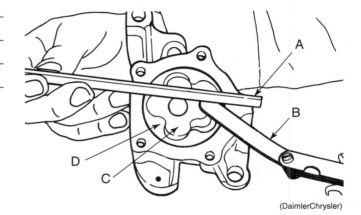

(DaimlerChrysler)

48. What is being checked in the following illustration?_____

_____

_____

Gears

Feeler
gauge

Pump body

(Buick)

49. A pressure relief valve may be part of the _____ or built   49. _____
into the _____.

_____

50. A pressure regulator can be cleaned with _____ cloth.   50. _____

ASE-Type Questions

1. All of the following cause heavy wear to occur at the top of   1. _____
the cylinder, *except:*
   (A) poor lubrication at the top of the cylinder.
   (B) pressure of the rings.
   (C) the cooling effect of the fuel.
   (D) combustion pressure.

2. All of the following statements about cylinders are true,   2. _____
except:
   (A) cylinder wear is greatest at the top of the ring travel.
   (B) if an oversize piston is installed in a standard
       size cylinder, piston slap will result.
   (C) if the glaze is not removed from the cylinder, the
       rings may never seat.
   (D) all cylinders in an engine should be bored to the
       same size.

3. Technician A says that new rings have minute tool marks   3. _____
around the ring-to-cylinder edge. Technician B says that a
cylinder should only be deglazed if it has some scoring.
Who is right?
   (A) A only.
   (B) B only.
   (C) Both A and B.
   (D) Neither A nor B.

4. Technician A says that after honing, the cylinder should be cleaned with hot, soapy water. Technician B says that after cleaning, cylinders should be wiped down with engine oil. Who is right?
    (A) A only.
    (B) B only.
    (C) Both A and B.
    (D) Neither A nor B.

3. _____

5. Technician A says that journal taper is calculated by computing the difference between diameter readings for both ends of the journal. Technician B says that connecting rod journals are more prone to taper than the main bearing journals. Who is right?
    (A) A only.
    (B) B only.
    (C) Both A and B.
    (D) Neither A nor B.

4. _____

6. Technician A says that all offset connecting rods should be discarded. Technician B says that all twisted connecting rods should be discarded. Who is correct?
    (A) A only.
    (B) B only.
    (C) Both A and B.
    (D) Neither A nor B.

6. _____

7. All of the following statements about crankshafts are true, *except:*
    (A) if a crankshaft is only slightly out of alignment, it may be possible to straighten it.
    (B) all insert sets have both upper and lower shells drilled for oil entry.
    (C) bearing clearance is best checked with Plastigage.
    (D) journals must be polished after they are ground.

7. _____

8. After a wick seal is installed, the excess should be _____.
    (A) driven down into the groove
    (B) allowed to stick up
    (C) placed across the block surface
    (D) trimmed flush with the block surface

8. _____

9. Technician A says that all ring grooves should be examined for burrs, dented edges, and side wear. Technician B says that particular attention should be given to the oil control ring grooves. Who is right?
    (A) A only.
    (B) B only.
    (C) Both A and B.
    (D) Neither A nor B.

9. _____

10. Technician A says that an oiled feeler gauge strip and a spring scale can be used to fit pistons in their cylinders. Technician B says that piston-to-cylinder clearance can be determined by carefully measuring both the piston and the cylinder. Who is right?

    (A) A only.
    (B) B only.
    (C) Both A and B.
    (D) Neither A nor B.

10. _____

11. Proper pin clearances depend on all of the following, *except:*

    (A) pin diameter.
    (B) piston material.
    (C) pin length.
    (D) piston operating temperature.

11. _____

12. All of the following statements about pistons and rings are true, *except:*

    (A) a ring compressor must be used to install the piston and rod assemblies.
    (B) pistons should be swapped between adjacent cylinders to equalize wear.
    (C) if the piston does not enter the cylinder with light tapping, something is wrong.
    (D) careless piston installation can damage the crankshaft journals.

12. _____

13. Plastigage is used to check _____ clearance.

    (A) connecting rod–bearing-to-crankshaft journal
    (B) piston to cylinder wall
    (C) compression ring to piston ring groove
    (D) piston pin to piston

13. _____

14. Some manufacturers recommend packing a rebuilt or new oil pump with _____ to make sure that it primes (draws in oil).

    (A) petroleum jelly
    (B) chassis lube
    (C) wheel bearing grease
    (D) camshaft lube

14. _____

15. During pump installation, all of the following steps should be taken, *except:*

    (A) disassemble, clean, and check the pressure relief valve.
    (B) check pickup tube and screen positioning.
    (C) stretch the relief valve spring to the specified length.
    (D) clean and attach any external lines.

15. _____

Name _____

Date _____ Period _____

Instructor _____

Score _____

Text pages 551–573

27

Valve Train and Intake Service

Objectives: After studying Chapter 27 in the textbook and completing this section of the workbook, you will be able to:

- Remove, check, and install a block-mounted camshaft.
- Remove, check, and install a cylinder head–mounted camshaft.
- Properly remove and install a vibration damper.
- Remove, check, and install valve lifters.
- Remove, check, and install rocker arms.
- Remove, check, and install push rods.
- Measure timing gear and chain wear.
- Remove and install camshaft gears and sprockets.
- Inspect a camshaft drive mechanism for problems.
- Service a front cover oil seal.
- Remove and install a timing belt.
- Clean and check intake manifolds and plenums.

Tech Talk: The importance of an engine's camshaft and valve train is often overlooked. Since the valves are directly involved in the entry of the air-fuel mixture into the cylinder, the development of compression, and the exhaust of burned gases, minor valve system defects can have a great effect on engine performance. Worn camshaft drive components can alter the valve timing and reduce engine performance. A jumped timing belt or chain can cause the pistons to hit the valves, damaging both. A worn camshaft will reduce the amount and duration of valve opening, lowering engine power. Worn cam bearings will leak and reduce engine oil pressure. Wear or careless repair of almost any part of the valve train will cause valve noise.

Review Questions

Instructions: Study Chapter 27 of the text and then answer the following questions in the spaces provided.

Camshaft Service

1. Worn camshaft bearings can lower _____ pressure and produce excessive throw off.

1. _____

2. Maximum camshaft-journal-to-bearing clearance will be _____.

2. _____

3. When cam lobes wear, they may not raise the _____ to the specified height.

3. _____

4. When installing two-piece cam bearings, make sure they are _____ properly in the saddles.

4. _____

5. What is being checked in the following illustration? _____

_____

_____

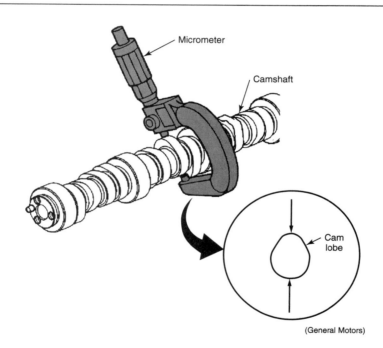

(General Motors)

Lifter Service

6. The lower portion of a used lifter body is often coated with gum or varnish, which makes _____ difficult.

6. _____

7. The lifter-to-cam-lobe surface should be _____ and free of cam wear, grooving, chipping, and galling.

7. _____

8. Due to the close _____, hydraulic lifters must be thoroughly cleaned and assembled in a spotless condition.

8. _____

9. Each lifter must have the correct _____ rate characteristic.

9. _____

10. Before installing hydraulic lifters, fill the lifters with _____.

10. _____

11. If the roller of a roller lifter shows evidence of damage, check the _____ for damage.

11. _____

12. A single defective roller lifter can be replaced without replacing the _____.

12. _____

Push Rod Service

13. Push rods should be _____ and both ends must be smooth.

13. _____

14. If a push rod carries oil, the _____ should be thoroughly cleaned.

14. _____

15. Push rod straightness can be checked with V-blocks and a(n) _____.

15. _____

Timing Gear Service

16. What is being checked in the following illustration?

_____

_____

Gears

Feeler gauge

(Sealed Power)

17. Some camshaft gears are _____ in place, while others are force fit on the shaft and must be pressed off.

17. _____

18. When installing a camshaft, turn the crankshaft so the timing mark faces the _____ of the front cam bearing.

18. _____

19. The camshaft gear timing mark should be aligned with the _____ timing mark.

19. _____

Servicing the Timing Belt and Sprocket

20. Before removing the timing belt, the technician will usually have to loosen the belt _____.

20. _____

21. Extra _____ are needed for dual overhead camshafts and timing belt–driven distributors.

21. _____

22. Identify the timing belt problems shown in the following illustration.

(A) _____

(B) _____

(C) _____

(D) _____

(E) _____

(F) _____

(G) _____

(H) _____

(I) _____

(J) _____

(DaimlerChrysler)

Servicing the Timing Chain and Sprockets

23. When installing a new timing chain, the technician should always use new _____.

23. _____

24. After installing the timing chain and oil slinger, check sprocket _____, chain slack, and camshaft endplay.

24. _____

25. The following illustration shows three methods of _____. These three methods are:

25. _____

(A) _____

(B) _____

(C) _____

(DaimlerChrysler)

Removing and Replacing Front Cover Oil Seal

26. Whenever the vibration damper is removed, the _____ should be replaced.

26. _____

27. Always install the timing cover seals so the seal lip faces _____.

27. _____

28. If the front cover is not positioned by dowel pins, a suitable _____ should be used to prevent oil seal leakage.

28. _____

29. Plastic intake manifolds should not be _____.
    - (A) placed in a hot tank
    - (B) sandblasted
    - (C) glass bead cleaned
    - (D) All of the above.

29. _____

30. Plastic manifolds may be damaged by common _____.

ASE-Type Questions

1. What is being checked in the following illustration?
    - (A) Camshaft lobe condition.
    - (B) Camshaft lobe lift.
    - (C) Camshaft journal bearing clearance.
    - (D) Camshaft journal bearing condition.

1. _____

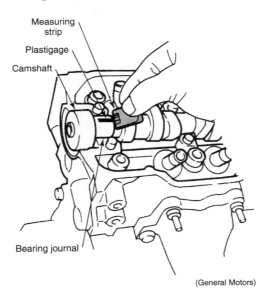

(General Motors)

2. Technician A says that cam lobe wear patterns should extend across the width of the cam lobe. Technician B says that grinding the lifters with slight crowns at their bases will keep the loading area away from the lobe edges. Who is right?
    - (A) A only.
    - (B) B only.
    - (C) Both A and B.
    - (D) Neither A nor B.

2. _____

3. All of the following statements about cleaning hydraulic lifters are true, *except:*
    (A) lifters can be soaked in solvent.
    (B) lifters should be kept in order.
    (C) allow the lifter components to air dry.
    (D) wipe all lifter parts with a lint-free cloth.

3. _____

4. Technician A says that excessive rocker arm-to-shaft clearance will restrict the flow of oil to the valve stems. Technician B says that excessive rocker-arm-to-shaft clearance will cause increased oil consumption. Who is right?
    (A) A only.
    (B) B only.
    (C) Both A and B.
    (D) Neither A nor B.

4. _____

5. The individual rocker shaft oil passages are generally positioned so they face the _____.
    (A) rocker arms
    (B) head
    (C) valve cover
    (D) rear of the engine

5. _____

6. Vibration dampers are being discussed. Technician A says that a vibration damper is properly removed by pulling on its outer rim. Technician B says that the insert between the inertial weight and the hub is made of rubber. Who is right?
    (A) A only.
    (B) B only.
    (C) Both A and B.
    (D) Neither A nor B.

6. _____

7. Which of the following tools can be used to check timing-gear backlash?
    (A) Feeler gauge.
    (B) Inside micrometer.
    (C) Outside micrometer.
    (D) Plastigauge.

7. _____

8. Technician A says that to prevent engine damage on some OHC engines, it is necessary to secure the camshaft with bolts or a special tool before removing the timing belt. Technician B says that timing belts should be turned inside out for inspection. Who is right?
    (A) A only.
    (B) B only.
    (C) Both A and B.
    (D) Neither A nor B.

8. _____

9. Which of the following timing-belt installation steps should be taken first?

    (A) Rotate the engine through two revolutions.

    (B) Install the belt-tensioner tool.

    (C) Align the sprocket timing marks.

    (D) Loosen the tensioner locknut.

9. _____

10. Technician A says that before removing the timing-chain assembly, the engine should be cranked until the timing marks on both sprockets face each other. Technician B says that after removing the timing chain parts, the engine should be cranked until the woodruff key on the crankshaft faces up. Who is right?

    (A) A only.

    (B) B only.

    (C) Both A and B.

    (D) Neither A nor B.

10. _____

Name _____

Date _____ Period _____

Instructor _____

Score _____

Text pages 575–585

28

Engine Assembly, Installation, and Break-In

Objectives: After studying Chapter 28 in the textbook and completing this section of the workbook, you will be able to:

- Summarize a typical sequence for assembling an engine.
- Describe how to install an engine in a car.
- Pressurize the engine lubrication system with oil.
- Make final checks before starting the engine.
- Operate an engine properly for safe break-in.

Tech Talk: Rebuilding an engine is a time-consuming and unforgiving job. A complete engine overhaul will take numerous hours and will require hundreds of different assembly operations. Only one small mistake, however, will cause all of that time to be wasted. You may have to remove and rebuild the engine over again. For this reason, you must use extreme care when assembling and installing an engine.

Review Questions

Instructions: Study Chapter 28 of the text and then answer the following questions in the spaces provided.

Typical Engine Assembly Sequence

As a general rule, engine parts should be installed in a specific order. For questions 1–12, place the following engine parts in the order that they should be installed. **Note:** Not all parts are listed.

_____ 1.   _____ 7.

_____ 2.   _____ 8.

_____ 3.   _____ 9.

_____ 4.   _____ 10.

_____ 5.   _____ 11.

_____ 6.   _____ 12.

(A) Camshaft.
(B) Crankshaft.
(C) Cylinder heads.
(D) Distributor.
(E) Exhaust manifold.
(F) Flywheel.

(G) Front cover.
(H) Oil pan.
(I) Piston/rod assemblies.
(J) Timing gears.
(K) Valve lifters.
(L) Wiring connectors.

13. Many overhaul jobs are ruined at the last minute by _____.

13. _____

14. When not actively working on an engine, the technician should keep it _____.

14. _____

Engine Installation

15. When installing an engine with the transmission in the car, the engine crankshaft and transmission input shaft center-lines must be _____.

15. _____

16. All wires, lines, and hoses should have been _____ so they can be reinstalled correctly.

16. _____

17. Remember that during the initial starting after a rebuild, the engine is operating without _____.

17. _____

18. List the checks you should make before starting an overhauled engine for the first time.

_____

_____

_____

_____

_____

_____

_____

Starting the Engine

19. The first _____ of operation of an overhauled engine is critical.

19. _____

20. Check for signs of fuel, oil, or coolant _____ as the engine warms up.

20. _____

21. When normal engine operation has been reached for the first time, turn off the engine and _____ the head and manifold bolts.

21. _____

22. Alternately accelerating and coasting during the first few miles of operation will help to seat the _____.

22. _____

23. The technician can break in the engine and make a road test in the shop using a(n) _____.

23. _____

Deliver the Vehicle to the Owner

24. Be sure to caution the owner to avoid sustained high-speed driving and heavy _____ during the first 200 to 300 miles of driving to provide proper break-in.

24. _____

25. After 2500 miles of break-in, engine oil consumption should drop to one quart every _____ miles or better.

25. _____

ASE-Type Questions

1. Technician A says that an oil pressurizer can be used to prime the oil pump and fill the filter, galleries, lifters, and bearings. Technician B says that an oil pressurizer can be used to check bearing clearance. Who is right?
    (A) A only
    (B) B only.
    (C) Both A and B.
    (D) Neither A nor B.

1. _____

2. An overhauled engine has been in storage. Which of the following should be performed before it is started?
    (A) Test compression.
    (B) Repressurize lubrication system.
    (C) Readjust valves.
    (D) Replace valve seals.

2. _____

3. Technician A says that an engine that has been properly overhauled will start readily when cranked by the starter. Technician B says that vehicles with electric fuel pumps may have to be cranked for about 60 seconds to allow the pump to fill the system. Who is right?
    (A) A only.
    (B) B only.
    (C) Both A and B.
    (D) Neither A nor B.

3. _____

4. As soon as the engine is started, idle the engine at a speed of around _____ rpm.
    (A) 600
    (B) 800
    (C) 1200
    (D) 2500

4. _____

5. All of the statements about the procedure shown in the following illustration are true, *except:*

    (A) the procedure should be performed before the engine is started after an overhaul.

    (B) the procedure should be performed after the engine has warmed up as part of break-in.

    (C) the procedure is not necessary on engines with mechanical lifters.

    (D) the procedure is not necessary on vehicles with self-adjusting valve trains.

5. _____

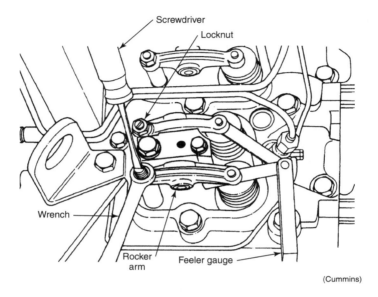

(Cummins)

Name _____

Date _____ Period _____

Instructor _____

Score _____

Text pages 587–605

29

Clutch and Flywheel Service

Objectives: After studying Chapter 29 in the textbook and completing this section of the workbook, you will be able to:

- Explain the construction, operation, and service of diaphragm spring clutches.
- Describe the construction, operation, and service of coil spring clutches.
- Adjust different types of clutch linkages.
- Summarize clutch break-in procedures.
- Diagnose clutch problems and suggest possible corrections.

Tech Talk: Manual clutches have not changed much over the years and neither have the causes of clutch problems. Many clutch failures are caused by improper driving techniques. For example, "riding the clutch" (resting foot on clutch pedal while driving) may cause premature throwout-bearing wear. Another poor driving technique is "slipping the clutch" (allowing clutch to slip for extended period when accelerating). This will speed clutch disc wear. When drivers complain of repeated clutch failures, question them about their driving habits.

Review Questions

Instructions: Study Chapter 29 of the text and then answer the following questions in the spaces provided.

1. To clean clutch parts, always use _____.
   (A) compressed air
   (B) a vacuum cleaning system
   (C) water
   (D) Both A and B.
   (E) Both B and C.

   1. _____

2. Why is it important to use the above method(s) to clean the clutch parts?

   _____

   _____

3. Before condemning the clutch, check the pedal _____.

3. _____

4. If the pressure plate will be reused, what should be done before it is removed from the flywheel?

_____

_____

5. Loosen each pressure plate bolt _____ turn(s) at a time.

5. _____

6. If the flywheel clutch contact area is scored or wavy, the flywheel must be _____ or _____.

6. _____

_____

7. Clutch parts that will be reused should be cleaned with a non-_____ cleaner, like brake cleaner and electrical parts cleaner.

7. _____

8. Never wash the _____ in any kind of solvent.

8. _____

9. The flywheel can be _____ to remove light glazing.

9. _____

10. Rebuilding a pressure plate requires a special fixture or a(n) _____ press.

10. _____

11. Pilot bushings can be removed using a special _____, or by packing the bearing recess with _____ and using a punch and hammer.

11. _____

_____

12. Pilot bearings should be installed with the _____ side facing away from the transmission.

12. _____

13. The following illustration shows clutch housing _____ runout being checked.

13. _____

Dial indicator

(DaimlerChrysler)

14. Oil leaking from the transmission or transaxle, flywheel bolts, or _____ can quickly ruin a new clutch disc.

14. _____

15. Label the parts of the clutch disc in the following illustration.

   (A) _____

   (B) _____

   (C) _____

   (D) _____

(Honda)

16. Do not tighten the pressure plate bolts until you have installed a(n) _____ in the clutch disc.

16. _____

17. The throwout bearing should be replaced whenever the clutch is _____.

17. _____

18. Some _____ throwout bearings are one-piece units.

18. _____

19. Do not press on the clutch pedal until _____ is complete.

19. _____

20. How is clutch pedal height usually determined? _____

   _____

   _____

21. What may cause a sticking clutch pedal? _____

   _____

   _____

22. When adjusting the clutch pedal's free travel, measure the distance the pedal moves from the_____ position to the point at which the release fingers are _____.

22. _____

   _____

23. If the fluid level is low in a hydraulic clutch linkage assembly, check the _____ cylinder, slave cylinder, hydraulic lines, and connections.

23. _____

24. After making about 20 starts with a new clutch assembly, recheck clutch pedal _____.

24. _____

25. Steam cleaning a clutch assembly may cause the parts to _____ together.

25. _____

ASE-Type Questions

1. Technician A says that the modern clutch consists of one friction disc installed between the flywheel and the pressure plate. Technician B says that clutch assemblies should be cleaned by blowing them clean with compressed air. Who is right?
    (A) A only.
    (B) B only.
    (C) Both A and B.
    (D) Neither A nor B.

1. _____

2. Checking flywheel runout requires a _____.
    (A) dial indicator
    (B) feeler gauge
    (C) micrometer
    (D) depth gauge

2. _____

3. If the flywheel surface is too deeply damaged to be sanded, it can be restored by a process called _____.
    (A) pinning
    (B) turning
    (C) grinding
    (D) spinning

3. _____

4. Which of the following is *not* a method of installing a ring gear on a flywheel?
    (A) Shrink fit.
    (B) Welds.
    (C) Bolts.
    (D) Staking.

4. _____

5. Technician A says that many pressure plates can be rebuilt. Technician B says that many technicians prefer to use a rebuilt or reconditioned pressure plate assembly rather than rebuild the old plate. Who is right?
    (A) A only.
    (B) B only.
    (C) Both A and B.
    (D) Neither A nor B.

5. _____

6. All of the following statements about replacing a clutch disc are true, *except:*
    (A) the clutch disc must be installed with the correct side facing the flywheel.
    (B) damaged cushion springs can be replaced.
    (C) the disc usually should be replaced when the clutch is taken apart.
    (D) an oil-soaked disc must be replaced.

6. _____

7. Technician A says that pilot bushings should be replaced only if they are obviously bad. Technician B says that pilot bushings should be replaced whenever the clutch is serviced. Who is right?
   - (A) A only.
   - (B) B only.
   - (C) Both A and B.
   - (D) Neither A nor B.

7. _____

8. Technician A says that the following illustration shows a method of checking for throwout bearing shaft endplay. Technician B says that the following illustration shows a method of checking for clutch housing face runout. Who is right?
   - (A) A only.
   - (B) B only.
   - (C) Both A and B.
   - (D) Neither A nor B.

8. _____

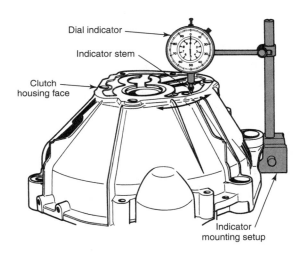

9. A clutch pedal does not return to the full up stop. All of the following could be causes, *except:*
   - (A) weak pedal return spring.
   - (B) excessive pressure plate spring tension.
   - (C) binding linkage.
   - (D) interference between the linkage and another part.

9. _____

10. Which of the following fluids should be used to refill a system using hydraulic clutch linkage?
   - (A) Brake fluid.
   - (B) Dexron automatic transmission fluid.
   - (C) Ethylene glycol antifreeze.
   - (D) Motor oil.

10. _____

Name _____

Date _____ Period _____

Instructor _____

Score _____

Text pages 607–626

30

Manual Transmission and Transaxle Service

Objectives: After studying Chapter 30 in the textbook and completing this section of the workbook, you will be able to:

- Explain manual transmission construction and operation.
- Explain manual transaxle construction and operation.
- Disassemble, check parts, and reassemble a manual transmission.
- Disassemble, check parts, and reassemble a manual transaxle.
- Diagnose manual transmission and transaxle problems.

Tech Talk: Although most vehicles have automatic transmissions or transaxles, many vehicles, especially economy cars and pickup trucks, have manual transmissions. A renewed interest in fuel economy and performance has made manual transmissions more popular in the last few years. Therefore, the technician must know how to service manual transmissions and transaxles.

Review Questions

Instructions: Study Chapter 30 of the text and then answer the following questions in the spaces provided.

Transmission and Transaxle Designs

1. The most obvious external difference between the transmission and transaxle is the presence of two _____ exiting the transaxle. In a transaxle, the transmission and _____ assemblies are installed in a single housing.

1. _____

_____

2. Label the working parts of the four-speed synchronized transmission shown in the following illustration.

(A) _____

(B) _____

(C) _____

(D) _____

(E) _____

(F) _____

(G) _____

(H) _____

(I) _____

(J) _____

(Ford)

Servicing Manual Transmissions and Transaxles

3. If the owner complains of hard shifting, gear clash, or jumping out of gear, check the _____ before road testing.

3. _____

4. Failure of the shift linkage to provide full engagement can result in _____.

4. _____

5. List three repairs that can be performed with the transmission or transaxle in the car.

_____

_____

6. Once a cross-and-roller U-joint is removed, _____ the loose roller bearings together to keep them from falling off.

6. _____

7. Before disassembling the transmission or transaxle, remove the top and side _____. Once this is done, slowly rock the gears while checking for damage. Inspect the synchronizer units for _____.

7. _____

_____

8. The technician must have the proper _____ for the transmission or transaxle being serviced.

8. _____

9. The transmission input shaft can often be removed by lowering _____ to the bottom of the case.

9. _____

10. Each synchronizer unit should be removed as _____.

10. _____

11. Which of the following *must* be replaced during a transmission overhaul?
    (A) Extension housing bushing.
    (B) Input shaft–bearing seal.
    (C) Cluster gear.
    (D) All of the above.

11. _____

12. What will happen to any metal particles that are allowed to remain in hard-to-reach places in the transmission case? _____

_____

_____

13. Check the clutch pilot-bearing end for _____ and _____.

13. _____

_____

14. When replacing a seal in an extension housing, drive the seal into the housing squarely to the proper _____.

14. _____

15. When inspecting the output shaft bearing surfaces, they should be smooth, with no evidence of _____.

15. _____

16. When an old part is unfit for service, check the new part against the old for _____.

16. _____

17. Label the parts of the synchronizer assembly in the following illustration.

(A) _____

(B) _____

(C) _____

(D) _____

(E) _____

(F)_____

(G) _____

18. On some transmissions and transaxles, a(n) _____ shaft may be needed to hold the counter gear bearings in place.

18. _____

19. The two types of lubricant commonly used in manual transaxles are _____ and _____.

19. _____

_____

20. What is being done in this illustration?   20. _____
   (A) Aligning shift linkage.
   (B) Drilling out a linkage pin.
   (C) Drilling out a counter gear roll pin.
   (D) Installing a detent ball.

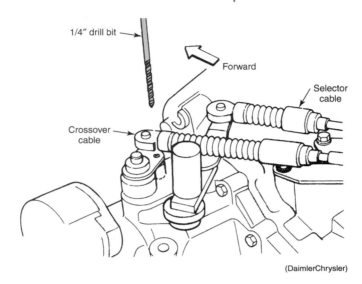

(DaimlerChrysler)

ASE-Type Questions

1. Which of the following is *most* likely to have an overdrive   1. _____
   gear?
   (A) Three-speed transmission.
   (B) Four-speed transmission.
   (C) Four-speed transaxle.
   (D) Five-speed transaxle.

2. Technician A says that whenever possible, a vehicle   2. _____
   should be road tested to help pinpoint transmission prob-
   lems. Technician B says that road tests should never
   include heavy acceleration or deceleration. Who is right?
   (A) A only.
   (B) B only.
   (C) Both A and B.
   (D) Neither A nor B.

3. All of the following statements about inspecting synchro-   3. _____
   nizers are true, *except:*
   (A) the inside of the blocking rings should show fine
       grooves.
   (B) the cone surfaces should show fine grooves.
   (C) the synchronizer teeth should be undamaged.
   (D) the synchronizer hub should be undamaged.

4. The endplay of various gears in a manual transmission or transaxle can be checked with a dial indicator or _____.
    (A) inside micrometer.
    (B) outside micrometer
    (C) proper size drill bit
    (D) feeler gauge.

4. _____

5. Technician A says that any type of oil can be used as a lubricant in a manual transaxle. Technician B says that the differential of some transaxles has a separate oil reservoir. Who is right?
    (A) A only.
    (B) B only.
    (C) Both A and B.
    (D) Neither A nor B.

5. _____

Name _____

Date _____ Period _____

Instructor _____

Score _____

Text pages 627–656

31

Automatic Transmission and Transaxle Service

Objectives: After studying Chapter 31 in the textbook and completing this section of the workbook, you will be able to:

- Explain automatic transmission and transaxle in-vehicle service and diagnosis.
- Explain towing procedures for vehicles with automatic transmissions and transaxles.
- List common automatic transmission and transaxle problems and corrections.
- Road test an automatic transmission or transaxle.
- Summarize the adjustment of automatic transmission and transaxle linkage.
- Explain automatic transmission and transaxle shift linkage and band adjustment.
- Summarize automatic transmission and transaxle removal and installation.

Tech Talk: Automatic transmissions and transaxles are complex devices that operate on various electronic, hydraulic, and mechanical principles. Modern automatic transmissions are partially or fully controlled by on-board computers. This means that the technician must have an understanding of mechanical, hydraulic, *and* electrical principles to properly service late-model automatic transmissions and transaxles. Nevertheless, many problems have simple solutions, and like any part on the vehicle, the transmission/transaxle assembly can be diagnosed using logical troubleshooting techniques. Answer the following questions to test your knowledge of automatic transmission and transaxle diagnosis and service.

Review Questions

Instructions: Study Chapter 31 of the text and then answer the following questions in the spaces provided.

Automatic Transmission/Transaxle

1. Automatic transmission gear ratios are obtained by the use of _____ gearsets.

1. _____

2. What are the three main types of holding members? _____

_____

_____

For questions 3–10, match the component on the left with its description on the right.

_____ 3. Main pressure regulator.

_____ 4. Manual valve.

_____ 5. Shift valves.

_____ 6. Throttle valve.

_____ 7. Governor valve.

_____ 8. Filter.

_____ 9. Accumulator.

_____ 10. Servos.

(A) Driven by output shaft.
(B) Redirect pressure to holding members.
(C) Controls overall system pressure.
(D) Operated by driver.
(E) Controls lockup clutch.
(F) Operated by linkage or modulator.
(G) Cushions shifts.
(H) Removes dirt and metal.
(I) Apply bands.

11. Electronic control systems eliminate the need for _____ and _____ valves.

11. _____

_____

In-Vehicle Service and Problem Diagnosis

12. A large percentage of transmission problems can be solved by _____.

12. _____

13. Vehicles with automatic transmissions cannot be started by _____.

13. _____

14. A vehicle with an automatic transmission should be _____ for only a short distance if the drive shaft is not removed.

14. _____

15. Always check transmission _____ before conducting any tests.

15. _____

16. What four things should the technician be sure of when checking the automatic transmission fluid level? _____

_____

_____

17. When examining the fluid level on the dipstick, check the oil for _____ and a(n) _____ smell.

17. _____

_____

18. If any water is present in the transmission fluid, the fluid will look _____.

18. _____

19. Identify the parts in the following illustration.

   (A) _____

   (B) _____

A

B

(DaimlerChrysler)

20. A pressure test is used to test the condition of the _____
    system.

20. _____

21. After adjusting the shift linkage, make sure that the _____
    pawl works correctly.

21. _____

22. Faulty throttle linkage adjustment will cause incorrect _____
    points and may result in transmission/transaxle failure.

22. _____

23. A leaking _____ can cause the transmission fluid level to
    drop without any visible leaks.

23. _____

24. Forced downshifts can be accomplished mechanically by
    a(n) _____ rod or cable or electrically by a switch and

    _____.

24. _____

    _____

25. If the neutral safety switch is operating correctly, the
    engine will crank with the shifter in _____ or _____ only.

25. _____

    _____

26. Do not attempt to diagnose transmission/transaxle prob-
    lems until you are sure the _____ is in good condition.

26. _____

27. Cracks in the transmission case can often be fixed with

    _____.

27. _____

28. Identify the parts indicated in the following illustration.

(A) _____

(B) _____

(C) _____

(D) _____

(E) _____

(F) _____

(G) _____

(H) _____

(I) _____

(J) _____

(K) _____

(L) _____

(Ford)

29. The lip of a transmission/transaxle seal should always face _____.

29. _____

30. How is the valve body assembly removed? _____

_____

_____

31. The removal of the valve body makes it possible to apply _____ pressure to check clutch and band operation.

31. _____

32. In the following illustration, air pressure is being applied to check _____ action.

32. _____

(DaimlerChrysler)

Transmission or Transaxle Removal

33. The transmission or transaxle and the _____ should always be removed from the vehicle together.

33. _____

34. To prevent damaging a new or rebuilt transmission with contaminated fluid, the oil cooler and lines should be _____.

34. _____

Transmission Installation

35. When installing a transmission, make sure the converter is mounted in the transmission front _____ to the full depth.

35. _____

ASE-Type Questions

1. Technician A says that most torque converters have a lockup clutch to increase fuel mileage. Technician B says that, if a lockup clutch applies in reverse or low, the engine will stall. Who is right?
   - (A) A only.
   - (B) B only.
   - (C) Both A and B.
   - (D) Neither A nor B.

1. _____

2. A sticking governor valve could cause which of the following?
   - (A) Low system pressures.
   - (B) Improper shift points.
   - (C) No movement in Drive.
   - (D) No movement in Reverse.

2. _____

3. The ECM controls all of the following transmission components, *except:*
   - (A) converter lockup clutch.
   - (B) system pressures.
   - (C) accumulator apply.
   - (D) shift points.

3. _____

4. Which of the following could cause fluid aeration?
   - (A) High fluid level.
   - (B) Incorrect type of fluid.
   - (C) Sludge in the hydraulic system.
   - (D) Old, worn out fluid.

4. _____

5. Technician A says that separating torque converter lockup from a gear change is sometimes difficult. Technician B says that a gear change will cause a smaller RPM drop than torque converter lockup. Who is right?
   - (A) A only.
   - (B) B only.
   - (C) Both A and B.
   - (D) Neither A nor B.

5. _____

6. Technician A says that the oil pan must be removed to change the transmission filter. Technician B says that the oil pan may have to be removed to make some band adjustments. Who is right?

    (A) A only.
    (B) B only.
    (C) Both A and B.
    (D) Neither A nor B.

6. _____

7. A stall test is useful for checking the condition of all the following, *except:*

    (A) bands.
    (B) disc clutches.
    (C) one-way clutches.
    (D) pressure regulators.

7. _____

8. All of the following statements about band adjustment are true, *except:*

    (A) some bands are adjusted by adding or subtracting shims inside the servo.
    (B) all bands can be adjusted without disassembling the transmission.
    (C) some bands are adjusted by turning an adjusting screw.
    (D) all bands should be adjusted to factory specifications.

8. _____

9. Following any type of band adjustment, the vehicle should be _____ tested.

    (A) pressure
    (B) leak
    (C) road
    (D) stall

9. _____

10. To change solenoids or pressure sensors, the _____ must be removed.

    (A) oil pan
    (B) valve body
    (C) transmission
    (D) servo cover(s)

10. _____

Name _____

Date _____ Period _____

Instructor _____

Score _____

Text pages 657–700

32

Axle and Driveline Service

Objectives: After studying Chapter 32 in the textbook and completing this section of the workbook, you will be able to:

- Explain the construction and operation of one- and two-piece Hotchkiss drive shafts.
- Service one- and two-piece drivelines.
- Describe cross-and-roller universal joints.
- Service cross-and-roller universal joints.
- Diagnose driveline and universal joint problems.
- Explain the construction and operation of front-wheel drive CV axles.
- Explain the construction and operation of CV joints.
- Service CV axles, joints, and boots.
- Explain the construction, operation, and service of axle housings.
- Compare drive axle types.
- Describe the construction, operation, and service of differentials.
- Diagnose differential and axle problems.

Tech Talk: Although many late-model cars are equipped with front-wheel drive systems, most new trucks and many older cars have rear-wheel drives. The technician must know how to service both types of drivelines. Study this chapter carefully.

Review Questions

Instructions: Study Chapter 32 of the text and then answer the following questions in the spaces provided.

Rear-Wheel Drive Shaft Service

1. The Hotchkiss drive consists of _____ or more pieces.

2. A two-piece propeller shaft requires the use of a(n) _____.

1. _____

2. _____

3. The cross rollers can be retained by _____ set into the yoke at the outer ends of the rollers.

3. _____

4. With a constant-velocity universal joint, both the input and the output sides of the joint _____ at the same _____ throughout the full 360° of rotation.

4. _____

_____

5. If the rear bearing caps are not retained on the cross with a thin strap, _____ them to prevent dropping the needle bearings.

5. _____

6. To remove a cross-and-roller U-joint, tap the _____ inward a small amount to free the snap ring. Place the yoke between the jaws of a heavy _____ so that the yoke is just free to move. Strike the yoke sharply with a lead, brass, or plastic hammer.

6. _____

_____

7. If the inside of the rollers and the trunnion bearing surfaces are free of _____ and _____, the parts may be reused.

7. _____

_____

8. During inspection, try the rollers on the trunnions to check for evidence of _____.

8. _____

9. If the bearing is the sealed type, pack the grease reservoirs at the ends of the _____.

9. _____

10. The constant velocity universal joint is literally _____ cross-and-bearing-cap joints attached to a center _____.

10. _____

_____

11. During propeller shaft installation, make sure the bearing cap _____ are underneath the locating _____ on the differential yoke before tightening the U-bolts.

11. _____

_____

12. If a car is being undercoated, keep the _____ and _____ covered.

12. _____

_____

13. The drive shaft may be checked for runout by using a(n) _____.

13. _____

14. In the following figure, the hose clamps are being used to correct drive-shaft _____.

14. _____

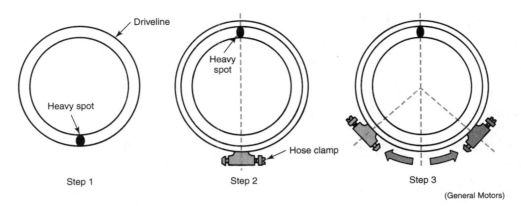

Step 1   Step 2   Step 3

(General Motors)

15. When universal joints are forced to operate at an angle   15. _____
    other than specified, they may cause _____.

16. Label the parts of the drive shaft shown in the following illustration.

    (A) _____

    (B) _____

    (C) _____

    (D) _____

    (E) _____

    (F) _____

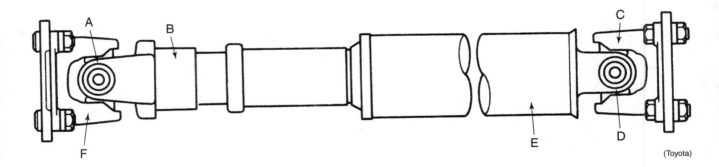

(Toyota)

17. Identify the parts of the universal joint shown in the following illustration.

    (A) _____

    (B) _____

    (C) _____

    (D) _____

    (E) _____

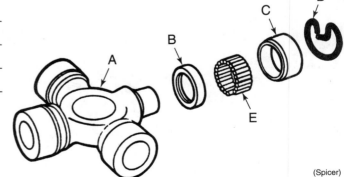

(Spicer)

Front-Wheel Drive CV-Axle Service

18. The two main types of CV joints are the _____ joint and the _____ joint.

18. _____

19. On some vehicles, the steering _____ must be removed before the CV axle can be removed.

19. _____

20. On most CV axles, the inner end of the axle is held to the transaxle by an internal _____.
    - (A) bolt and nut
    - (B) snap ring
    - (C) C-lock
    - (D) All of the above, depending on the manufacturer.

20. _____

21. Identify the parts in the following illustration.
    - (A) _____
    - (B) _____
    - (C) _____
    - (D) _____
    - (E) _____
    - (F) _____
    - (G) _____

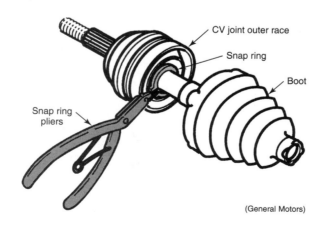

(DaimlerChrysler)

22. In the following illustration, the snap ring is holding the _____ to the CV-joint assembly.
    - (A) spindle
    - (B) axle
    - (C) boot
    - (D) boot clamp

22. _____

CV joint outer race

Snap ring

Boot

Snap ring pliers

(General Motors)

23. Identify the parts in the following illustration.

   (A) _____

   (B) _____

   (C) _____

   (D) _____

   (E) _____

(Saginaw Division of General Motors)

Servicing Axle Shafts

24. The rear axles used with independent rear suspensions resemble small _____.

   24. _____

25. To pull an axle, remove the wheel and the _____, which may have retaining fasteners. Then, remove the nuts from the bearing retainer. Attach a slide-type puller to the axle _____. With a few sharp blows, remove the axle.

   25. _____

   _____

26. On some rear axle applications, the axle is retained with a(n) _____ located in the differential assembly.

   26. _____

27. A section of broken axle may often be retrieved from the housing with a(n) _____, a(n) _____, or other special tool.

   27. _____

   _____

28. Following axle removal, always install new _____ to prevent oil leakage.

   28. _____

29. When installing an axle, pass the splined end of the axle through the _____ very carefully to avoid damage.

   29. _____

30. Explain how to measure axle endplay. _____

   _____

   _____

31. Identify the parts of the axle assembly in the following illustration.

   (A) _____

   (B) _____

   (C) _____

   (D) _____

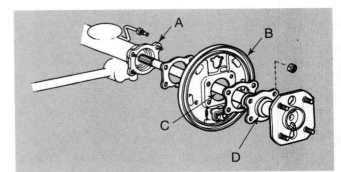

(Toyota)

Ring and Pinion, Differential Unit Service

32. The ring and pinion used on modern rear-wheel drive vehicles are _____ gears.

32. _____

33. When testing a drive axle, check the action during drive, _____, float, and _____ conditions.

33. _____

_____

34. Various vehicle parts can cause noises that may be mistaken for rear-axle noises, including _____.
    (A) tires
    (B) front wheel bearings
    (C) engine
    (D) All of the above.

34. _____

35. What sounds do bearings tend to produce? _____

_____

_____

36. Defective _____ bearings produce a continuous growl that is the same in all pull conditions.

36. _____

37. When will differential pinion and side gear noise be noticeable? _____

_____

_____

38. Before removing the differential case side bearing caps, make certain each _____ and _____ is marked.

38. _____

_____

39. When new side bearings are used, use new _____ also.

39. _____

40. When either a new ring gear or pinion drive is required, they should be replaced as a(n) _____.

40. _____

41. Ring gear fasteners should be tightened in a(n) _____ order, forming a(n) _____ pattern.

41. _____

_____

42. Define *gear ratio.* _____

_____

_____

43. Define *pinion gear depth.* _____

_____

_____

44. If the original ring and pinion, inner bearing, and carrier will be used, the original _____ will give the proper pinion depth.

44. _____

45. To prevent the pinion gear from moving away from the _____ under load, the pinion bearings must be properly _____.

45. _____

_____

46. A(n) _____ is used to check gear backlash.

46. _____

47. What is being measured in the following illustration? _____

_____

_____

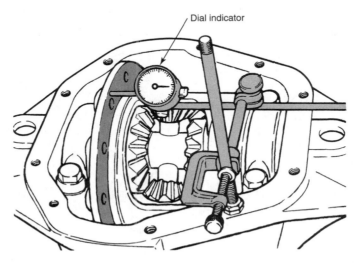

48. Define *limited slip differential*.

_____

_____

49. In the following illustration, the technician is checking the condition of a(n) _____.

49. _____

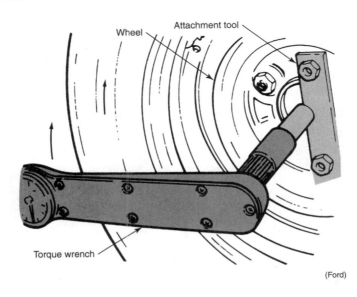

(Ford)

50. When checking the fluid level on a cold differential assembly, the level should be up to _____ below the filler hole.

50. _____

ASE-Type Questions

1. Rear-wheel drive shaft components are being discussed. Technician A says that the slip yoke allows lengthwise movement between the transmission and the rear axle housing. Technician B says that, when a center support bearing is used, a slip yoke is not needed. Who is right?
   (A) A only.
   (B) B only.
   (C) Both A and B.
   (D) Neither A nor B.

1. _____

2. A rear-wheel drive vehicle vibrates at cruising speeds. Which of the following is the *least* likely cause of the problem?
   (A) Loose universal joint.
   (B) Undercoat on drive shaft.
   (C) Driveline angles too large.
   (D) Driveline angles too small.

2. _____

3. All of the following statements about CV joint service are true, *except:*
   (A) the Rzeppa joint-cage must be tilted to remove the balls.
   (B) if a CV joint will not come apart, it can be tapped lightly with a hammer.
   (C) the boot straps should be removed after the CV joint is disassembled.
   (D) tripod joints will slide apart.

3, _____

4. A semifloating rear-wheel drive axle does all of the following, *except:*
   (A) retains the wheel.
   (B) supports the wheel.
   (C) drives the wheel.
   (D) dampens vibration in the driveline.

4. _____

5. When a rear axle shaft is held in the housing by a retainer plate, the axle seal between the bearing and rear axle assembly is installed in the _____.
   (A) retainer plate
   (B) axle housing
   (C) bearing assembly
   (D) axle flange

5. _____

6. Technician A says that to remove an axle with a C-lock, the differential drive pinion should be removed. Technician B says that to remove an axle with a C-lock, the differential pinion shaft should be removed. Who is right?
    (A) A only.
    (B) B only.
    (C) Both A and B.
    (D) Neither A nor B.

6. _____

7. Technician A says that sounds produced by gears will usually vary in pitch. Technician B says that gear noises will sound the same under all pull conditions. Who is right?
    (A) A only.
    (B) B only
    (C) Both A and B.
    (D) Neither A nor B.

7. _____

8. A clanking sound during acceleration or deceleration may be caused by all of the following, *except:*
    (A) worn universal joints.
    (B) insufficient ring and pinion backlash.
    (C) a worn transmission.
    (D) a worn drive pinion shaft.

8. _____

9. The following illustration shows a technician performing which of the following procedures?
    (A) Checking ring and pinion backlash.
    (B) Checking tooth contact pattern.
    (C) Removing the differential carrier bearings.
    (D) Removing the pinion lock screw.

9. _____

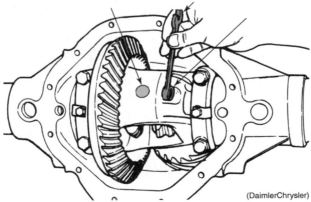

(DaimlerChrysler)

10. Technician A says that a heel contact pattern is caused by excessive backlash. Technician B says that excessive backlash is corrected by moving the ring toward the pinion. Who is right?
    (A) A only
    (B) B only
    (C) Both A and B.
    (D) Neither A nor B.

10. _____

Name _____

Date _____ Period _____

Instructor _____

Score _____

Text pages 701–722

33

Four-Wheel Drive Service

Objectives: After studying Chapter 33 in the textbook and completing this section of the workbook, you will be able to:

- Explain four-wheel drive construction and operation.
- Define part-time and full-time four-wheel drive.
- Diagnose mechanical and electronic transfer case problems.
- Disassemble, check parts, and reassemble a transfer case.
- Disassemble, check parts, and reassemble a typical locking hub assembly.
- Diagnose four-wheel drive problems.

Tech Talk: Once only used by the military, forest rangers, and hunters, four-wheel drive vehicles are now commonly used for commuting and shopping. The advantages of four-wheel drive vehicles on wet and icy roads have prompted their popularity. It is common to find four-wheel drive vehicles that have never been taken off the pavement. The automotive technician must know how to diagnose and repair these vehicles. This chapter will improve your knowledge of four-wheel drive service operations.

Review Questions

Instructions: Study Chapter 33 of the text and then answer the following questions in the spaces provided.

Four-Wheel Drive Components

1. All four-wheel drive vehicles incorporate a transfer case between the transmission _____ shaft and the drive shafts.

1. _____

Transfer Case

2. Identify the parts of the transmission and transfer case assembly in the following illustration.

(A) _____

(B) _____

(C) _____

(D) _____

(E) _____

(F) _____

(G) _____

(H) _____

(I) _____

(J) _____

(K) _____

(L) _____

(M) _____

(N) _____

(O) _____

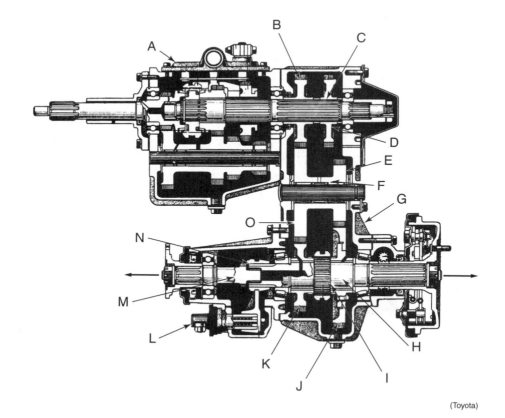

(Toyota)

3. If a part-time four-wheel drive vehicle is operated on hard,
dry surfaces, _____ will cause damage to the unit.

3. _____

4. Full-time four-wheel drive vehicles make use of a(n) _____ assembly or a(n) viscous clutch.

4. _____

5. Full-time four-wheel drive units often use a(n) _____ drive instead of gears.

5. _____

6. A full-time four-wheel drive _____ mechanism should never be engaged on dry pavement.

6. _____

7. Identify the parts of the transfer case assembly in the following illustration.

(A)_____ (I) _____

(B)_____ (J) _____

(C) _____ (K)_____

(D) _____ (L) _____

(E)_____ (M) _____

(F)_____ (N) _____

(G) _____ (O) _____

(H) _____ (P)_____

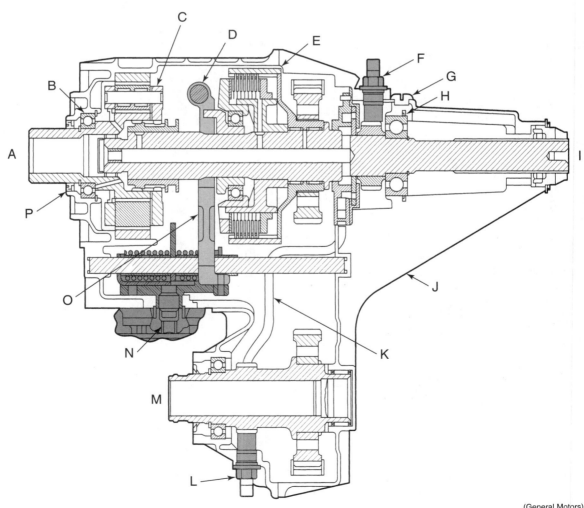

(General Motors)

Transfer Case Service

8. To begin diagnosis of a four-wheel drive mechanism, _____ the vehicle after checking the lubricant level.

8. _____

9. The following figure shows a speed sensor being removed. After this unit is removed, check the tip for _____.

9. _____

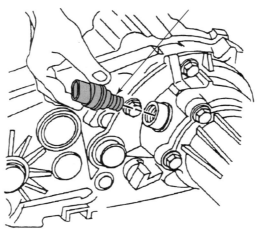

(General Motors)

10. A technician suspects a transfer case has an electronic control-system problem. What should the technician do first?
    (A) Check ground straps.
    (B) Retrieve trouble codes.
    (C) Check sensor connectors.
    (D) Check fuses.

10. _____

11. Transfer case overhaul is similar to that of a(n) _____.

11. _____

12. If a support member must be removed, support the engine and transmission with a(n) _____.

12. _____

13. If all dirt and metal particles are not removed from the transfer case, they will be loosened by the _____ and find their way into moving parts.

13. _____

14. Use a(n) _____ hammer when parts must be hammered loose.

14. _____

15. A leaking viscous coupling will show up as globules of _____ in the transfer case lubricant.

15. _____

16. All suspect transfer case parts should be replaced because _____.
    (A) removing and disassembling the transfer case is difficult
    (B) every part will fail eventually
    (C) some parts are no longer available
    (D) replacement parts are lighter and stronger than original equipment

16. _____

17. Any grease used to hold internal parts in place must _____ in the lubricant used in the transfer case.

17. _____

18. Label the parts of the transfer case lubrication pump shown in the following illustration.
    (A) _____
    (B) _____
    (C) _____
    (D) _____
    (E) _____

(Borg-Warner)

19. In the following illustration, the dial indicator is being used to check output shaft _____.

19. _____

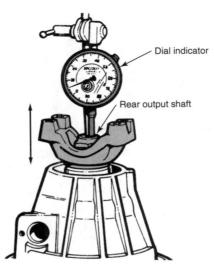

(Range Rover)

Locking Hubs

20. Locking hubs are much more likely than other parts to be damaged from the effects of _____.
    - (A) dirt and water
    - (B) low case oil level
    - (C) maladjusted linkage
    - (D) All of the above.

20. _____

21. Most locking hub components are held in place by a(n) _____.

21. _____

22. It is important to check all locking hub _____ and _____ for damage that would allow foreign material to enter the hub.

22. _____

_____

Vacuum Motor Service

23. If the shift collar of a vacuum motor sticks, the front axle may be _____ at all times.

23. _____

24. Label the parts of the vacuum motor and shift fork assembly shown in the following illustration.
    - (A) _____
    - (B) _____
    - (C) _____
    - (D) _____
    - (E) _____

(DaimlerChrysler)

25. When reassembling the vacuum motor, make sure that the shift _____ engages the shift collar.

25. _____

ASE-Type Questions

1. Technician A says that the transfer case is needed to supply output shaft torque to the front and rear drive shafts. Technician B says that all four-wheel drive vehicles have automatic transmissions. Who is right?
    - (A) A only.
    - (B) B only.
    - (C) Both A and B.
    - (D) Neither A nor B.

1. _____

2. A failed control computer may cause all of the following   2. _____
transfer case problems, *except:*
  (A) constantly illuminated MIL.
  (B) MIL not on at any time.
  (C) incorrect shifts.
  (D) no forward movement.

3. Which of the following wiring problems is *least* likely to   3. _____
cause an incorrect sensor input?
  (A) Corroded sensor output wire.
  (B) Sensor input wire touching frame.
  (C) Corroded ground wire connector.
  (D) Ground wire touching frame.

4. Technician A says that drive shafts should be marked   4. _____
before removal. Technician B says that pilot bolts can be
used to facilitate transfer case removal. Who is right?
  (A) A only.
  (B) B only.
  (C) Both A and B.
  (D) Neither A nor B.

5. All of the following statements about transfer case service   5. _____
are true, *except:*
  (A) transfer cases may have mechanical and elec-
      tronic parts.
  (B) a front axle problem may be caused by a vacuum
      leak.
  (C) a leaking viscous coupling should be leak tested
      and refilled.
  (D) sometimes a scan tool is needed to diagnose a
      transfer case.

Name _____

Date _____ Period _____

Instructor _____

Score _____

Text pages 723–765

34

Brake Service

Objectives: After studying Chapter 34 in the textbook and completing this section of the workbook, you will be able to:

- Explain drum brake construction, operation, and service.
- Summarize disc brake construction, operation, and service.
- Explain the operation and service of power brakes.
- Explain the operation and service of anti-lock brake system.
- Describe master cylinder operation, construction, and service.
- Diagnose brake hydraulic system problems.
- Diagnose brake friction system problems.
- Diagnose power brake system problems.

Tech Talk: The principles of automotive brake system operation have not changed much over the years. The same basic parts are used to produce friction for slowing and stopping the car. Recent trends to increase fuel economy, however, have made auto manufacturers use lighter materials. Aluminum and plastic parts are now being used in the brake system. Keep this in mind during service. These parts may be damaged more easily than conventional cast iron components.

Review Questions

Instructions: Study Chapter 34 of the text and then answer the following questions in the spaces provided.

Brake Inspection

1. Brake friction materials may contain _____, which is a known carcinogen.

1. _____

2. During system inspection, the pedal should be firm, with no _____ feel.

2. _____

3. A periodic wheel cylinder and brake shoe inspection generally involves pulling a(n) _____ and _____.

3. _____

_____

4. Caliper pistons should show no signs of fluid _____.

4. _____

5. Identify the brake system inspection points in the following illustration.

(A) _____

(B) _____

(C) _____

(D) _____

(E) _____

(F) _____

(G) _____

(H) _____

(I) _____

(J) _____

(K) _____

(L) _____

(M) _____

(N) _____

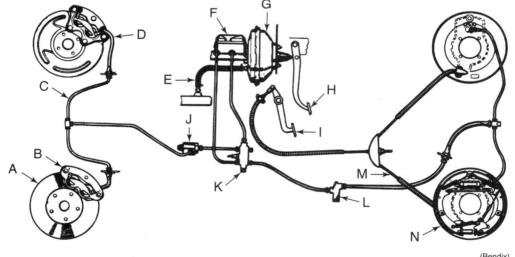

(Bendix)

Brake Hydraulic System Operation

6. The brake hydraulic system uses a master cylinder to develop _____.

6. _____

7. Never mix regular and _____ brake fluids.

7. _____

8. When the brake pedal is released, the brake shoe _____ force fluid to flow backward into the master cylinder reservoir.

8. _____

9. For years, the hydraulic brake system had one serious flaw; a system leak could cause _____.

9. _____

10. The double piston (also called the dual piston or split system) master cylinder was developed so that the _____ for the front and rear wheels could be completely separated.

10. _____

11. With the dual system, failure of either the front or rear system will be evident by a sudden increase in brake _____.

11. _____

12. A quick-take-up master cylinder is used with _____ disc brake calipers.

12. _____

13. Brake fluid level should be around _____″ (_____mm) from the top of the reservoir.

13. _____

_____

14. If the wheel cylinder has a special glassy, rolled finish, it should not be _____.

14. _____

15. If an aluminum master cylinder has any scratches or other imperfections, it should be _____.

15. _____

16. The cylinder, cups, and pistons must be coated with _____ before assembly.

16. _____

17. Define *brake pedal free travel.* _____

_____

_____

18. In the following illustration, the distance that the master cylinder push rod protrudes from the vacuum booster is being checked. Label the parts indicated.

(A) _____

(B) _____

(C) _____

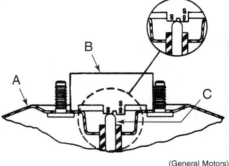

(General Motors)

19. Identify the tools and parts in the following illustrations.

    (A) _____

    (B) _____

    (C) _____

    (D) _____

    (E) _____

(Neihoff)

20. Disc brakes, as the name implies, use a heavy _____ instead of a conventional brake drum.

20. _____

21. Disc brakes are highly resistant to _____.

21. _____

22. The floating caliper often uses only one _____.

22. _____

23. Caliper bore size must not be increased by more than _____″ (_____mm).

23. _____

    _____

24. A caliper cylinder bore in good condition can often be cleaned by _____.

24. _____

25. A proportioning valve is used in the brake system with _____ brakes in the front and _____ brakes in the rear.

25. _____

    _____

26. A proportioning valve controls (and finally limits) _____ to the rear wheels. This reduces the possibility of rear-wheel _____ during heavy braking.

26. _____

    _____

27. A height-sensing proportioning valve compensates for changes in the vehicle's _____, which affect the vehicle's height.

27. _____

28. Explain the function of the metering valve. _____

_____

_____

_____

29. Explain the purpose of the pressure differential switch. _____

_____

_____

30. Label the parts of the pressure differential warning switch in the following illustration.

(A) _____

(B) _____

(C) _____

(D) _____

(E) _____

(General Motors)

31. Define *combination valve.* _____

_____

_____

32. Label the parts of the combination valve in the following illustration.

(A) _____

(B) _____

(C) _____

(D) _____

(E) _____

(Bendix)

33. Define *brake bleeding.* _____

_____

_____

Brake Friction Member Service

34. The _____ must be disengaged to remove the rear drums.

34. _____

35. To loosen brake shoes in a self-adjusting system, hold the adjuster lever free while loosening the _____.

35. _____

36. Before attempting to remove the shoes, study the _____.   This will help you during reassembly.

36. _____

37. When changing disc brake pads, use a(n) _____ to force pistons back into the bore.

37. _____

38. Make sure the pads are _____ with the rotor surface.

38. _____

39. After installing new brake linings or pads, it is important that they be given the proper _____.

39. _____

Drum and Disc Service

40. Drum scoring, bell-mouth and barrel wear may be removed by _____.

40. _____

41. Any drum measuring more than 0.010″ (0.254 mm) out-of-round or more than 0.005″ (0.127 mm) taper should be _____ by turning or grinding.

41. _____

42. A turned brake drum finish can be slightly more resistant to _____ and _____ than a ground surface.

42. _____

_____

43. New brake drums are usually given a protective coating to guard against rust. Remove the coating and clean the braking surface with _____.

43. _____

44. As long as the disc is smooth, rotor scoring up to _____″ (_____ mm) deep is permissible.

44. _____

_____

45. Before checking a disc for lateral runout, set the _____ clearance to just remove any endplay.

45. _____

46. Maximum rotor runout should not exceed _____″ (_____ mm).

46. _____

_____

47. When wear has reduced disc thickness beyond recommended limits, the disc should be _____.

47. _____

Parking Brake Adjustments

48. To adjust a rear wheel–type parking brake, apply the parking brake about three notches (about 1 3/4″ or 44.5 mm) travel. Adjust the equalizer _____ until a slight drag is noticeable at the rear wheels.

48. _____

Power Brake Service

49. What is the purpose of a power booster? _____

_____

_____

50. If a vacuum booster has been exhausted (all vacuum used up) and the brake pedal is depressed, what should happen when the engine is started?

_____

_____

ASE-Type Questions

1. A thorough brake inspection must include checks for all of the following, *except:*
    - (A) leaking brake lines.
    - (B) loose wheel bearings.
    - (C) worn steering parts.
    - (D) incorrect wheel balance.

    1. _____

2. Once removed from the vehicle, brake drums should be checked for all of the following, *except:*
    - (A) out-of-round.
    - (B) taper.
    - (C) scoring.
    - (D) carbon buildup.

    2. _____

3. Disc brake inspection is being discussed. Technician A says that disc brake pads should be checked for taper. Technician B says that disc brake rotors should be checked for scoring and parallelism. Who is right?
    - (A) A only.
    - (B) B only.
    - (C) Both A and B.
    - (D) Neither A nor B.

    3. _____

4. Which of the following can be used to wash wheel cylinder parts?
    - (A) Kerosene.
    - (B) Diesel fuel.
    - (C) Denatured alcohol.
    - (D) Transmission fluid.

    4. _____

5. Technician A says that rebuilt wheel cylinders generally last longer than new cylinders. Technician B says that it is often more expensive to rebuild a wheel cylinder than to replace it. Who is right?
    - (A) A only.
    - (B) B only.
    - (C) Both A and B.
    - (D) Neither A nor B.

    5. _____

6. Technician A says that compression fittings may be used to splice brake lines. Technician B says that double-lap flares and I.S.O.-type flares can be used interchangeably. Who is correct?
   - (A) A only.
   - (B) B only.
   - (C) Both A and B.
   - (D) Neither A nor B.

6. _____

7. Technician A says that double-wrapped steel tubing should be used for brake lines. Technician B says that copper tubing can be used for brake lines. Who is right?
   - (A) A only.
   - (B) B only.
   - (C) Both A and B.
   - (D) Neither A nor B.

7. _____

8. All of the following statements about brake bleeding are true, *except:*
   - (A) pressure bleeding is slower than manual bleeding.
   - (B) always start bleeding at the wheel farthest from the master cylinder.
   - (C) bench bleeding the master cylinder saves time.
   - (D) the metering valve must be blocked open when bleeding disc brakes.

8. _____

9. Technician A says that brakes with self-adjusting shoes need only an initial adjustment following installation of new shoes. Technician B says that a special gauge is used to adjust the self-adjusting shoes to the brake drum diameter. Who is right?
   - (A) A only.
   - (B) B only.
   - (C) Both A and B.
   - (D) Neither A nor B.

9. _____

10. A rear brake drum is deeply scored and is 0.010″ (0.254 mm) smaller than its discard diameter. Technician A says that the drum should be discarded. Technician B says that the drum should be turned no more than 0.005″ (0.128 mm). Who is right?
    - (A) A only.
    - (B) B only.
    - (C) Both A and B.
    - (D) Neither A nor B.

10. _____

11. On-car brake lathes are used for which of the following
reasons?
    (A) Ensuring that the drum is turned to a 50
        microinch finish.
    (B) Ensuring that the rotor is turned without thick-
        ness variations.
    (C) Ensuring that the rotor is turned without warping.
    (D) To save time.

11. _____

12. Technician A says that one type of rear wheel parking-
brake assembly contains a small drum brake inside of the
main brake rotor. Technician B says that a screw-actuated
rear disc parking brake usually requires only cable adjustment. Who is right?
    (A) A only
    (B) B only
    (C) Both A and B.
    (D) Neither A nor B.

12. _____

13. All of the following statements about vacuum power brake
boosters are true, *except:*
    (A) loss of vacuum in the booster when the engine is
        off indicates an internal leak.
    (B) vacuum at the booster vacuum inlet should
        match engine vacuum.
    (C) if engine vacuum is high, the booster may not
        produce enough boost.
    (D) some boosters have an inlet air filter.

13. _____

14. On most Hydroboost systems, the _____ provides
hydraulic pressure to assist the master cylinder.
    (A) power steering pump
    (B) oil pump
    (C) automatic transmission
    (D) wheel cylinder

14. _____

15. When the vehicle brake pedal is depressed, it gradually
goes to the floor. Which of the following is the *least* likely
cause of this problem?
    (A) Leaking internal master cylinder seals.
    (B) Leaking wheel cylinder.
    (C) Leaking power brake booster.
    (D) Leaking metering valve

15. _____

Name _____

Date _____ Period _____

Instructor _____

Score _____

Text pages 767–784

35

Anti-Lock Brake and Traction Control System Service

Objectives: After studying Chapter 35 in the textbook and completing this section of the workbook, you will be able to:

- Explain anti-lock brake system operation.
- Identify anti-lock brake system components.
- Diagnose anti-lock brake system problems.
- Explain traction control system operation.
- Identify traction control system components.
- Diagnose traction control system problems.

Tech Talk: Anti-lock brakes are becoming increasingly common on new vehicles. Anti-lock brakes reduce braking distances, decrease the chance of skidding on wet and icy surfaces, make the vehicle more controllable, and reduce tire flat-spotting during hard braking. The basic principles of anti-lock brake system operation are essentially the same for all systems, but individual systems and components vary widely. The technician must carefully study this chapter to obtain the information needed to service anti-lock brake systems.

Review Questions

Instructions: Study Chapter 35 of the text and then answer the following questions in the spaces provided.

Anti-Lock Brake Systems

1. Identify the ABS system components shown in the illustration on the following page.

(A) _____

(B) _____

(C) _____

(D) _____

(E) _____

(F) _____

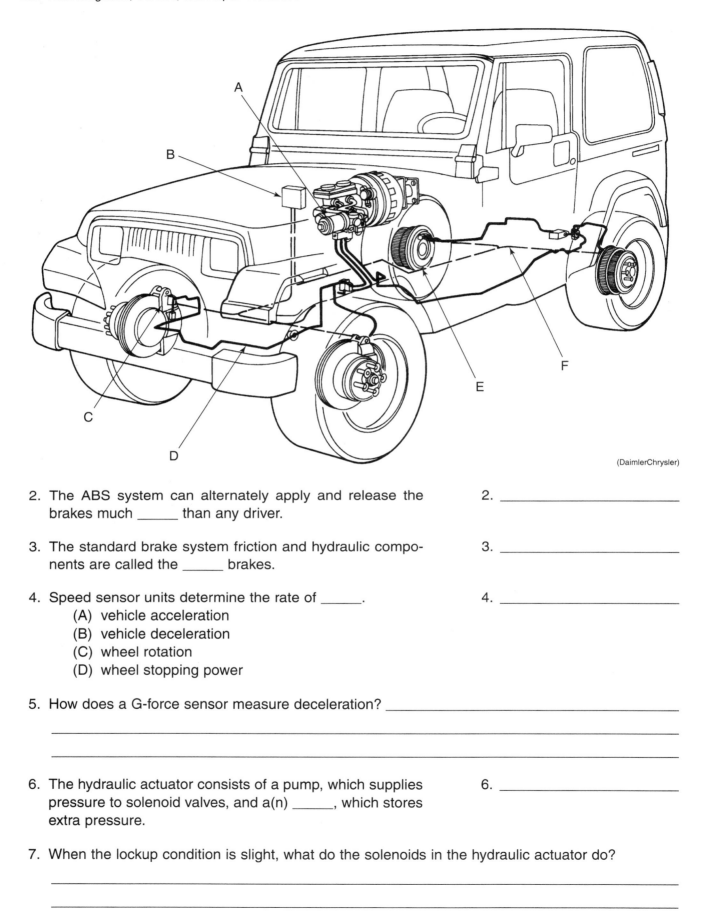

(DaimlerChrysler)

2. The ABS system can alternately apply and release the brakes much _____ than any driver.

2. _____

3. The standard brake system friction and hydraulic components are called the _____ brakes.

3. _____

4. Speed sensor units determine the rate of _____.
   - (A) vehicle acceleration
   - (B) vehicle deceleration
   - (C) wheel rotation
   - (D) wheel stopping power

4. _____

5. How does a G-force sensor measure deceleration? _____

_____

_____

6. The hydraulic actuator consists of a pump, which supplies pressure to solenoid valves, and a(n) _____, which stores extra pressure.

6. _____

7. When the lockup condition is slight, what do the solenoids in the hydraulic actuator do?

_____

_____

8. When the lockup condition is severe, what do the solenoids in the hydraulic actuator do?

_____

_____

9. Pressure in the piston-operated hydraulic actuator is produced by pistons operated by small electric motors
   through _____ gears.

9. _____

Anti-Lock Brake System Maintenance

10. Maintenance of the ABS system consists of checking the
    wheel speed sensors and rotors for _____ and _____.

10. _____

_____

11. How do procedures for checking and adding fluid to
    an ABS master cylinder differ from those for non-ABS
    systems?
    (A) The master cylinder reservoir is under pressure.
    (B) The reservoir cap is held by a series of set
        screws.
    (C) The reservoir does not contain brake fluid.
    (D) There is no difference.

11. _____

Troubleshooting Anti-Lock Brake Systems

12. A scan tool can be used to retrieve ABS _____.

12. _____

13. Wheel speed sensors can be checked with what kind of
    meter?
    (A) Ohmmeter.
    (B) Ammeter.
    (C) Voltmeter.
    (D) Oscilloscope.

13. _____

14. Identify the following ABS system tools and components.

(A) _____

(B) _____

(C) _____

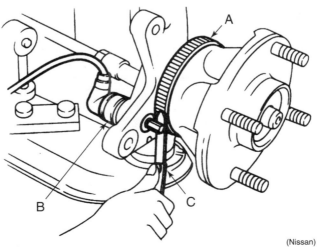

(Nissan)

15. The previous illustration shows a speed sensor _____ being checked.

15. _____

16. Some scan tools can operate the control module by inputting _____ signals.

16. _____

Replacing ABS Parts

17. Most of the ABS hydraulic system and _____ are similar to those on units without ABS.

17. _____

18. Before repairing any part of the hydraulic system, the system should be _____.

18. _____

19. Pumping the brake pedal at least _____ times will discharge the accumulator.

19. _____

20. A standard (non-ABS) brake hose may _____ if it is used on an ABS system.

20. _____

21. If a G-force sensor is not mounted in exactly the same position as the old sensor, what will happen?

_____

_____

_____

22. After an ABS hydraulic system has been repaired, the light may remain on for a few minutes, until the system has _____.

22. _____

Traction Control Systems

23. When it is operating, the traction control will apply the _____ on a wheel that is spinning at a faster rate than the others.

23. _____

24. A traction control light is used to tell the operator when the traction control system is _____.

24. _____

25. List two brake service procedures that are unaffected by the presence of ABS or traction controls.

_____

_____

ASE-Type Questions

1. Wheel speed sensors can be mounted in all of the following places, *except:*
    (A) brake rotor.
    (B) rear axle.
    (C) transmission input shaft.
    (D) transmission output shaft.

1. _____

2. The pedal travel switch is used to alert the control module
   that pedal _____ is becoming excessive.
   - (A) travel
   - (B) pulsation
   - (C) pressure
   - (D) drop rate

2. _____

3. No brake pedal pulsation is felt during hard braking on a
   vehicle with an ABS system. Technician A says to check
   the brake fluid level first. Technician B says that pulsation
   will only be felt if the vehicle is moving at a speed lower
   than 6 mph (9.7 km/h). Who is right?
   - (A) A only.
   - (B) B only.
   - (C) Both A and B.
   - (D) Neither A nor B.

3. _____

4. Technician A says that the ABS electrical system operates
   directly from the battery, with no fuse protection.
   Technician B says that problems with the charging system
   can affect ABS operation. Who is right?
   - (A) A only.
   - (B) B only.
   - (C) Both A and B.
   - (D) Neither A nor B.

4. _____

5. All of the following statements about ABS troubleshooting
   are true, *except:*
   - (A) pulsation of the brake pedal under hard braking
     is not a sign of an ABS problem.
   - (B) low brake fluid levels in the master cylinder can
     turn on the ABS warning light.
   - (C) worn foundation brakes can cause ABS pressure
     problems.
   - (D) mismatched tires will not affect ABS operation.

5. _____

6. Erratic speed sensor readings can be caused by all of the
   following, *except:*
   - (A) a sensor that runs too close to a strong magnetic
     field.
   - (B) a warped drum or rotor.
   - (C) low tire pressure.
   - (D) loose wheel bearings.

6. _____

7. All of the following are ways to check an ABS system
   hydraulic actuator, *except:*
       (A) check for resistance with an ohmmeter.
       (B) check for voltage drop with a waveform meter.
       (C) check pressures with a pressure gauge.
       (D) disassemble the unit and look for debris and
           wear.

7. _____

8. All of the following are methods of installing a new ABS
   rotor to the brake disc or axle, *except:*
       (A) the rotor is bolted in place.
       (B) the rotor is part of the axle or disc brake rotor
           and the whole assembly is replaced.
       (C) the rotor is hammered into place, and then the
           overlapping metal is peened.
       (D) the rotor is pressed onto the axle or disc brake
           rotor.

8. _____

9. Technician A says that traction control systems operate by
   increasing engine power. Technician B says that traction
   control systems share some parts with the ABS system.
   Who is right?
       (A) A only.
       (B) B only.
       (C) Both A and B.
       (D) Neither A nor B.

9. _____

10. The traction control system may adjust vehicle perfor-
    mance in all of the following ways, *except:*
        (A) retarding ignition timing.
        (B) reducing engine speed.
        (C) applying brakes.
        (D) stiffening suspension.

10. _____

Name _____

Date _____ Period _____

Instructor _____

Score _____

Text pages 785–810

36

Suspension System Service

Objectives: After studying Chapter 36 in the textbook and completing this section of the workbook, you will be able to:

- Explain the construction, operation, and service of conventional front suspensions.
- Explain the construction, operation, and service of conventional rear suspensions.
- Describe the function of coil springs, torsion bars, and leaf springs.
- Describe the function of load-carrying and following ball joints.
- Explain the construction, operation, and service of MacPherson strut suspensions.
- Describe the function of control arms, strut rods, and stabilizer bars.
- Describe the function of shock absorbers and MacPherson strut dampers.
- Summarize the operating principles and service of front and rear suspensions.
- Diagnose problems in suspension systems.

Tech Talk: There have been many suspension system design advances in the last 20 years. Most late-model cars use MacPherson strut front suspension systems. However, many cars and most pickup trucks and SUVs have conventional suspension systems, with upper and lower control arms and either coil springs or torsion bars. It is important that you understand modern suspension technology. Study this chapter thoroughly!

Review Questions

Instructions: Study Chapter 36 of the text and then answer the following questions in the spaces provided.

Front Suspension Systems

1. Most modern front suspension systems use coil springs. A few vehicles use _____ bars.

1. _____

2. Identify the parts of the suspension system in the following illustration.

(A) _____

(B) _____

(C) _____

(D) _____

(E) _____

(F) _____

(G) _____

A
B
C
D
E
F
G

(Ford)

3. To remove the coil spring, a spring _____ may be necessary.

3. _____

4. Identify the parts of the suspension system in the following illustration.

(A)_____   (F) _____

(B)_____   (G) _____

(C) _____   (H) _____

(D) _____   (I) _____

(E)_____

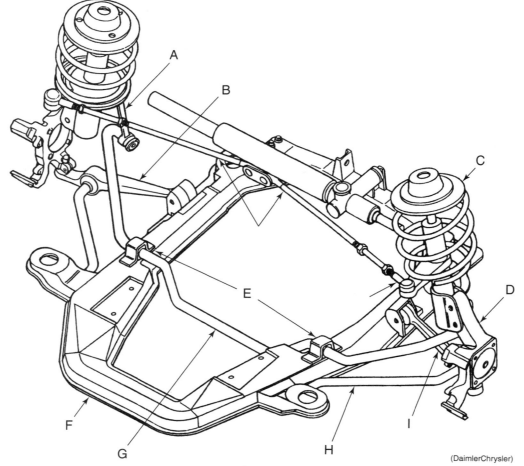

(DaimlerChrysler)

5. Shocks in good condition will allow about _____ free bounce(s).

5. _____

6. Inspect each shock absorber for signs of _____.

6. _____

7. List three conditions that may result from excessive ball joint wear.

_____

_____

8. The most rapid wearing ball joint is the _____ joint.

8. _____

9. Define *axial movement.* _____

_____

_____

10. The following figure shows the _____ ball joint being checked for wear.

10. _____

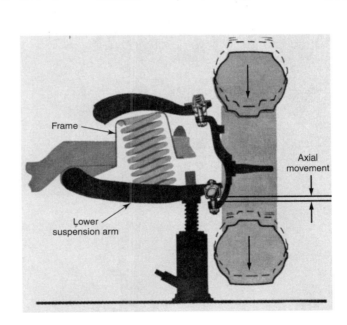

For questions 11–17, place the following ball joint removal steps in the proper order. (*Note:* not all steps are given.)

_____ 11.
_____ 12.
_____ 13.
_____ 14.
_____ 15.
_____ 16.
_____ 17.

(A) Loosen the stud nut several turns.
(B) Remove the cotter key from the stud nut.
(C) Place a jack under the lower control arm.
(D) Strike the steering spindle sharply with a hammer to break the taper.
(E) Remove the stud nut and lower the control arm.
(F) Remove rivets or bolts holding the ball joint to the control arm.
(G) Use a special tool to apply pressure to the ball joint stud.

18. The _____ must be removed to replace the control arm bushings.

18. _____

19. When one control arm bushing is worn, always replace _____.

19. _____

20. For removal, the control arm bushings are usually _____ out.

20. _____

21. Strut rods may be threaded for _____ adjustment.

21. _____

Rear Suspension Systems

22. When coil springs are used, _____ must be used to provide proper rear axle housing alignment.

22. _____

23. Some independent rear suspensions make use of a single leaf spring mounted in a(n) _____ position.

23. _____

24. When working with leaf springs, always allow the _____ of the vehicle to rest on bushings before torquing the shackle bolts.

24. _____

25. Identify the parts of the rear suspension system in the following illustration.

(A) _____

(B) _____

(C) _____

(D) _____

(Toyota)

Automatic Level Control

26. Most automatic level control systems are operated by _____ pressure.

26. _____

27. Identify the automatic level control system parts in the following illustration.

(A) _____

(B) _____

(C) _____

(D) _____

(E) _____

(F) _____

(G) _____

(H) _____

(I) _____

(J) _____

(K) _____

(L) _____

(M) _____

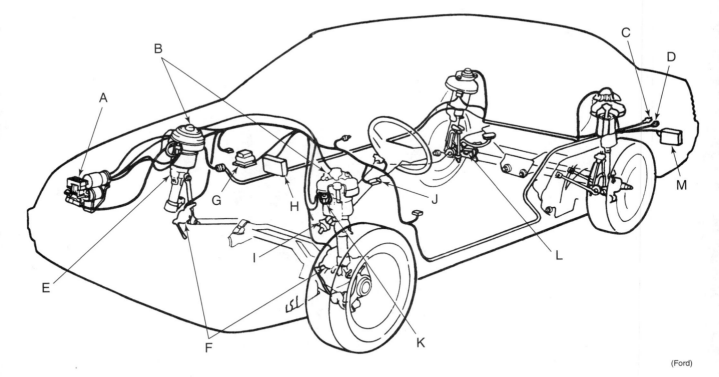

(Ford)

Computerized Ride Control Systems

28. The ride control system may override the driver settings during hard _____.

28. _____

29. The ride control module is generally located in the vehicle's _____.

29. _____

30. Actuators control the flow of hydraulic fluid in the adjustable _____.

30. _____

ASE-Type Questions

1. Technician A says that coil springs must be compressed with a special tool to remove them. Technician B says that torsion bar tension must be removed by removing the lower ball joint at the spindle. Who is right?
   - (A) A only.
   - (B) B only.
   - (C) Both A and B.
   - (D) Neither A nor B.

1. _____

2. All of the following statements about the following illustration are true, *except:*
   - (A) this tool is used to disassemble MacPherson strut assemblies.
   - (B) the strut cartridge cannot be replaced without using this tool.
   - (C) when the spring is compressed in this way, it must be replaced.
   - (D) all fasteners must be in place before the spring is released.

2. _____

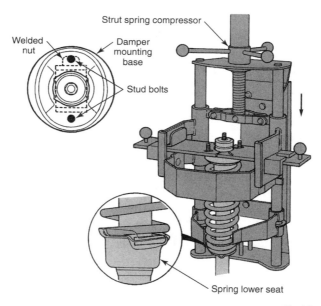

(Honda)

3. A rattling noise is heard when the one corner of the vehicle is pushed up and down. Which shock absorber defect is the *most* likely cause of this noise?
   - (A) External leaks.
   - (B) Internal leaks.
   - (C) Incorrect damper rate.
   - (D) Missing rubber bushing.

   3. _____

4. Technician A says that the load-carrying ball joints carry most of the vehicle weight. Technician B says that load-carrying ball joints are always tension loaded. Who is right?
   - (A) A only.
   - (B) B only.
   - (C) Both A and B.
   - (D) Neither A nor B.

   4. _____

5. To check ball joints for wear, they must be properly _____.
   - (A) loaded.
   - (B) unloaded.
   - (C) lubricated.
   - (D) compressed.

   5. _____

6. Control arm bushing replacement is being discussed. Technician A says that some control arms contain only one bushing. Technician B says that, if the control arm has two bushings, only the defective bushing should be replaced. Who is right?
   - (A) A only.
   - (B) B only.
   - (C) Both A and B.
   - (D) Neither A nor B.

   6. _____

7. Modern vehicles use all of the following rear suspension systems, *except:*
   - (A) coil spring.
   - (B) leaf spring.
   - (C) torsion bar.
   - (D) MacPherson strut.

   7. _____

8. The most common leaf spring failure is defective _____.
   - (A) springs
   - (B) body mountings
   - (C) attaching bolts
   - (D) bushings

   8. _____

9. Ride control input sensors include all of the following,
   except:
   - (A) steering.
   - (B) transmission gear.
   - (C) brake.
   - (D) acceleration.

9. _____

10. Like other computer control systems, the ride control
    system can perform all of the following, *except:*
    - (A) process inputs
    - (B) store trouble codes
    - (C) illuminate a dashboard warning light
    - (D) recalibrate input sensors.

10. _____

Name _____

Date _____ Period _____

Instructor _____

Score _____

Text pages 811–840

37

Steering System Service

Objectives: After studying Chapter 37 in the textbook and completing this section of the workbook, you will be able to:

- Explain the differences between conventional and rack-and-pinion steering systems.
- Identify the components of conventional steering systems.
- Identify the components of rack-and-pinion steering systems.
- Diagnose problems in conventional and rack-and-pinion steering systems.
- Summarize the construction, operation, and service of steering linkage components.

Tech Talk: The introduction of the rack-and-pinion steering system was among the design advances in steering systems in the last 20 years. Although rack-and-pinion steering systems are used in most late-model vehicles, many large cars and light trucks continue to use the conventional parallelogram linkage steering system. The technician must, therefore, be familiar with the service procedures for both types of systems. To gain this knowledge, study this chapter thoroughly and apply the information to your work.

Review Questions

Instructions: Study Chapter 37 of the text and then answer the following questions in the spaces provided.

Steering Column and Steering Wheel

1. Worn or dry steering shaft bearings can cause _____.
   - (A) roughness
   - (B) binding
   - (C) squeaking
   - (D) All of the above.

1. _____

2. What service task is being performed in this illustration?
   2. _____
   (A) Removing a steering wheel.
   (B) Installing a steering wheel.
   (C) Adjusting steering wheel position.
   (D) Adjusting steering shaft bearing preload.

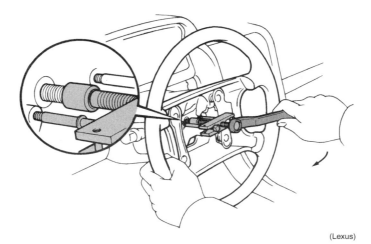

(Lexus)

3. The manual steering gear must turn the _____ of the steering wheel into side-to-side motion at the steering linkage.
   3. _____

4. The gearbox is attached to the frame and is usually connected to the steering shaft by a shock-absorbing _____ joint.
   4. _____

5. Name the two adjustments found on most steering gears. _____
   _____
   _____

6. In the following illustration, the technician is adjusting _____.
   6. _____

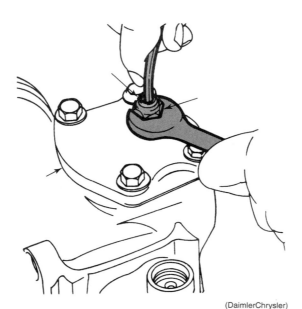

(DaimlerChrysler)

Conventional Power Steering Service

7. Define *integral power cylinder.* _____

   _____

   _____

8. The basic adjustments for inline power steering systems   8. _____
   include _____.
   - (A) worm-to-rack-piston preload
   - (B) pitman shaft depth
   - (C) thrust bearing alignment
   - (D) All of the above.

9. With respect to handling, the most critical inline power   9. _____
   steering adjustment is the _____.

10. When reinstalling a steering gear, check for proper _____   10. _____
    with steering shaft. This will prevent binding and prema-
    ture wear.

Rack-and-Pinion Steering Gears

11. On a rack-and-pinion steering system, the sector shaft, or   11. _____
    pinion, is connected directly to the _____.

12. Ideally, a rack-and-pinion gear should be _____ if it is   12. _____
    defective.

13. In the following illustration, the technician is taking steps   13. _____
    necessary to remove the _____ from the rack assembly.

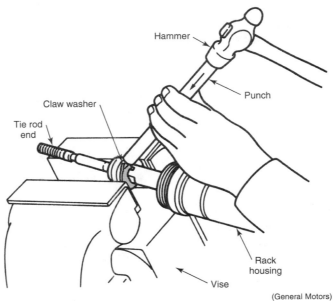

(General Motors)

14. The power rack-and-pinion uses a(n) _____ control valve   14. _____
    to regulate hydraulic flow.

Power Steering Pumps and Hoses

15. The three main types of power steering pumps are the vane, slipper, and _____.

15. _____

16. Never pry on the pump reservoir or _____ when adjusting belt tension.

16. _____

17. Modern power steering systems take special fluid, not _____ fluid, as much older systems did.

17. _____

18. If power steering pressure specifications call for checking the pressure at a certain temperature, a thermometer should be placed in the power steering _____.

18. _____

19. Which of the following parts have to be removed to properly bleed a power steering system?
    (A) High-pressure hose.
    (B) Return hose.
    (C) Reservoir.
    (D) None of the above.

19. _____

20. When removing a pump pulley, always use a(n) _____.

20. _____

21. Always install high-pressure hose on the pump's _____ circuit.

21. _____

22. A power steering pressure-sensing switch can be used to control the _____ or turn the air conditioner on or off.

22. _____

Steering Linkage

23. When inspecting steering linkage, check the ball sockets for looseness by _____ the center link and tie rods.

23. _____

24. In conventional systems, _____ are threaded into the inner and outer tie rod ends.

24. _____

25. What part is being removed in the following illustration?_____

_____

_____

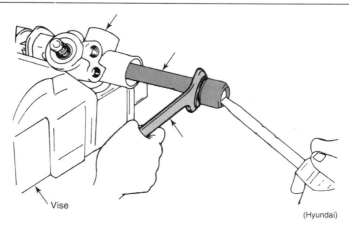

Vise

(Hyundai)

ASE-Type Questions

1. Technician A says that before servicing the steering wheel, the air bag (if used) must be disabled. Technician B says that most steering wheels can be removed by prying gently around the hub. Who is right?
   - (A) A only.
   - (B) B only.
   - (C) Both A and B.
   - (D) Neither A nor B.

1. _____

2. Which of the following does *not* have to be removed to overhaul a conventional steering gear?
   - (A) Pittman arm.
   - (B) Idler arm.
   - (C) Sector shaft.
   - (D) Worm shaft.

2. _____

3. Technician A says that the rack-and-pinion steering system has more parts than the conventional steering system. Technician B says that the rack-and-pinion system is a more direct linkage arrangement than the conventional system. Who is right?
   - (A) A only.
   - (B) B only.
   - (C) Both A and B.
   - (D) Neither A nor B.

3. _____

4. A loose or worn power steering pump belt will squeal when the wheels are _____.
   - (A) sharply turned
   - (B) slightly turned
   - (C) returned from a turn
   - (D) in the straight-ahead position

4. _____

5. When bleeding the power steering, which of the following steps should be done first? (Not all steps are listed.)
   - (A) Shut off the engine and allow the vehicle to idle for two minutes.
   - (B) Refill the reservoir if necessary.
   - (C) Start the engine and allow it to run for one minute.
   - (D) Road test the vehicle.

5. _____

6. All of the following statements are true, *except:*
   - (A) pitman arms are used on conventional steering systems only.
   - (B) idler arms are used on conventional and rack-and-pinion steering systems.
   - (C) tie rod ends are used on all steering systems.
   - (D) adjuster sleeves on conventional linkage are threaded into the inner and outer tie rod ends.

6. _____

7. When inspecting steering linkage, all of the following can be considered defects, *except:*
    - (A) loose fasteners.
    - (B) bent parts.
    - (C) leaking grease seals.
    - (D) undercoating on parts.

7. _____

8. Technician A says that a special tool or hammer can be used to break a taper joint. Technician B says that cotter pins can be reused if they are in good shape. Who is right?
    - (A) A only.
    - (B) B only.
    - (C) Both A and B.
    - (D) Neither A nor B.

8. _____

9. Technician A says that the following illustration shows a tie rod end being checked for looseness. Technician B says that the following illustration shows a tie rod end being replaced. Who is right?
    - (A) A only.
    - (B) B only.
    - (C) Both A and B.
    - (D) Neither A nor B.

9. _____

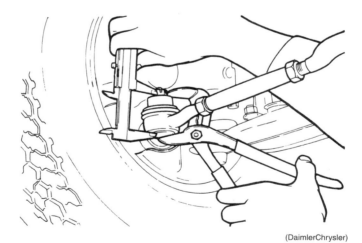

(DaimlerChrysler)

10. If an electronic assist steering system fails, which of the following will happen?
    - (A) Complete loss of steering control.
    - (B) Loss of assist in one direction only.
    - (C) Extremely loose steering when turning at low speeds.
    - (D) Steering will operate like a non-electronic system.

10. _____

Name _____

Date _____ Period _____

Instructor _____

Score _____

Text pages 841–862

38

Wheel and Tire Service

Objectives: After studying Chapter 38 in the textbook and completing this section of the workbook, you will be able to:

- Describe the construction, operation, and service of wheel bearings.
- Explain tire and wheel construction and service.
- Summarize tire size, type, and quality ratings.
- Diagnose common wheel bearing- and tire-related problems.

Tech Talk: Wheel and tire components are the final connection between the vehicle and the road. Bearings must be properly lubricated, installed and adjusted. Air pressure in the tires is crucial to proper handling and ride, and to prevent blowouts at cruising speeds. Tires must be rotated on schedule to reduce wear.

Review Questions

Instructions: Study Chapter 38 of the text and then answer the following questions in the spaces provided.

Wheel Bearings

1. The tapered roller bearing consists of the rollers and _____ and the inner and outer races.

1. _____

2. When removing a wheel _____, pry a little at a time, being careful to avoid bending it out of shape.

2. _____

3. Some wheel covers are held in place by _____.

3. _____

4. If the vehicle has front disc brakes, the _____ must be removed to remove the bearings.

4. _____

5. Each wheel bearing must be filled with the specified wheel bearing grease by using a(n) _____ or by packing by hand.

5. _____

6. If the slot in the nut does not line up with the hole in the spindle, the nut should be _____ just enough to make the holes line up.

6. _____

7. Describe the steps in the spindle nut adjustment procedure shown in the following illustration.

(A) _____

(B) _____

(C) _____

A   B   C

(Ford)

8. Identify the bearing-related parts shown in the following illustration.

(A) _____

(B) _____

(C) _____

(D) _____

(DaimlerChrysler)

9. Most sealed front bearings must be _____ out of the steering knuckle.

9. _____

10. Regardless of the technique used for bending, always use a(n) _____ cotter pin.

10. _____

11. If the steering knuckle is removed, the front end will need _____ after it is reinstalled.

11. _____

12. Most sealed bearings used on rear wheel, _____ axles are replaced as a unit.

12. _____

Vehicle Wheel Rims

13. Rim size is determined by what three measurements? _____

_____

_____

14. Why is it important to tighten the lug nuts to the proper torque, even if the rim is steel?

_____

_____

15. Describe the process for installing a wheel lug shown in the following illustration.

_____

_____

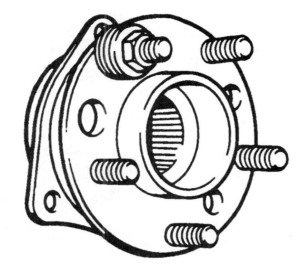

Vehicle Tires

16. List four cord materials. _____

_____

_____

Use the following tire rating information to answer questions 17–21.

P205/75HR15

17. The letter P indicates that the tire is designed for a(n) _____.

17. _____

18. The number 205 represents the tire's _____.

18. _____

19. The number 75 is the _____.

19. _____

20. The letter H is the tire's _____ rating.

20. _____

21. The number 15 is the _____ in inches.

21. _____

22. A tire with a DOT temperature resistance rating of B will resist heat generation better than one graded _____.

22. _____

23. A tire with a DOT traction grade of A has _____ traction than tires graded B or C.

23. _____

24. A tire with a DOT tread wear rating of 225 will give approximately _____ percent more mileage than a tire with a rating of 100.

24. _____

25. A space saver tire is for _____ use only.

25. _____

26. What is the purpose of a tire interchange chart?

____

____

27. When a vehicle is driven, the _____ of the tires increases. 27. ______________

28. Identify the tire inflation conditions in the following illustration.

(A) ______________

(B) ______________

(C) ______________

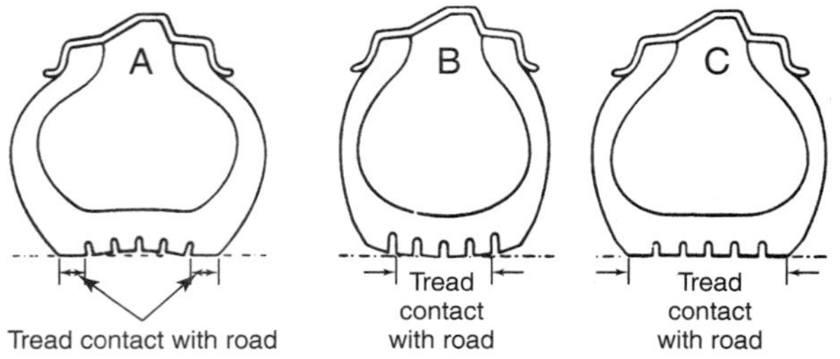

A
B
C
Tread contact with road
Tread contact with road
Tread contact with road

(Rubber Mfg. Association)

29. Tires should be rotated every _____ miles (_____km). 29. ______________

____

30. In the following illustration, draw the lines and arrows showing how the tires should be rotated.

Front
Front
Conventional tire rotation
Rotating 5 tires
Rotating 4 tires
Left
Right
Left
Right
Front
Front
Radial tire rotation
Rotating 5 tires
Rotating 4 tires
Left
Right
Left
Right

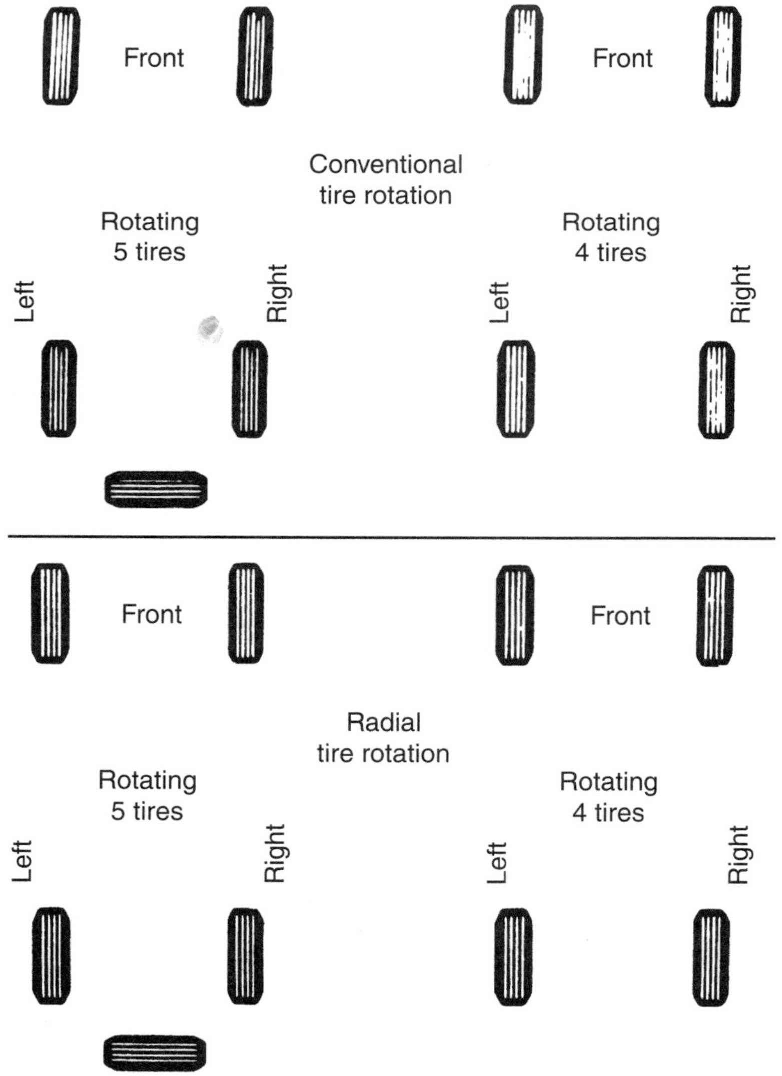

(DaimlerChrysler)

31. To prevent tire bead damage, always use rubber _____ on the bead surfaces.

31. _____

32. If a rim has a dirty sealing area, it should be _____. If a rim has any splits or cracks, it should be _____.

32. _____

_____

33. Repairing a tire without _____ it is no longer recommended.

33. _____

34. A tire with _____ can be repaired.
    (A) tread separation
    (B) an embedded nail
    (C) sidewall damage
    (D) a damaged bead wire

34. _____

35. To be in static balance, the wheel's mass (weight) must be evenly distributed around the _____.

35. _____

36. To be in dynamic balance, the centerline of the mass (weight) must be in the same _____ as the centerline of the wheel.

36. _____

37. A wheel and tire assembly with excessive _____ or _____ runout cannot be balanced.

37. _____

_____

38. Runout may be checked with a(n) _____.

38. _____

39. When balancing the rear wheels on the car, remember that when one wheel is on the floor, the other is free to _____.

39. _____

40. When using an on-car balancer on cars equipped with the limited-slip differential, raise _____ to prevent the car from running off the jack.

40. _____

ASE-Type Questions

1. All of the following statements about bearings are true, *except:*
    (A) tapered roller bearings can be cleaned and greased.
    (B) tapered roller bearings can be adjusted.
    (C) flat roller and ball bearings can be adjusted.
    (D) rear-wheel drive vehicles have tapered roller bearings on the front wheels.

1. _____

2. Technician A says that a bearing showing the slightest sign of wear should be replaced. Technician B says that bearing rollers and races should always be replaced together. Who is right?
   (A) A only.
   (B) B only.
   (C) Both A and B.
   (D) Neither A nor B.

2. _____

3. Improper wheel bearing adjustment *cannot* cause which of the following?
   (A) wheel shake
   (B) tie rod wear
   (C) poor brake performance
   (D) poor steering

3. _____

4. Technician A says that some sealed bearings must be pressed out of the steering knuckle. Technician B says a cotter pin may be reused if it is in good shape. Who is right?
   (A) A only.
   (B) B only.
   (C) Both A and B.
   (D) Neither A nor B.

4. _____

5. Wheel rims are made of all of the following, *except:*
   (A) steel.
   (B) aluminum.
   (C) cast iron.
   (D) graphite.

5. _____

6. Technician A says that wheel lug nuts should be tightened with an impact wrench and torque sticks. Technician B says that if necessary, wheel lug nuts can be tightened with a torque wrench. Who is right?
   (A) A only.
   (B) B only.
   (C) Both A and B.
   (D) Neither A nor B.

6. _____

7. Today, the most commonly used type of tire is the _____ type.
   (A) bias
   (B) radial
   (C) belted bias
   (D) double-belted bias

7. _____

8. Technician A says that an underinflated tire will wear in the center. Technician B says that an underinflated tire will generate excessive heat. Who is right?
   - (A) A only.
   - (B) B only.
   - (C) Both A and B.
   - (D) Neither A nor B.

8. _____

9. All of the following could cause feathered edges on front tires, *except:*
   - (A) excessive toe in.
   - (B) excessive toe out.
   - (C) hard cornering.
   - (D) a bent wheel rim.

9. _____

10. Technician A says that at cruising speeds, even a slight tire imbalance can cause the tire and wheel to hop up and down. Technician B says that at cruising speeds, a slight tire unbalance can cause the tire and wheel to shake from side to side. Who is right?
    - (A) A only.
    - (B) B only.
    - (C) Both A and B.
    - (D) Neither A nor B.

10. _____

Name _____

Date _____ Period _____

Instructor _____

Score _____

Text pages 863–884

39

Wheel Alignment

Objectives: After studying Chapter 39 in the textbook and completing this section of the workbook, you will be able to:

- Explain the importance of wheel alignment.
- Describe the purposes of major alignment settings.
- Identify adjustable and nonadjustable alignment settings.
- Summarize wheel alignment procedures.
- Explain the difference between two-wheel and four-wheel alignment.
- Diagnose common alignment-related problems.

Tech Talk: Alignment is critical to vehicle handling and tire wear. Modern vehicles use lighter components and are, therefore, easier to knock out of alignment. Many late-model vehicles have front-wheel drive. Rear-wheel alignment is also adjustable on these vehicles. These new alignment factors make it imperative that you study this chapter thoroughly.

Review Questions

Instructions: Study Chapter 39 of the text and then answer the following questions in the spaces provided.

Defining Wheel Alignment

1. Define *wheel alignment*. _____

_____

_____

2. Label the caster-related items shown in the following illustration.

(A) _____

(B) _____

(C) _____

(D) _____

(E) _____

(Perfect Equipment Corp.)

3. Negative camber occurs when the _____ of the wheel is tilted inward.

3. _____

4. Makers often recommend setting _____ more degrees of positive camber in the left wheel to compensate for road crown.

4. _____

5. Steering axis inclination is formed by tilting the _____ ball joint or strut mount inward.

5. _____

6. _____ is the relative position of the front and rear of a tire in relation to the tire on the other side of the vehicle.

6. _____

7. Rear-wheel drive vehicles are usually toed _____.

7. _____

8. When a car rounds a corner, the inner front wheel is forced to follow a smaller _____ than the outer wheel.

8. _____

9. Although toe on turns specifications vary, they usually call for a(n) _____° difference between the inner- and outer-wheel turning angles.

9. _____

For questions 10–13, match the alignment angle with its purpose.

_____ 10. Positive caster.

_____ 11. Positive camber, left wheel.

_____ 12. Toe.

_____ 13. Toe-out on turns.

(A) Reduces tire slip.
(B) Makes turning easier.
(C) Compensates for road crown.
(D) Assists the wheels in maintaining a straight-ahead position.
(E) Allows the tires to run parallel when driving straight ahead.

14. Wheel _____ is the ability of the rear wheels to follow directly in the path of the front wheels.

14. _____

15. Why was the two-wheel alignment the only type performed in the past?

_____

_____

16. What is the purpose of turning plates? _____

_____

_____

Performing Wheel Alignments

17. During prealignment checks, check the passenger compartment and trunk for excessive _____.

17. _____

18. If the engine has _____, start the engine before centering the steering wheel.

18. _____

19. Rear toe and camber are set by _____ or threaded rod.

19. _____

20. If the vehicle has four-wheel steering, the rear gearbox must be _____ with a special tool before camber and toe are set.

20. _____

21. Some rear end alignments can be adjusted with _____.

21. _____

22. Never attempt to make non-factory adjustments by _____ or welding parts.

22. _____

Wheel Alignment

23. The following illustration shows the alignment tool being attached to the wheel rim. Label the parts indicated.

   (A) _____

   (B) _____

   (C) _____

   (D) _____

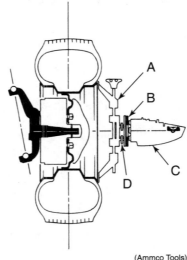

(Ammco Tools)

24. The following illustration shows a method of adjusting caster. Label the indicated parts.

(A) _____

(B) _____

(C) _____

(D) _____

(Hunter Engineering Co.)

25. The following illustration shows a method of adjusting the camber. Label the indicated parts.

(A) _____

(B) _____

(C) _____

(Hunter Engineering Co.)

ASE-Type Questions

1. Improper alignment can cause all of the following, *except:*
    (A) rapid tire wear.
    (B) wandering.
    (C) hard steering.
    (D) vibration.

1. _____

2. Improper caster angle can cause all of the following, *except:*
    (A) high-speed instability.
    (B) tire wear.
    (C) wandering.
    (D) hard steering.

2. _____

3. Camber angles, whether negative or positive, are usually quite small, averaging around _____.
   - (A) 1°
   - (B) 5°
   - (C) 10°
   - (D) 15°

3. _____

4. Technician A says that steering axis inclination (SAI) is adjustable on most vehicles. Technician B says that common causes of incorrect SAI include a bent spindle, frame, or strut. Who is right?
   - (A) A only.
   - (B) B only.
   - (C) Both A and B.
   - (D) Neither A nor B.

4. _____

5. Technician A says that before beginning the alignment, the vehicle should be checked for worn parts. Technician B says that a suspected alignment problem can often be traced to tires, brakes, or other unrelated parts. Who is right?
   - (A) A only.
   - (B) B only.
   - (C) Both A and B.
   - (D) Neither A nor B.

5. _____

6. Check all of the following before performing an alignment, *except:*
   - (A) tire condition.
   - (B) suspension parts.
   - (C) steering linkage.
   - (D) steering system hydraulic pressure.

6. _____

7. Technician A says that curb weight is obtained with a full tank of gas and the driver in the vehicle. Technician B says that toe should always be set first. Who is right?
   - (A) A only.
   - (B) B only.
   - (C) Both A and B.
   - (D) Neither A nor B.

7. _____

8. A four-wheel alignment has been performed on a front-wheel drive sedan. Which of the following was *not* set?
   - (A) Front camber.
   - (B) Front toe.
   - (C) Rear camber.
   - (D) Rear caster.

8. _____

9. All of the following statements about alignment are true, *except:*
   (A) a four-wheel alignment should always begin with the front wheels.
   (B) some imports are aligned with weight in the vehicle.
   (C) wheel rims always have some runout.
   (D) never attempt to align a vehicle with mismatched tires.

9. _____

10. Technician A says that adjustable alignment angles include caster, camber, and toe-out on turns. Technician B says that nonadjustable alignment angles include steering axis inclination and toe-in. Who is right?
    (A) A only.
    (B) B only.
    (C) Both A and B.
    (D) Neither A nor B.

10. _____

Name _____

Date _____ Period _____

Instructor _____

Score _____

Text pages 885–926

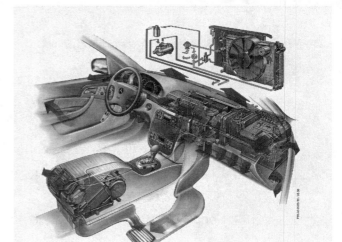

40

Air Conditioning and Heater Service

Objectives: After studying Chapter 40 in the textbook and completing this section of the workbook, you will be able to:

- Explain basic refrigeration theory.
- Identify major air conditioning system components.
- List the safety rules for air conditioning service.
- Inspect an air conditioning system for problems.
- Recover, evacuate, and recharge an air conditioning system.
- Service air conditioning and heater parts.
- Install air conditioning system parts.
- Service refrigerant oil.
- Diagnose air conditioning and heating problems.

Tech Talk: Automotive air conditioning is almost universal: it is installed in over 90% of new cars, trucks, and SUVs. An automotive technician will often be asked to diagnose and repair an air conditioning system. Understanding the material in this chapter is necessary to service air conditioners and avoid equipment damage, personal injury, and environmental damage.

Review Questions

Instructions: Study Chapter 40 of the text and then answer the following questions in the spaces provided.

Principles and Major Components of Air Conditioning

1. Basic air conditioning is a process in which air entering the vehicle is cooled, cleaned, and _____.

1. _____

For Questions 2–6, match the air conditioner refrigerant system component with its purpose.

_____ 2. Controls flow of refrigerant into the evaporator based on evaporator temperature.

_____ 3. Controls flow of refrigerant but cannot open or close.

_____ 4. Removes heat from the refrigerant.

_____ 5. Absorbs heat from the air.

_____ 6. Keeps liquid from entering the compressor.

(A) Accumulator.
(B) Expansion valve.
(C) Condenser.
(D) Receiver-drier.
(E) Orifice tube.
(F) Evaporator.

7. If evaporator temperature drops below 32°F (0°C), condensation on the evaporator core will _____.

7. _____

8. Define *Schrader valve.* _____

_____

_____

9. Temperature sensors are sometimes called _____.

9. _____

10. A(n) _____ sensor prevents compressor clutch engagement when the outside air temperature is too low.

10. _____

Air Conditioning System Service

11. Always wear _____ when working on an air conditioner.

11. _____

12. Refrigerant should never be discharged directly into the _____.

12. _____

13. If refrigerant vapor contacts an open flame, it will produce _____ gas.

13. _____

14. The center manifold connection is _____ to both valves.

14. _____

Leak Detection

15. The air conditioning system must be at least partially _____ before leak testing can be done.

15. _____

16. Modern electronic leak detectors are extremely sensitive and can detect a leak as small as _____ of refrigerant per year.

16. _____

17. A(n) _____ solution can locate only large leaks.

17. _____

Refrigerant Recovery/Recycling

18. Most recovery and recycling machines clean and remove _____ from the refrigerant to allow its immediate reuse.

18. _____

19. Never attempt to recycle _____ refrigerant.

19. _____

Evacuating the System

20. An air conditioning system cannot be drawn down to 29″ (737 mm) at sea level. What are some possible causes? _____

_____

_____

21. After shutting off the valve in the vacuum line and stopping   21. _____
the vacuum pump, the system should hold vacuum with no
more than a 2″ (51 mm) drop in a(n) _____-minute period.

Refrigerant Charging

22. List three items that must be checked before recharging the system.

_____

_____

23. Unless specifically recommended, never allow liquid   23. _____
refrigerant to enter the _____ side of the system.

Component Replacement

24. Always use two wrenches when loosening a(n) _____.   24. _____

25. When servicing lines, fittings, or components, always use   25. _____
new _____.

26. The compressor clutch can be removed without _____ the   26. _____
system.

27. _____ flushing is done with the flushing agent blown   27. _____
through the individual component.

28. Adapting an R-12 system to operate on R-134a is called   28. _____
_____. This may become necessary as the supply of R-12

_____.   _____

Heater System

29. If the heater shut-off valve sticks in the closed position,   29. _____
which of the following will happen?
    (A) The heater will not warm the vehicle interior in
        winter.
    (B) The heater will overheat the vehicle interior in
        summer.
    (C) The engine will overheat.
    (D) The engine will not reach operating temperature.

30. If the heater core is suspected of leaking, conduct a(n)   30. _____
_____ system pressure test.

31. If the blower motor is inoperative, check the _____ first.

31. _____

32. A defective _____ or burned _____ are common causes when the blower has some speeds but not others.

32. _____

_____

Air Distribution and Control System Service

33. The blend door controls the temperature of the incoming air by blending air from the _____ and the heater core.

33. _____

34. A(n) _____ air distribution system allows the driver and passenger to separately control airflow and temperature.

34. _____

35. What can happen to the intake ductwork at the front cowl?

_____

_____

ASE-Type Questions

1. If the restrictor at the evaporator is not variable, the _____ is turned on and off to control refrigerant flow.
    (A) compressor
    (B) blower
    (C) condenser cooling fan
    (D) engine cooling fan

1. _____

2. All of the following statements about refrigerants are true, *except:*
    (A) newer air conditioning systems use R-134a refrigerant.
    (B) older air conditioning systems use a refrigerant called R-12.
    (C) some refrigerant blends contain R-134a or R-22.
    (D) drop-in refrigerants can be used in any air conditioner.

2. _____

3. Technician A says the sight glass is used to check the refrigerant for bubbles or foam. Technician B says that the sight glass cannot be used to check refrigerant level on an R-134a system. Who is right?
    (A) A only.
    (B) B only.
    (C) Both A and B.
    (D) Neither A nor B.

3. _____

4. All of the following are steps in preventing refrigeration   4. _____
   system contamination, *except:*
   - (A) as soon as a line or part is disconnected, it
     should be capped.
   - (B) leave the system open overnight to allow mois-
     ture to evaporate out of the lines.
   - (C) refrigerant lines should be close to room temper-
     ature before they are disconnected.
   - (D) the receiver-drier should be connected into the
     system last.

5. Technician A says that compressor oil containers should   5. _____
   be capped. Technician B says that the system should be
   evacuated before recharging. Who is right?
   - (A) A only.
   - (B) B only.
   - (C) Both A and B.
   - (D) Neither A nor B.

6. Screw-on service fittings are found only on _____ systems.   6. _____
   - (A) R-12
   - (B) R-134a
   - (C) blended refrigerant
   - (D) SNAP-approved refrigerant

7. Technician A says that the pressure cycling switch is the   7. _____
   most common clutch control switch. Technician B says that
   the thermostatic temperature cycling switch is used on
   many aftermarket air conditioners. Who is right?
   - (A) A only.
   - (B) B only.
   - (C) Both A and B.
   - (D) Neither A nor B.

8. An air conditioning system appears to have a low refriger-   8. _____
   ant charge. Which of the following is the *least* likely cause?
   - (A) Different refrigerants mixed.
   - (B) Refrigerant leak.
   - (C) System improperly charged.
   - (D) Relief valve opened.

9. When can air conditioning refrigerant be legally dis-   9. _____
   charged into the atmosphere?
   - (A) When recovery equipment is not available.
   - (B) When recovering the refrigerant would contami-
     nate the recovery equipment.
   - (C) When the refrigerant is contaminated with mois-
     ture or dirt.
   - (D) At no time.

10. Technician A says that system evacuation is done to remove air and moisture. Technician B says that the evacuation process requires a compressor. Who is right?
    (A) A only.
    (B) B only.
    (C) Both A and B.
    (D) Neither A nor B.

10. _____

11. Which of the following is the *least* likely way that oil can be lost from a refrigeration system?
    (A) a leak in the system
    (B) evaporator icing
    (C) replacing a system part
    (D) discharging too fast

11. _____

12. All of the following statements about refrigerant oils are true, *except:*
    (A) refrigerant oils will absorb moisture.
    (B) oil can be injected into a charged system.
    (C) POE oils can be used in R-134a systems only.
    (D) mineral oils should be used with R-12 only.

12. _____

13. Technician A says that in most vehicles, the heater system relies on engine coolant to provide heat. Technician B says that the job of the heater shut-off valve is to control the flow of air through the heater core. Who is right?
    (A) A only.
    (B) B only.
    (C) Both A and B.
    (D) Neither A nor B.

13. _____

14. The air conditioner and heater can be operated by the vehicle occupants through all of the following, *except:*
    (A) vacuum switches.
    (B) electrical switches.
    (C) cables.
    (D) expansion valves.

14. _____

15. Which of the following problems will be caused by misaligned air conditioning system ductwork?
    (A) Improper air conditioning system pressures.
    (B) Improper airflow inside of the vehicle.
    (C) Evaporator icing.
    (D) Loss of refrigerant through the high-pressure relief valve.

15. _____

Name _____

Date _____ Period _____

Instructor _____

Score _____

Text pages 927–932

41

Repair Orders and Cost Estimates

Objectives: After studying Chapter 41 in the textbook and completing this section of the workbook, you will be able to:

- Explain the purpose of repair orders.
- Fill out a repair order.
- Explain the purpose of cost estimates.
- Complete a cost estimate.

Tech Talk: You may be able to specialize within the automobile repair business and avoid many repair operations that you are not interested in performing. However, one job you will have to do—no matter how disinterested you are—is the paperwork. Filling out forms is a part of modern life, and you must know how to fill out repair orders and cost estimates. This chapter will prepare you for these inevitable paperwork duties.

Review Questions

Instructions: Study Chapter 41 of the text and then answer the following questions in the spaces provided.

Repair Orders

1. Many repair orders are set up to record both _____ and _____ work.

1. _____

   _____

2. The customer will not be billed for parts and labor covered under _____.

2. _____

3. When the technician begins the repair, he or she will write or punch in the _____ time. When repairs are completed, he or she will write or punch in the _____ time.

3. _____

   _____

4. When parts are used, the part _____, description, and price must be listed on the repair order.

4. _____

For questions 5–10, calculate the labor bill for a 3.5 hour job at the hourly labor rate listed.

5. $21.00

5. _____

6. $25.00

6. _____

7. $29.50

7. _____

8. $35.00

8. _____

9. $47.50

9. _____

10. $55.00

10. _____

11. When a shop buys a part for $35.00 and the price to the customer is $42.00, the markup is _____ percent.

11. _____

12. A deposit on an old part is called the _____ charge.

12. _____

13. Give some examples of outside service costs. _____

_____

_____

14. Rags, solvents, and lubricants are examples of _____.

14. _____

15. A disposal fee may be charged for such things as used _____ or scrap _____.

15. _____

_____

16. Sales _____ are levied almost everywhere.

16. _____

For questions 17–21, calculate the final bill for the repair costs listed. For this exercise, the local sales tax is 5% on parts and labor.

17. $75.00

17. _____

18. $120.00

18. _____

19. $205.00

19. _____

20. $275.50

20. _____

21. $311.00

21. _____

Cost Estimates

22. Should the technician assume that the customer will want needed repairs done?

22. _____

23. When listing parts and outside services on an estimate, do not forget to add the normal _____.

23. _____

24. List two things that should be added to the basic parts and labor sections of the cost estimate?

_____

_____

25. If a technician discovers that additional parts or labor are required to complete a job for which an estimate was given, what should the technician do?_____

_____

_____

ASE-Type Questions

1. Which of the following would *not* be on a repair order?
   - (A) Vehicle identification number (VIN).
   - (B) Manufacturers of individual vehicle components.
   - (C) Vehicle owner and address.
   - (D) Vehicle body style.

1. _____

2. A shop is basing its labor charges on a current flat rate manual. According to the manual, a water pump replacement should take 2.5 hours. However, one of the pump bolts is rounded off, and removing the bolt increases the total time for the job to 3.0 hours. The labor rate is $30.00 per hour. What will the shop charge the customer for labor?
   - (A) $30.00
   - (B) $60.00
   - (C) $75.00
   - (D) $90.00

2. _____

3. Technician A says that if the shop forgets to charge sales tax to the customer, it is not liable for the tax. Technician B says that sales taxes do not apply to some exempt customers. Who is right?
   - (A) A only.
   - (B) B only.
   - (C) Both A and B.
   - (D) Neither A nor B.

3. _____

4. If a flat rate manual is not available, the shop can obtain parts prices by calling _____.
   - (A) local part outlets
   - (B) other repair shops
   - (C) vehicle manufacturers
   - (D) the information division of the local library

4. _____

5. A flat rate book is being used to calculate the cost of fuel pump replacement. The power steering pump must be removed to gain access to the pump fasteners. Technician A says that the labor for removing the powering steering pump must be added to the cost estimate. Technician B says that the labor for removing the power steering pump is already figured into the time for a fuel pump replacement. Who is right?

    (A) A only.
    (B) B only.
    (C) Both A and B.
    (D) Neither A nor B.

5. _____

Name _____

Date _____ Period _____

Instructor _____

Score _____

Text pages 933–941

42

ASE Certification

Objectives: After studying Chapter 42 in the textbook and completing this section of the workbook, you will be able to:

* Explain why technician certification is necessary.
* Explain the process of registering for ASE tests.
* Explain how to take the ASE tests.
* Identify typical ASE test questions.
* Explain what is done with ASE test results.

Tech Talk: ASE tests are a fact of life for the modern automotive technician. A technician without any ASE certifications will increasingly be looked on as an unskilled laborer. To get the best jobs and stay competitive, you must take and pass the ASE tests. Tests are given in several languages all over the United States, Canada, Mexico, and some other countries in the Western Hemisphere. This chapter will help prepare you for ASE certification.

Review Questions

Instructions: Study Chapter 42 of the text and then answer the following questions in the spaces provided.

Reasons for ASE Tests

1. ASE was established in 1975 to provide a(n) _____ process for automobile technicians.

   1. _____

2. Define *standardized test.* _____

   _____

   _____

Applying for the ASE Tests

3. ASE tests are given _____ each year.

   3. _____

4. It has recently become possible to register for ASE test using _____.

4. _____

5. In addition to English, ASE tests are available in _____ and _____.

5. _____

_____

Taking the ASE Tests

6. All ASE test questions are multiple-choice questions with _____ possible answers.

6. _____

7. What three areas of knowledge do the ASE tests measure? _____

_____

_____

8. A test question with statements from Technician A and Technician B is a(n) _____ part question.

8. _____

9. In most cases, checking your answers more than _____ is unnecessary.

9. _____

10. An incomplete sentence question asks the technician to select one of _____ choices to correctly complete a sentence.

10. _____

Test Results

11. All ASE test scores are _____ and are provided to the _____ only.

11. _____

_____

12. A technician's score report says that he or she made a score of 37 on a particular test. The score report says that the total number of test questions was 50 and 36 correct answers were needed to pass. Did the technician pass this test?
    (A) The technician passed the test.
    (B) The technician failed the test.
    (C) The technician failed the test, but qualified for a retest.
    (D) The technician failed the test and did not qualify for a retest.

12. _____

13. How many times can you take a certification test?
    (A) Once.
    (B) Twice.
    (C) Depends on test score.
    (D) As many times as needed.

13. _____

Other ASE Tests

14. Engine machinists can be certified in _____ skill areas.

14. _____

Recertification Tests

15. Recertification tests emphasize _____ in automotive service.

15. _____

ASE-Type Questions

1. Which of the following may *not* be substituted for all or part of the 2 years work experience required to become certified?
    - (A) Training courses.
    - (B) Apprenticeship courses.
    - (C) A commitment to complete an apprenticeship course.
    - (D) Time spent on similar work.

1. _____

2. Technician A says that the technician must include payment for the required fees when sending in the application form. Technician B says that the technician should send in the application forms as early as possible to be able to take the test at the test center of choice. Who is right?
    - (A) A only.
    - (B) B only.
    - (C) Both A and B.
    - (D) Neither A nor B.

2. _____

3. If you have passed an ASE test, ASE will send you your test report and all of the following, *except:*
    - (A) wallet card.
    - (B) certificate.
    - (C) pocket card.
    - (D) list of potential employers.

3. _____

4. An alternative fuels test is used to test for proficiency in _____ fuels.
    - (A) diesel
    - (B) alcohol
    - (C) CNG
    - (D) propane/butane

4. _____

5. The certified technician must take a recertification test every _____ years to keep his or her certification.
    - (A) 2
    - (B) 5
    - (C) 10
    - (D) 15

5. _____

Job 1

Perform Safety and Environmental Inspections

Introduction

Thousands of technicians are injured or killed every year while on the job. Most of these technicians were breaking basic safety rules when their accidents occurred. The technicians that survived learned to respect safety precautions the painful way. Hopefully, you will learn how to avoid accidents the easy way, by studying and abiding by shop safety rules.

Environmental protection is the law. You must follow all applicable environmental laws or risk heavy fines. In addition, proper waste disposal and recycling will save money.

Objective

Given access to an auto shop, you will be able to locate the shop's fire extinguishers, fire exit, safety glasses, and other shop safety equipment. You will also learn the general safety rules of an auto shop. You will learn the methods of preventing environmental damage through environmentally friendly work procedures.

Materials and Equipment

* Floor jack.
* Jack stand.
* Access to the shop.

Instructions

As you read the job instructions, answer the questions and perform the tasks. Print your answers neatly and use complete sentences. Ask your instructor for help as needed. A careful study of Chapter 2, *Safety and Environmental Protection*, will prepare you for some of the procedures in this job.

Warning

Before performing this job, review all pertinent safety information in the text and discuss safety procedures with your instructor.

Procedure

Eye Safety

1. Eye protection (safety glasses or goggles) should be worn during any operation that could injure your eyes. See **Figure 1-1.**

Completed ❑

Name _____

List five common tasks that require the use of safety goggles. _____

_____

_____

_____

_____

(Hunter Engineering Company)

Figure 1-1: This technician is wearing safety goggles to protect his eyes from flying debris.

2. Where are the safety glasses and goggles   Completed ❑
   kept in your shop?

_____

_____

_____

_____

Material Safety Data Sheets

3. Walk through the shop and familiarize   Completed ❑
   yourself with the chemicals stored there.

4. Find the material safety data sheets for the   Completed ❑
   chemicals in the shop. Look up information
   about the chemicals stored in the shop.

Name _____

Were any chemicals improperly stored according to the material safety data sheet? _____ Were material safety data sheets missing for any of the chemicals in the shop? _____ If so, which ones? _____

Fire Safety

5. Walk around the shop and locate all of the fire extinguishers, the fire exit, and fire alarms. This will help you during an emergency.

Completed ❑

6. To help prevent an emergency, memorize these important fire prevention tips:
   - Always take actions to prevent a fire.
   - Store gasoline and oily rags in safety cans.
   - Wipe up spilled gasoline and oil immediately.
   - Hold a rag around the fitting when removing a car's fuel line.

Completed ❑

7. How many fire extinguishers and alarms are there in your shop? _____

Completed ❑

8. Where are the fire extinguishers located?

Completed ❑

_____

_____

_____

_____

9. Where are the fire alarms?

Completed ❑

_____

_____

_____

_____

10. How do you leave the shop in case of a fire?

Completed ❑

_____

_____

_____

_____

Name _____

Electrical Safety

11. Check the shop for unsafe electrical condi-   Completed ❑
    tions, such as damaged electrical cords and
    overloaded outlets. List the details of any
    unsafe electrical conditions found.

    _____

    _____

    _____

    _____

    _____

12. Check that all electrically operated tools   Completed ❑
    and equipment have grounding plugs. If any
    tools or equipment have had the grounding
    plug removed, list them here.

    _____

    _____

    _____

    _____

    _____

Compressed Air Safety

13. Describe the dangers posed by compressed air.   Completed ❑

    _____

    _____

    _____

    _____

    _____

14. What is the air pressure setting on the shop   Completed ❑
    compressor? _____

Shop Cleanliness

15. Check the shop floor for unsafe conditions,   Completed ❑
    including spills and trip hazards. List any
    unsafe conditions.

    _____

    _____

    _____

    _____

Name _____

16. Check the shop tools for cleanliness and organization. List any ways that tool storage can be improved.

Completed ❑

_____

_____

_____

_____

Clothing Safety

17. Examine your clothing. List and explain any changes that would make your clothing safer or better suited for the shop.

Completed ❑

_____

_____

_____

Carbon Monoxide

18. Locate the exhaust hoses in the shop. Where are they located?

Completed ❑

_____

_____

_____

19. Locate the controls to operate the exhaust fans. Where are they located?

Completed ❑

_____

_____

_____

Grinder and Drill Press Safety

20. Go to the electric grinder and inspect it closely. Locate the power switch. Observe the position of the tool rest and face shield. Also, check the condition of the grinding wheel. Is the electric grinder in the shop safe? Explain.

Completed ❑

_____

_____

_____

_____

Name _____

21. Locate and inspect the operation of the shop's drill press. Find the on/off button, feed lever, chuck, and other components. List the safety precautions associated with operating a drill press.

Completed ❑

_____

_____

_____

_____

Floor Jack and Jack Stand Safety

22. Check out a floor jack and a set of jack stands.

Completed ❑

23. Without lifting a car, practice operating a floor jack. Close the valve on the jack handle. Pump the handle up and down to raise the jack. Then, lower the jack slowly. It is important that you know how to control the lowering action of the jack.

Completed ❑

24. In what direction must you turn the jack handle valve to raise the jack? To lower the jack?

Completed ❑

_____

_____

_____

_____

STOP Warning: Ask your instructor for permission before beginning the next step. Your instructor may need to demonstrate the procedures to the class.

25. After getting your instructor's approval, place the jack under a proper lift point on the car (frame, rear axle housing, suspension arm, or reinforced section of the unibody). If in doubt about where to position the jack, refer to a service manual for the particular car. Instructions will usually be given in one of the front sections of the manual.

Completed ❑

26. To raise the car, place the transmission in Neutral and release the emergency brake. This will allow the car to roll as the jack

Completed ❑

Name _____

goes up. If the car cannot roll and the small wheels on the jack catch in the shop floor, the car could slide off the jack.

27. As soon as the car is high enough, place jack stands under the suggested lift points. Lower the car onto the stands slowly. Check that they are safe. Then, remove the floor jack and block the wheels. It should now be safe to work under the car.

Completed ❏

28. Raise the car. Remove the jack stands. Lower the car and return the equipment to the proper storage area.

Completed ❏

29. Where did you position the floor jack when raising the car?

Completed ❏

_____

_____

30. Where did you position the jack stands when securing the car?

Completed ❏

_____

_____

Environmental Protection

31. Carefully observe all areas of the shop, paying particular attention to the creation and storage of waste.

Completed ❏

32. Locate and list the types of solid waste produced. How are solid wastes disposed of?

Completed ❏

_____

_____

_____

_____

33. Locate and list the types of liquid waste produced. How are liquid wastes disposed of?

Completed ❏

_____

_____

_____

_____

Name _____

34. Locate and list types of gases or airborne particles produced. How are they kept from entering the atmosphere?

Completed ❏

_____

_____

_____

35. From the previous lists, identify the types of solid and liquid waste that could be recycled. Identify the materials that could be returned for a core deposit.

Completed ❏

_____

_____

_____

36. Do any Environmental Protection Agency (EPA) regulations apply to the wastes generated in the shop? _____. If yes, briefly summarize them.

Completed ❏

_____

_____

_____

37. Do any local and state regulations apply to the wastes generated by the shop? _____. If yes, briefly summarize them.

Completed ❏

_____

_____

_____

38. List any of the shop's waste disposal practices that require improvement. Explain what improvements could be made.

Completed ❏

_____

_____

_____

39. Have your instructor check your work and sign this worksheet.

Completed ❏

Instructor's signature: _____

Job 2

Access and Use
Service Information

Introduction

Service information contains detailed directions for working on vehicles. Service information can refer to service manuals, troubleshooting charts, or schematics. Today, these materials are available in printed manuals, in electronic files, or from the Internet.

Service information is an essential reference tool of the automotive technician. When a technician must perform an unfamiliar or difficult repair, service information outlines what must be done. It also provides specifications, capacities, and other useful information.

Objective

Supplied with service and diagnosis manuals, a computer, electronic files, and Internet access, you will find and use service information.

Materials and Equipment

- Service manual(s).
- Troubleshooting chart(s).
- Access to a computer.
- Electronic files (discs or CDs).
- Access to the Internet.
- Internet website or e-mail addresses.

Instructions

As you read the job instructions, answer the questions and perform the tasks. Print your answers neatly and use complete sentences. Use the service manual to look up and list the information requested in the job. Use the shop computer to access electronic files and locate website or e-mail addresses. Ask your instructor for help as needed.

Warning

Before performing this job, review all pertinent safety information in the text and discuss safety procedures with your instructor.

Procedure

---

1. Ask your instructor for the following vehicle information. You may be able to look up data on a car of your choice, or

Completed ❑

Name _____

your instructor may assign a particular make and model of car. Record this information in the following space. It will be used for the remainder of the job.

- Vehicle make: _____
- Vehicle model: _____
- Check the appropriate line: Automobile ___ Light Truck ___ SUV ___ Other ___
- Year: _____
- Engine size: _____
- Transmission type: _____

Index and Contents Sections

2. Obtain a service manual that covers the vehicle selected. Read the front section explaining how to use the manual. This information is normally inside the front cover or on first few pages of the manual.

Completed ❏

3. What is the title of the service manual you are using?

Completed ❏

_____

4. What model years does it cover? _____

Completed ❏

5. Does your manual contain a table of contents? _____

Completed ❏

6. Does your manual contain an index? Where is it located? Are there small contents or index pages at the beginning of each repair section?

Completed ❏

_____

_____

_____

Finding Page Numbers

7. How are the pages numbered? Is this different from the way the pages are numbered in your textbook?

Completed ❏

_____

_____

_____

Name _____

8. On what page is the section covering the repair of the engine? (The page number should be given in either the contents or the index.) _____

Completed ❑

9. List the service manual page numbers that explain the service and repair of the following engine parts. Also, read the instructions covering each area of repair.

Completed ❑

Area of Repair	Page Number
(A) Engine assembly	_____
(B) Valve lifters	_____
(C) Connecting rods	_____
(D) Main bearings	_____
(E) Valve guides	_____
(F) Timing chain, gears, or belt	_____
(G) Pistons	_____
(H) Camshaft	_____

Bolt Torque Specifications

10. Normally provided in the specifications section, list the torque specifications for the following engine components.

Completed ❑

Component	Torque Specifications
(A) Cylinder heads	_____
(B) Main bearings	_____
(C) Intake manifold	_____
(D) Connecting rod bolts	_____
(E) Flywheel bolts	_____
(F) Exhaust manifold	_____

Maintenance Tune-up Specifications

11. Look up and list the following maintenance tune-up information for the engine.

Completed ❑

Tune-up Item	Specification
(A) Spark plug type	_____
(B) Compression pressure	_____
(C) Spark plug gap	_____

12. In the service manual for your engine, locate a simplified top-view illustration of the engine vacuum hose routing similar to **Figure 2-1.** Note how the illustration shows the vacuum lines traveling from the source of vacuum (usually the intake manifold) to

Completed ❑

Name _____

the vacuum-operated devices. This type of illustration may be needed when tracing a vacuum leak or when the intake manifold or other parts must be removed.

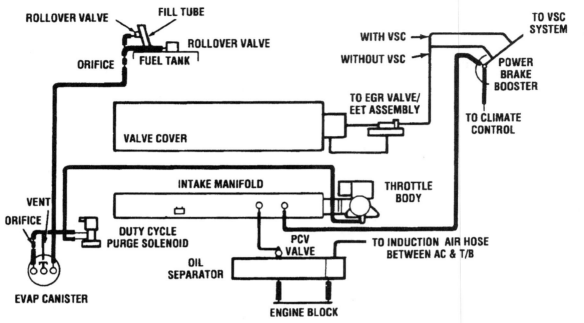

(DaimlerChrysler)

Figure 2-1: A simple vacuum diagram can be very helpful when tracking a vacuum leak.

13. Draw your engine illustration in **Figure 2-2.** Use a ruler to draw a close approximation of the engine as it is shown in the service manual vacuum routing diagram. Then draw lines representing the vacuum lines from intake manifold or vacuum pump to the vacuum operated devices.

Completed ❑

Figure 2-2

Name _____

Troubleshooting Charts

14. Locate a troubleshooting chart or diagnosis information in a service manual or a separate diagnosis manual.   Completed ❑

    • Type of chart: _____

    • Page number: _____

15. Using the information found in the previous step, explain one vehicle problem and its possible causes and corrections.   Completed ❑

    _____

    _____

    _____

    _____

Review Other Service Manuals

16. When you have spare time, thumb through different service manuals. Inspect the illustrations and repair procedures for various components. Also, compare the organization of one service manual to another.   Completed ❑

Accessing Service Information from Electronic Files

17. Access the shop computer or another computer designated by your instructor.   Completed ❑

18. Start the computer and ensure that it boots up properly. If the computer is already running, skip this step. If necessary, familiarize yourself with the computer keyboard and mouse. How can you tell that the computer has started correctly?   Completed ❑

    _____

    _____

    _____

    _____

19. Ask your instructor to select a service procedure that can be accessed by using the computer.   Completed ❑

20. Locate the floppy disks or CDs that contain the service procedure assigned by your instructor.   Completed ❑

Name _____

21. Insert the first disk or CD into the   Completed ❑
    computer.

22. Use the computer as necessary to call up the   Completed ❑
    main menu of the needed files.

23. Use the computer to print out the selected   Completed ❑
    file or portion of a file. Attach the printout
    to this job sheet. If a printer is not available,
    summarize the information in the space
    provided.

_____

_____

_____

_____

_____

_____

_____

_____

_____

_____

_____

_____

24. Return the discs or CDs to your instructor,   Completed ❑
    or to storage as necessary.

**Note: If the next portion of the job cannot be
performed, proceed to step 32.**

Accessing Service Information from the Internet

25. Ask your instructor for the following   Completed ❑
    information.

    Your Internet service provider: _____

    The service procedure or information being sought: _____

    _____

    _____

    The e-mail address or website of the contact: _____

Name _____

26. Use the computer to access your school's Internet provider.

Completed ❑

Note: Your instructor may assign a different Internet task. Follow your instructor's directions.

---

27. Use the computer to call up the website or contact the e-mail address, locate the needed information, and print it out.

Completed ❑

28. Attach the printout to this job sheet. If a printer is not available, summarize the information in the space provided.

Completed ❑

_____

_____

_____

_____

_____

_____

29. Exit the Internet site and return to the main Internet menu using the keyboard and/or mouse.

Completed ❑

30. Exit the Internet.

Completed ❑

31. Properly shut down the computer if directed to do so by your instructor.

Completed ❑

32. Have your instructor check your work and sign this worksheet.

Completed ❑

33. Have your instructor check your work and sign this worksheet.

Completed ❑

Instructor's signature: _____

Name _____ Date _____

Score _____ Instructor _____

Job 3

Use Precision Measuring Tools

Introduction

Precision measuring tools are frequently used during auto service and repair. These tools are needed to check various auto parts for wear. If a part is worn beyond manufacturer's specifications, it must be repaired or replaced. During this job, you will use a micrometer, sliding caliper, and dial indicator to measure several shop units. Once these basic measuring tools have been mastered, you can easily transfer this knowledge and skill to other measuring tasks. Other measuring devices will be covered in later jobs.

Objective

You will use the tools and materials below in a series of precision measuring exercises.

Materials and Equipment

- Instructor prepared set of flat feeler gauges (numbered but with sizes removed).
- Sliding caliper.
- Dial indicator.
- Two large screwdrivers.
- Six engine valves.
- Six valve spring shims.
- 0"–1" outside micrometer, or an equivalent metric micrometer.
- Differential assembly.

Instructions

As you read the job instructions, answer the questions and perform the tasks. Print your answers neatly and use complete sentences. Make sure that you have completed Chapter 4, *Precision Tools and Test Equipment*, in both the textbook and workbook before working on this job.

Warning

Before performing this job, review all pertinent safety information in the text and discuss safety procedures with your instructor.

Procedure

1. As a quick review, identify the parts of the micrometer in **Figure 3-1.** Also, use the space provided to describe the four steps in reading a micrometer.

Completed ❑

Name _____

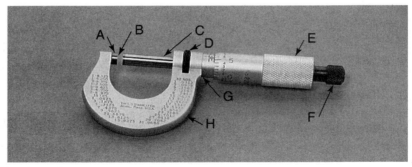

Figure 3-1: A typical outside micrometer.

(A) _____

(B) _____

(C) _____

(D) _____

(E) _____

(F) _____

(G) _____

(H) _____

Step 1. Read: _____

_____

_____

_____

_____

Step 2. Count: _____

_____

_____

_____

Step 3. Count: _____

_____

_____

_____

Step 4. Add: _____

_____

_____

_____

_____

Name _____

2. Inspect the flat feeler gauge set. Ten of the blades should have the factory blade sizes removed. These blades should be renumbered (engraved, labeled with tape, or notched) from one to ten.

Completed ❏

3. Before measuring the unsized blades, practice measuring a few of the sized blades. You can then get feedback about measuring accuracy by comparing your mike readings to the actual sizes written on the blades.

Completed ❏

4. Using the micrometer, measure the thickness of the ten unsized feeler gauge blades and record your measurements in the following chart. See **Figure 3-2.** If needed, ask your instructor for help getting started.

Completed ❏

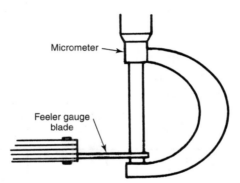

Figure 3-2: Measure the feeler gauge blades with an outside micrometer.

Feeler Gauge Blade Number	1	2	3	4	5	6	7	8	9	10
Your "Mike" Reading										

5. Inspect the six engine valves. Check that they are numbered from one to six.

Completed ❏

6. Measure the valve stems using both the micrometer and the sliding caliper, **Figure 3-3.** Try to measure the largest (least worn) and the smallest (most worn) part of each valve stem. Only measure on the operational, machined, or shiny portion of the stem. Record your results in the following chart.

Completed ❏

Name _____

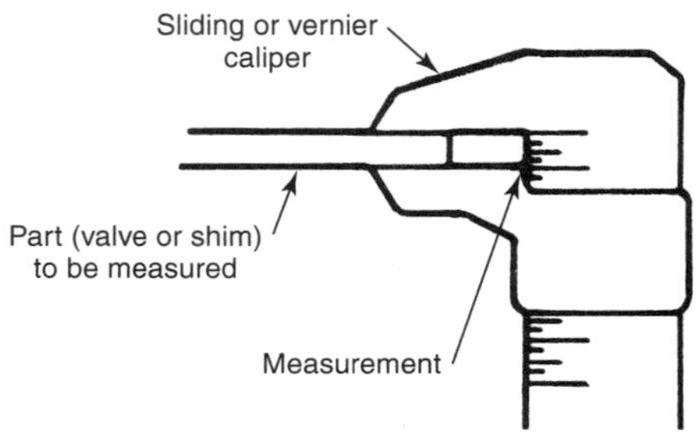

Figure 3-3: Measure the feeler valve stems with a sliding or vernier caliper.

Valve Number	1	2	3	4	5	6	
Caliper Reading							Largest Reading
							Smallest Reading
"Mike" Reading							Largest Reading
							Smallest Reading

7. Repeat this same measuring and recording procedure on the six valve spring shims.

Completed ❑

Shim Number	1	2	3	4	5	6	
Caliper Reading							Largest Reading
							Smallest Reading
"Mike" Reading							Largest Reading
							Smallest Reading

8. Inspect the dial indicator and mounting attachments. Be sure all parts are present.

Completed ❑

9. Ask your instructor which differential unit will be used in this section of the job. Preferably, you will have a bench-mounted differential.

Completed ❑

10. Mount the dial indicator as shown in the following illustration, **Figure 3-4.** The indicator stem should contact the smooth, back side of the ring gear. The base of the indicator should be clamped to or placed on the outer edge of the carrier.

Completed ❑

Name _____

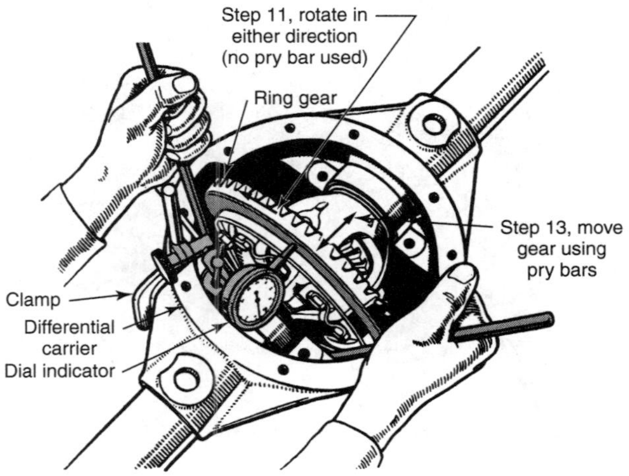

Figure 3-4: In step 11, rotate or turn the ring gear as if it were operating. The amount of runout (wobble) will register on the dial. In step 13, you must pry sideways as shown above.

11. Now, adjust the indicator dial to read zero. The face of the indicator should turn and adjust to zero. Slowly rotate the ring gear while watching the movement of the indicator needle. The amount of needle movement indicates the runout (wobble) of the ring gear in thousandths of an inch.

Completed ❑

12. How much runout is there on the ring gear?

_____

Completed ❑

13. Without changing the indicator mounting, use screwdrivers or small pry bars to move the ring gear sideways. Pry one way and then the other while watching the indicator needle. This will measure gear endplay.

Completed ❑

14. How much sideplay or endplay is there in the case bearings? _____

Completed ❑

15. To measure the gear backlash, mount the indicator as shown in **Figure 3-5.** The indicator stem should contact one of the ring gear teeth. Hold the pinion gear solid to prevent any movement while moving (wiggling) the ring gear back and forth gently. Do not turn the ring gear, however. The amount of dial indicator needle movement indicates gear backlash.

Completed ❑

Name _____

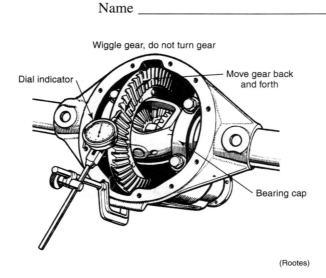

(Rootes)

Figure 3-5: Use a dial indicator to measure the amount of ring and pinion backlash.

16. What is the amount of backlash between   Completed ❑
    the differential gears? _____

17. Have your instructor check your work and   Completed ❑
    sign this worksheet.

    Instructor's signature: _____

Job 4

Use an Automotive Multimeter

Introduction

Due to the advanced electronic technology in today's vehicles, multimeters are becoming a necessary tool in auto service and repair. Electronic devices are used in almost every system of today's vehicles. A multimeter is a combination of many testers in a single case. The ohmmeter portion of a multimeter measures the resistance to current flow in a circuit or component. The voltmeter portion measures electrical pressure, and the ammeter measures the electron flow in a circuit. By measuring actual electrical values in a circuit and comparing them to specifications, a technician can quickly determine the source of electrical and electronic problems. In this job, we will cover the ohmmeter, voltmeter, and ammeter features of a multimeter.

Objective

You will use the ohmmeter, voltmeter, and ammeter portions of a multimeter to perform various tests.

Materials and Equipment

- Multimeter.
- Piece of wire.
- Piece of rubber vacuum hose.
- Resistor spark plug wire.
- Alternator diode.
- 12-volt battery installed in a working vehicle.
- Simple electrical circuit in a working vehicle.

Instructions

Study the information in the test equipment section of textbook Chapter 4, *Precision Tools and Test Equipment*. Ask your instructor for the location of the equipment to be used in this job and for any added details, such as safety precautions and optional steps. Only connect your multimeter to the components that are mentioned. Always make the proper connections, especially when making resistor checks, since the ohmmeter could be damaged if it is connected to a live circuit. As you read the job instructions, answer the questions and perform the tasks. Print your answers neatly and use complete sentences.

Warning

Before performing this job, review all pertinent safety information in the text and discuss safety procedures with your instructor.

Name _____

Procedure

Ohmmeter Section

1. Turn the multimeter power switch on, and check that the multimeter operates properly. If the multimeter does not work properly, ask your instructor for advice. The multimeter battery may need replacement.

   Completed ❑

2. Attach the test leads to the proper sockets on the multimeter. Typical multimeter sockets are shown in **Figure 4-1.**

   Completed ❑

(Fluke)

Figure 4-1: This is a typical multimeter.

3. Touch the two test leads together. The scale should read zero ohms.

   Completed ❑

4. Separate the test leads. The scale should read infinity or very high ohms. Carefully note that the multimeter scale has changed. With the leads together, the reading was in single ohms. When the test leads were separated, the scale changes to K (thousands) or M (millions) of ohms. You have actually measured the resistance of a known conductor (the meter test leads = no resistance) and a known nonconductor (surrounding air separating untouched test leads = infinite resistance).

   Completed ❑

Name _____

5. Measure the resistance of the following objects with the multimeter. Record your results in the space provided.

Completed ❑

- Piece of wire: _____ (ohms, K ohms, M ohms)
- Piece of rubber vacuum hose: _____ (ohms, K ohms, M ohms)

6. Explain the difference in your resistance readings.

Completed ❑

_____

_____

7. Measure the resistance between the terminals of the spark plug wire. See **Figure 4-2.** Record your results in the space provided.

Completed ❑

- Spark plug wire: _____ (ohms, K ohms, M ohms)

Figure 4-2: Note the arrangement for testing spark plug wires.

8. Is the spark plug wire usable? Explain.

Completed ❑

_____

_____

9. Locate the alternator diode and connect the ohmmeter leads in both directions. Record your readings for both directions. Remember that a diode is an electrical check valve. It will allow current to flow in one direction but not the other.

Completed ❑

- First reading: _____
- With leads reversed: _____

Name _____

10. Is the diode usable? Explain.   Completed ❑

_____

_____

Voltmeter Section

11. Set the multimeter to measure 15 dc volts.   Completed ❑
    A voltmeter, as with any meter, must
    always be set to measure slightly above the
    expected maximum value.

12. Attach the test leads to the proper sockets   Completed ❑
    on the multimeter.

13. Open the hood of the test vehicle and   Completed ❑
    connect the test leads across the battery
    posts, as shown in **Figure 4-3.**

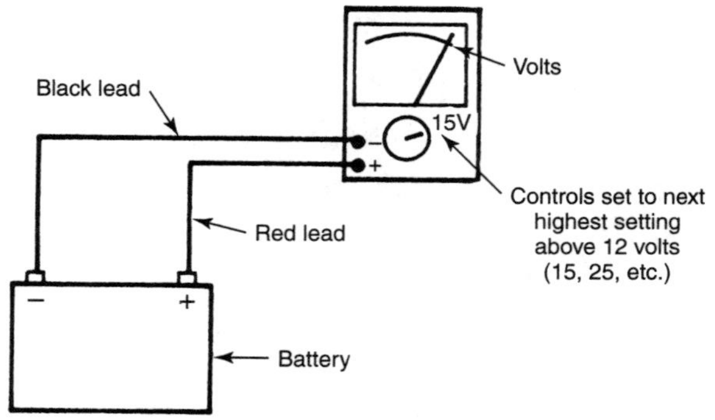

Figure 4-3: Measure battery voltage as shown.

STOP Warning: Do not reverse the leads. Meter
damage could occur. The red meter lead
should go to positive, and the black lead
should go to negative.

14. What is the voltage of your battery? A   Completed ❑
    charged 12-volt battery should read approx-
    imately 12.6 volts. Reading: _____

15. Start the vehicle engine and measure the   Completed ❑
    amount of voltage across the battery posts.
    The voltage should be higher than 12.6 volts.
    Reading: _____

16. Stop the vehicle engine.   Completed ❑

Name _____

17. How much did the voltage increase?   Completed ❑
    - Voltage in Step 15: _____
    - Subtract voltage from Step 14: _____
    - Voltage difference: _____
    - What does this tell you? _____

18. Place one terminal lead on a battery post   Completed ❑
    and one lead on battery terminal. See
    Figure 4-4.

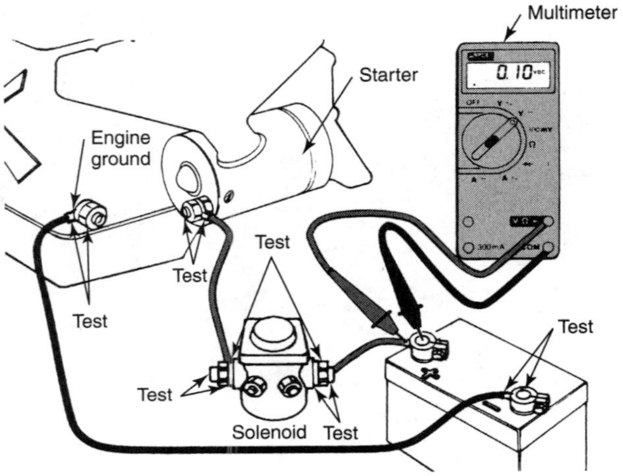

Figure 4-4: If the connection has a high resistance, current will flow through the meter.

19. Have an assistant crank the engine as you   Completed ❑
    observe the multimeter. Reading: _____

20. The voltmeter should read less than 0.5 volt.   Completed ❑
    A higher reading indicates high resistance
    in the post-to-terminal connection (elec-
    tricity is attempting to flow through the
    meter instead of the connection). This is
    called voltage drop. Voltage drop is an indi-
    cator of circuit resistance. A high voltage
    drop indicates component resistance. A low
    voltage drop indicates a low resistance
    connection. Based on this information,
    what can you say about the vehicle that you
    are testing? _____

Ammeter Section

21. Set the multimeter to measure amperage.   Completed ❑
    Ask your instructor to pick a circuit for you
    to measure. Set the meter to measure slightly
    above the expected maximum value.

Name _____

 Caution: If the multimeter cannot measure 10A or more, do not use it for this test. Check with your instructor before proceeding.

22. Attach the test leads to the proper sockets on the multimeter, or attach the inductive amps probe. See **Figure 4-5.**

Completed ❑

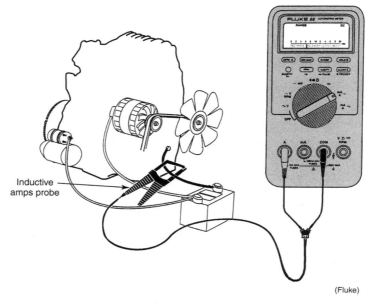

(Fluke)

Figure 4-5: An inductive amps probe.

23. Select a vehicle circuit to measure. Ask your instructor if you are unsure as to which circuit to use. A light circuit is usually the easiest and safest circuit to test.

Completed ❑

24. Attach the test leads in series with the circuit, or place the inductive clamp over a wire in the circuit to be tested.

Completed ❑

 Caution: Never connect an ammeter in parallel with a circuit. A short circuit may be produced, causing an electrical fire.

25. Before turning on the power, double-check your meter connections and settings.

Completed ❑

26. Turn the circuit controls to the On position and measure the current in your circuit. Record the multimeter readings for the On and Off switch positions.

Completed ❑

Name _____

- On: _____
- Off: _____
- *Optional:* Use the ohmmeter part of the multimeter to determine the resistance in the circuit. Next, divide the battery voltage by the resistance to get the amperage. Does the calculated amperage match the recorded amperage? If not, can you explain?

_____

_____

27. Return all tools and equipment to storage.   Completed ❑

28. Have your instructor check your work and sign this worksheet.   Completed ❑

Instructor's signature: _____

Job 5

Use a Scan Tool and a Waveform Meter

Introduction

It is nearly impossible to diagnose modern vehicles without scan tools and waveform meters. Older testers cannot check the electronic devices used in today's vehicles. Scan tools communicate directly with vehicle computers, allowing technicians to quickly access trouble codes and determine problem areas. Waveform meters represent electrical activity in the component as a wave pattern.

Objective

Using the materials listed below, you will correctly operate scan tools and waveform meters.

Materials and Equipment

- Scan tool.
- Waveform meter.
- Vehicle to be tested.

Instructions

Read over the test equipment section in Chapter 4, *Precision Tools and Test Equipment*, in your textbook. As you read the job instructions, answer the questions and perform the tasks. Print your answers neatly and use complete sentences. Ask your instructor for the location of the equipment to be used and for any additional information. Only hook up the circuits that are described in the job. Always double-check connections before turning on the power.

Warning

Before performing this job, review all pertinent safety information in the text and discuss safety procedures with your instructor.

Procedure

Using the Scan Tool

1. Obtain a scan tool compatible with the vehicle to be used for this job.

   Completed ❑

2. Study the controls on your scan tool. Be sure that you know how to use it. Refer to the manual if you are unsure about how to operate the particular scan tool.

   Completed ❑

Name _____

3. Ensure that the ignition switch is in the Off position.   Completed ❑

4. Attach the scan tool to the vehicle's diagnostic connector. Most diagnostic connectors are located under the dashboard as shown in **Figure 5-1.**   Completed ❑

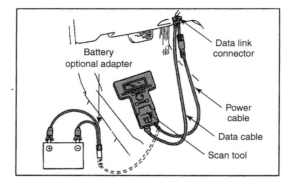

Figure 5-1: Most diagnostic connectors are located under the dashboard.

5. Turn the ignition switch to the On position (do not start the engine).   Completed ❑

6. Use the scan tool menu to retrieve trouble codes and list them here.   Completed ❑

_____

_____

7. Using the correct factory service manual, determine the vehicle system problems indicated by the trouble codes.   Completed ❑

_____

_____

8. Clear the existing codes using the procedure in the factory service manual. Remove the scan tool and store it securely.   Completed ❑

9. Start the vehicle and drive it for several miles or allow it to idle for several minutes. This will allow you to separate the hard codes from the intermittent codes.   Completed ❑

10. Stop the engine.   Completed ❑

11. Repeat Steps 4–6 to retrieve trouble codes.   Completed ❑

Name _____

12. Compare the trouble codes (if any) with those found in Step 6.   Completed ❑

   • List all hard codes: _____

   • List all intermittent codes: _____

   • What do you think is causing trouble codes to reset? _____

   _____

   _____

   _____

   _____

13. Use the scan tool to obtain other information about engine condition (this will vary with scan tool capabilities). Record the readings for as many of the following as you can.   Completed ❑

   • Engine rpm: _____

   • Air fuel ratio: _____

   • Ignition timing: _____

   • Coolant temperature: _____

   • Incoming air temperature: _____

   • Manifold air pressure (MAP): _____

14. Use the scan tool to clear all trouble codes.   Completed ❑

Using the Waveform Meter

15. Locate the waveform meter.   Completed ❑

16. Study the controls on the waveform meter. Refer to the manual if you are unsure about how to operate the waveform meter.   Completed ❑

17. Ensure that the ignition switch is in the Off position.   Completed ❑

18. Locate a fuel injector on the test vehicle.   Completed ❑

Note: Your instructor may direct you to check the waveform patterns on another vehicle device. Follow your instructor's directions to make the test.

---

19. Remove the fuel injector connector. Use jumper wires or a specially prepared connector to attach the waveform meter   Completed ❑

Name _____

leads in a way that does not damage the fuel injector wiring. Do not allow the injector wiring to ground against the engine.

20. Set the waveform meter to the proper range (usually dc volts).

Completed ❑

21. Start the engine and observe the fuel injector waveform. Draw the waveform in the following space.

Completed ❑

22. Compare the waveform with the waveforms shown in **Figure 5-2.**

Completed ❑

- Does your drawing of the waveform match one of the waveforms in Figure 5-2? _____
- Can you identify the type of waveform?

_____

_____

23. Turn off the engine.

Completed ❑

24. Return the scan tool and waveform meter to storage.

Completed ❑

25. Have your instructor check your work and sign this worksheet.

Completed ❑

Instructor's signature: _____

Name _____

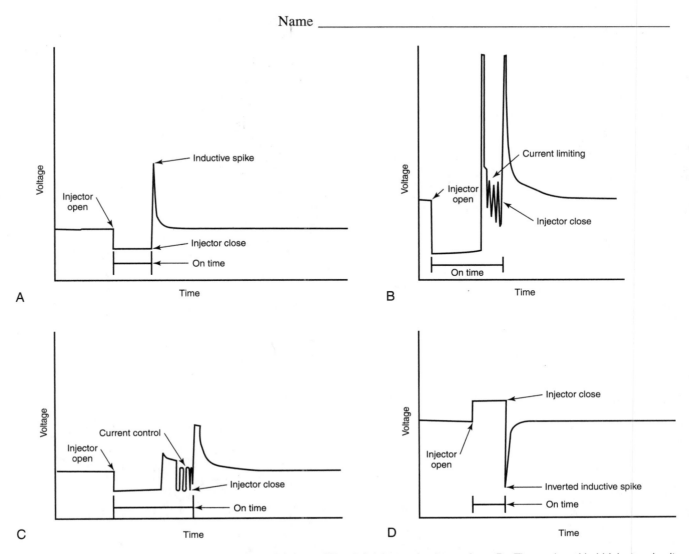

Figure 5-2: Common injector circuit waveforms. A—The saturated switch injector circuit waveform. B—The peak-and-hold injector circuit waveform. C—The Bosch-type peak-and-hold injector circuit waveform. D—The PNP injector circuit waveform.

Job 6

Perform Preventive Maintenance

Introduction

Preventive maintenance can prevent costly breakdowns. Periodic lubrication is one of the most important and common maintenance operations performed on a vehicle. Changing the engine oil and filter and lubricating high-friction points at prescribed intervals prolongs the useful life of an automobile. If a vehicle's fluid levels are not checked and topped off regularly, the useful life of the vehicle will be reduced. When a technician checks a fluid level (engine oil, coolant, power steering fluid, differential lubricant), it is a form of preventive maintenance. Anytime a low fluid level, dirty oil, or other problem is found, it tells the technician something about the condition of the vehicle. For example, an excessively low fluid level may indicate a serious leak. Then, the technician would know to inspect that system to find the source of the problem.

Objective

Given a vehicle and the listed tools, you will perform an engine oil and filter change, lubrication of the suspension and steering systems, and a check of the vehicle's fluids.

Materials and Equipment

- Basic hand tools.
- Vehicle in need of service.
- Oil filter wrench.
- Grease gun.
- Oil drain pan.
- Correct type and grade of oil.
- Other vehicle fluids as needed.
- Shop towels.
- Antifreeze tester.
- Eye protection.

Instructions

Review the information in textbook Chapter 11, *Preventive Maintenance*. Ask your instructor for details of the job. Your instructor may want you to perform the job on an actual vehicle or demonstrate the procedure on a shop vehicle or stand-mounted engine.

As you read the job instructions, answer the questions and perform the tasks. Print your answers neatly and use complete sentences. Ask your instructor for help as needed.

Warning

Before performing this job, review all pertinent safety information in the text and discuss safety procedures with your instructor.

Name _____

Procedure

Draining Oil

1. During an actual oil change, the engine should be warmed to operating temperature. Any dirt or contaminants will be picked up, suspended, and drained out of the engine with the warm oil.

Completed ❑

2. Normally, you will have to raise the vehicle on a lift or with a jack and then secure it on jack stands. The vehicle should be level to allow all of the oil to drain from the pan.

Completed ❑

> ⚠ **Caution: To prevent accidental starting during the procedure, remove the key from the ignition switch.**

3. Place the oil drain pan under the engine oil pan. Check that you are not looking at the transmission pan. Put the drain pan slightly to one side of the engine pan, as in **Figure 6-1.**

Completed ❑

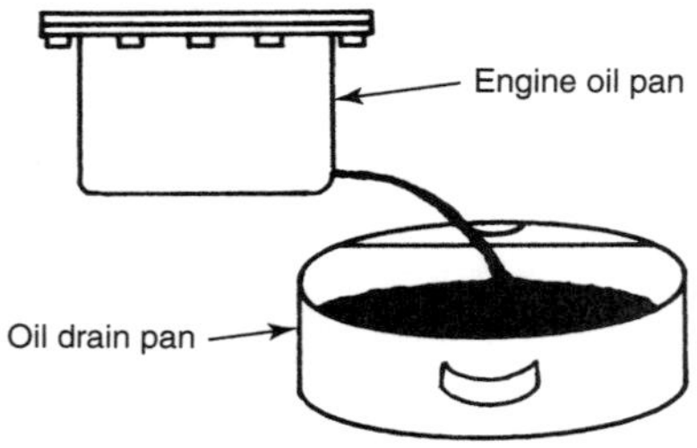

Figure 6-1: Be careful. The oil drained from the pan will be hot.

4. With a box-end wrench of the correct size, turn the oil drain plug counterclockwise and remove it, **Figure 6-2.**

Completed ❑

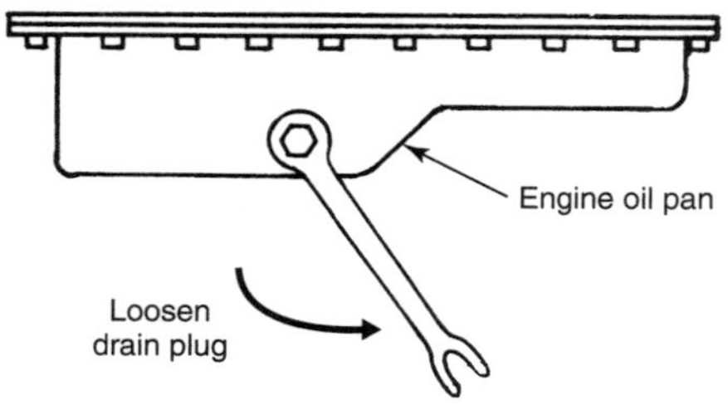

Figure 6-2: An oil pan drain plug will strip easily. Turn the plug counterclockwise to remove it.

Name _____

 Caution: Keep your arm out of the way of the hot oil as it flows from the engine oil pan.

5. While the oil is draining for three or four minutes, inspect the drain plug. The plastic washer or seal should be uncracked and unsplit or it will leak. Check the threads for damage and wear. If this is an actual oil change, replace a damaged washer seal or drain plug.

Completed ❑

6. What is the condition of the drain plug and seal? _____

Completed ❑

7. Being extremely careful not to cross thread, overtighten, and strip its threads, install and snug the oil drain plug. The drain plug only needs to be tight enough to slightly compress the plug seal. Overtightening will cause part damage and leakage.

Completed ❑

8. Move the oil drain pan under the engine oil filter. Loosen the oil filter using the filter wrench, **Figure 6-3.** Spin the filter the rest of the way off, be careful not to let hot oil run down your arm.

Completed ❑

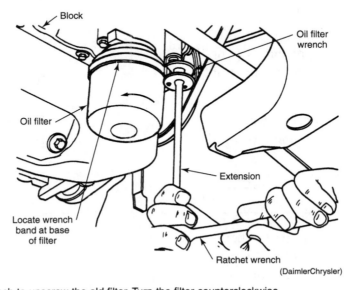

Figure 6-3: Use an oil filter wrench to unscrew the old filter. Turn the filter counterclockwise.

Name _____

Replacing the Oil Filter

9. Wipe off the mounting base for the oil filter to remove any dirt and contaminated oil, **Figure 6-4.** Also, check that the old filter seal is *not* stuck on the engine.

Completed ❑

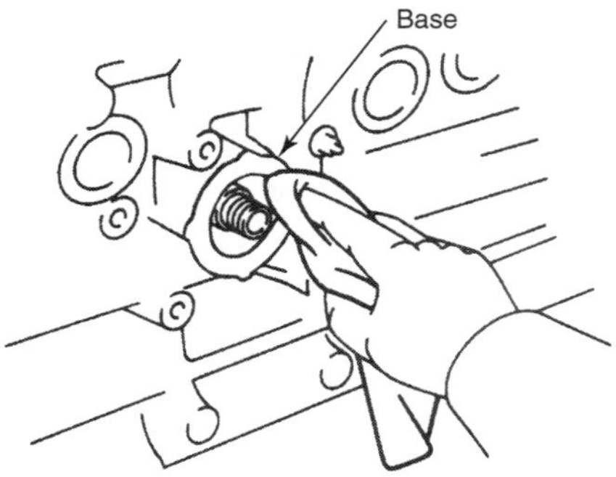

Figure 6-4: After removing the oil filter, clean off the filter mounting base to help prevent leakage.

10. Make certain that the new filter is a proper replacement. Make sure the rubber O-ring seals are identical. The diameters of each should measure the same.

Completed ❑

11. What is the measured diameter of the filter seal? _____

Completed ❑

12. What is the measured diameter of the threaded hole in the filter? _____

Completed ❑

13. If the filter fits on the engine upright, fill the new filter with oil. This prevents a temporary lack of oil pressure while the empty filter is filling.

Completed ❑

14. Describe the method of attachment and the angle at which the filter mounts to the engine.

Completed ❑

_____

_____

15. Wipe some clean engine oil on the new oil filter rubber seal. This will help in the proper tightening of the filter, reducing the chance of leaks. See **Figure 6-5.**

Completed ❑

Name _____

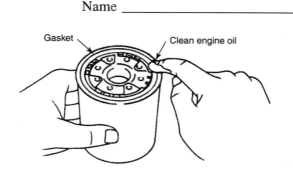
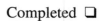

Figure 6-5: Wipe some oil on the seal of the new filter. This will allow proper tightening of the filter.

16. Screw on and hand-tighten the new oil filter, being careful not to cross thread it. (If this is just an exercise, reuse the old filter.) Your hands and the filter should be clean and dry. Look at **Figure 6-6.** After the seal makes contact with the engine base, use a rag or towel to help turn the filter an additional 1/2 to 3/4 turn.

Completed ❏

Figure 6-6: With your hands and the outside of the filter free of oil, tighten the filter by hand. After the seal touches the base, turn the filter an additional 1/2 to 3/4 of a turn.

17. Avoid tightening the filter with an oil filter wrench or the rubber seal may be smashed, causing a serious oil leak.

Completed ❏

Lubricating the Suspension and Steering Systems

18. If you have your instructor's OK and a front-end teaching unit or automobile, you can perform a grease job. Locate and lubricate the grease fittings on the upper and lower ball joints.

Completed ❏

 Caution: Do not overfill the rubber boots or they may rupture. As soon as you see them swell a little, stop!

Name _____

19. Look for other components needing lubrication. Sometimes tie rod ends, idler arms,
and universal joints can be lubricated.

Completed ❏

20. If the vehicle has never been lubed, the
small hex head screws will have to be
removed so that fittings can be installed.
Figure 6-7 shows the most common location of grease fittings.

Completed ❏

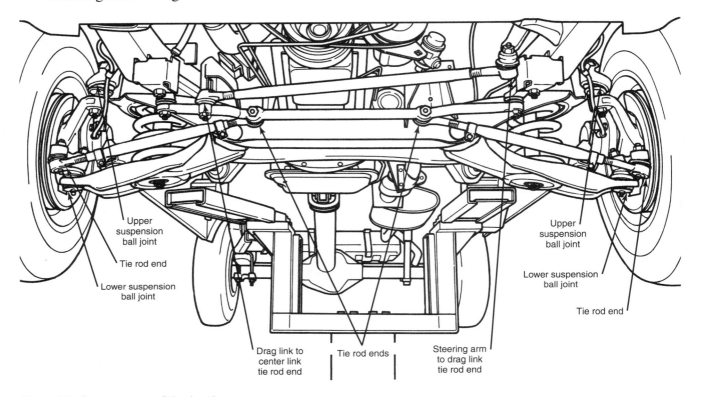

Figure 6-7: Common grease fitting locations.

21. How many grease fittings did you find?

_____

Completed ❏

22. Where were they located?

_____

_____

_____

_____

Completed ❏

23. Wipe up any spilled oil or grease and empty
the oil drain pan.

Completed ❏

Name _____

> **Note: If this is a learning exercise on a shop-owned engine, your instructor may want you to reinstall the same oil in the engine.**

Refilling with Oil

24. Lower the vehicle to the ground and remove the filler cap. It is usually on a valve cover.   Completed ❏

25. Make sure that you have the right type, weight, and quantity of engine oil. Use the type of oil recommended by the auto manu-facturer. Engine oil capacity can vary from four quarts in small gasoline engines to seven quarts in large diesel engines. Refer to the specifications given in the owner's manual or service manual.   Completed ❏

26. What type, weight, and brand of oil should be installed in this engine?   Completed ❏

    _____

    _____

27. Remove the oil container cap and remove the seal. Pour the oil into the filler opening in the engine. Use a funnel if the oil filler is located in an awkward position. Repeat this operation until the engine is filled to the proper level.   Completed ❏

28. How much oil does the engine require?   Completed ❏

    _____

29. Replace the filler cap and wipe off any oil that you might have dripped on the engine, workbench, or floor.   Completed ❏

30. Start the engine and watch the oil pressure indicator light or gauge. The gauge should begin to register almost immediately, and the oil warning light should go out within 15 or 20 seconds. If the oil pressure does not register, or if the warning light does not go out in the allotted time, shut off the engine and find the problem.   Completed ❏

31. How long did it take for the engine to develop oil pressure? _____   Completed ❏

Name _____

32. Shut off the engine.   Completed ❑

33. Allow the engine to sit for several minutes   Completed ❑
    and then recheck oil level.

34. If oil level is low, add enough oil to bring   Completed ❑
    the level up to the full mark on the dipstick.

Lubricating Other Units

35. Let the engine run for about 5 minutes   Completed ❑
    while you check for leaks under the engine.
    Lubricate other high friction and wear
    points such as door hinges and latches.

36. Did you find any leaks? _____   Completed ❑

37. If this is a service of a customer vehicle,   Completed ❑
    place a small amount of grease between the
    parts that rub on the hood hinges and hood
    latch. Squirt a small amount of oil on the
    door hinges and rub a little non-staining
    lubricant on the door latches and posts.

38. If this is an actual oil change, fill out a   Completed ❑
    service label including the date and mileage
    when the service was performed. Some
    labels are installed on the edge of the
    driver's door above the latch, while others
    are placed on the upper left side of the
    windshield.

> **Note: The following checks should be made whenever the engine oil is changed.**

Checking Brake Fluid Level

39. Locate the brake master-cylinder reservoir.   Completed ❑
    It is normally directly attached to the
    master cylinder, which is often bolted to the
    firewall on the driver's side of the vehicle.
    Some vehicles have a separate brake fluid
    reservoir located away from the master
    cylinder.

40. Clean and remove the master cylinder   Completed ❑
    reservoir cover. Most modern caps are

Name _____

threaded, and can be turned 1/4–1/2 turn to remove them. If spring clips are used, pry them off with a screwdriver. A wrench is needed if the cover is bolted in place.

41. Inspect the level and condition of the brake fluid. Typically, the brake fluid should *not* be more than 1/4″ (6.4 mm) down in the reservoir. See **Figure 6-8.**

Completed ❑

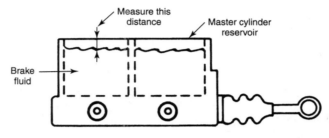

Figure 6-8: Brake fluid must be kept at the specified level. Always refer to a service manual for the proper specifications.

STOP Warning: Keep oil, dirt, and grease out of the brake fluid. Oil and grease can ruin the rubber parts in a brake system.

42. Is the master cylinder full or low? Explain.

Completed ❑

_____

_____

_____

43. When the brake fluid becomes low in a short period of time, inspect the entire brake system for leaks. Check the wheel cylinders, brake lines, and back of the master cylinder for wetness. Brake fluid will show up as a dark, damp spot. Inspect your vehicle for leaks. If you find evidence of leaks, write down the locations and the likely causes.

Completed ❑

_____

_____

44. To improve your troubleshooting ability, memorize the smell of brake fluid. Then, if fluid leakage is found, you can quickly smell and identify the leak.

Completed ❑

Name _____

45. If you spill brake fluid, wipe it up immediately. Brake fluid can ruin the paint job on a vehicle. It dissolves paint in seconds.

Completed ❑

Checking Battery Condition

46. Inspect the battery. The top of the battery should be clean and dry. Moisture on the battery case top can cause battery leakage (current shorts from one cell to another across dirt on the battery). Also, check the condition of the battery terminals. They should be clean, and tight. Corroded terminals may keep the engine from cranking properly.

Completed ❑

47. Describe the outer condition and cleanliness of the battery case and terminals. Could you find any other problems?

Completed ❑

_____

_____

_____

_____

Checking Engine Coolant

48. Determine whether the vehicle has a closed or an open cooling system, **Figure 6-9.** A closed system will have a plastic reservoir tank on one side of the radiator. The radiator cap may also be labeled "Do not open."

Completed ❑

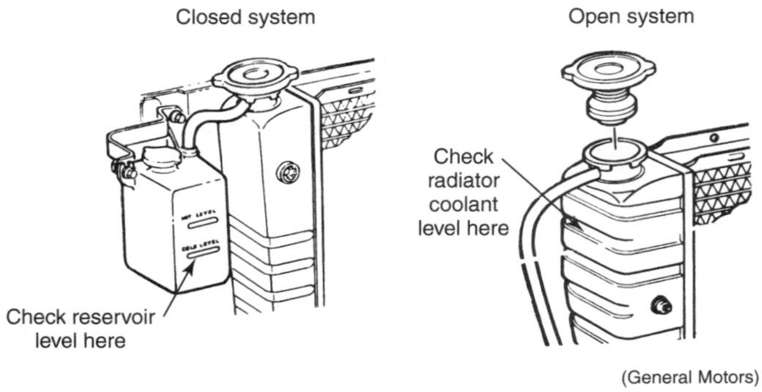

Closed system   Open system

Check radiator coolant level here

Check reservoir level here

(General Motors)

Figure 6-9: Note the difference between closed and open cooling systems. You do not have to remove the radiator cap to check the coolant level in a closed system.

Name _____

- To check the coolant level in a closed system, inspect the amount of coolant in the plastic reservoir tank. Compare the level with the marks on the side of the tank.
- To check the coolant level in an open system, first make sure the radiator is cool. If the radiator is warm, do not remove the cap. Hot coolant could blow into your face, causing serious injury. The coolant in an open system should be about one inch down in the radiator, Figure 6-9.
- If either type cooling system needs coolant, add a 50-50 mixture of antifreeze and water.

49. How was the level of coolant in the cooling system? _____   Completed ❑

50. Find an antifreeze tester. Draw coolant into the tester. Use the directions with the particular tester to determine the freeze protection of the coolant.   Completed ❑

51. Also, inspect the coolant for rust or discoloration. After prolonged use, antifreeze can break down and become very corrosive. It can cause rapid rust formation and damage to the cooling system. For this reason, antifreeze must be drained and replaced at recommended intervals.   Completed ❑

52. Inspect the radiator for signs of leakage. A leak in a cooling system will usually be easy to see. The area around the leak will often be a bright, rust color or the color of the antifreeze.   Completed ❑

53. Is the coolant rusty or could you find any leaks? Explain.   Completed ❑

_____

_____

54. To increase your troubleshooting skills, memorize the smell of antifreeze. Then, when a leak is found, the smell will tell you whether it is antifreeze or another fluid.   Completed ❑

55. How much freeze-up protection does the coolant provide for the system? _____   Completed ❑

Name

Checking Power Steering Fluid

56. Locate the power steering pump. It is normally on the lower front of the engine. Some newer vehicles may have a separate reservoir located away from the pump. With the engine off, remove the power steering pump cap and wipe off the dipstick. Reinsert dipstick and pull it back out. Holding the dipstick over a shop rag, inspect the level of fluid. The fluid should be between the Full and Add marks, as in **Figure 6-10.** If needed, add enough recommended power steering fluid to fill to the Add mark. Do not add too much, or fluid will blow out the pump after engine is started.

Completed ❑

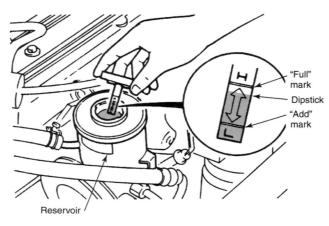

Figure 6-10: The power steering–fluid dipstick is normally attached to the pump cap. Add the recommended fluid as needed.

57. If the power steering fluid level is excessively low, check the power steering system for leaks. Look under the pump, around line fittings and any other component containing fluid.

Completed ❑

58. Was the power steering fluid at the proper level? If it was not, what is the most likely cause?

Completed ❑

_____

_____

_____

_____

Name _____

Checking Automatic Transmission or Transaxle Fluid

59. To check the automatic transmission or transaxle fluid level, start the engine and bring it up to operating temperature. Move the transmission selector through the gears. Apply the parking brake, shift the transmission into Park, and block the wheels. Leave the engine running.

Completed ❑

60. Remove and wipe off the transmission dipstick. It will usually be at the rear of the engine on one side. Reinsert the dipstick and pull it back out. Hold the dipstick over a rag while you check the fluid level. Refer to **Figure 6-11.** Smell the fluid. If the fluid smells burned, then the bands or clutches inside the transmission may be worn and damaged. Observe the condition of the fluid. A dark color indicates overheated fluid and possible internal damage. Milky looking fluid indicates water contamination.

Completed ❑

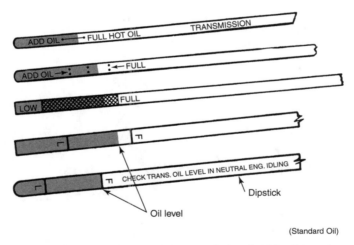

(Standard Oil)

Figure 6-11: Note the different types of markings on these automatic transmission dipsticks. Some show one pint from Add to Full. Others show one quart from Add to Full. Add only enough fluid to reach the Full mark.

61. Describe the condition and level of fluid in the automatic transmission. Does fluid need to be added?

Completed ❑

_____

_____

_____

_____

Name _____

62. When adding fluid to an automatic trans-   Completed ❑
    mission, make sure you have the correct
    type of fluid. Different units require
    different fluids. Always install the type
    fluid recommended by the manufacturer.
    Do not overfill the transmission. Overfilling
    can cause fluid foaming, poor transmission
    operation, and seal leakage.

Checking Differential Lubricant

63. Ask your instructor for approval before   Completed ❑
    completing the last section of this job. If
    the vehicle has front-wheel drive with a
    transverse-mounted (sideways) engine, it
    probably does not have a separate differen-
    tial unit.

64. After getting your instructor's approval,   Completed ❑
    raise the vehicle on a lift or raise it with a
    floor jack and secure it on jack stands.

65. Remove the differential filler plug, not the   Completed ❑
    drain plug. The filler plug will be on the
    front or rear of the housing, about halfway
    up on the differential. This is pictured in
    Figure 6-12.

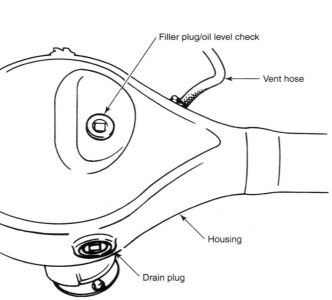

Figure 6-12: Note the locations of the filler and drain plugs in this differential.

Name _____

66. Stick your finger in the filler hole to check the lubricant level. The differential fluid should be approximately 1/2″ (12.7 mm) below the filler hole. This can vary with some vehicles, so always refer to a service manual for an exact specification.

Completed ❏

67. If the differential fluid level is low, add just enough recommended differential lubricant to meet factory recommendations. Memorize the smell of differential fluid so that you can quickly diagnose leaks.

Completed ❏

> ⚠ **Caution: If the vehicle has a positive-traction differential, do not add conventional differential oil. This will cause chattering on turns. Most positive-traction differentials will have an information tag attached to the fill plug.**

68. Was the lubricant in the differential low? If so, look for leaks at the differential seals. Consult your instructor if any leaks are found. _____

Completed ❏

Checking Manual Transmission Lubricant

69. Check the lubricant in a manual transmission. If the vehicle has an automatic transmission or transaxle, you will need another vehicle or a training unit.

Completed ❏

70. Remove the filler plug on the side of the transmission, **Figure 6-13.** Insert your finger into the hole. The lubricant should be slightly below or almost even with the filler hole. How was the lubricant level in the transmission? _____

Completed ❏

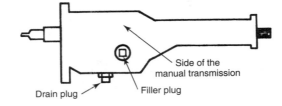

Side of the manual transmission

Drain plug

Filler plug

Figure 6-13: The filler plug is located on one side of a manual transmission. The drain plug will be on the bottom. Remove only the filler plug when checking the fluid level.

Name _____

71. If needed, add the recommended gear oil for the particular transmission. Learn to identify the smell of gear oil.   Completed ❑

72. Check the windshield washer solvent. It will be in a plastic reservoir on one side of the engine compartment. The solvent should be almost even with the Full mark on the side of the container. Add the recommended type of solution to the windshield washer tank. If water is added, it could freeze in cold weather and damage the windshield washer reservoir and pump.   Completed ❑

73. Have your instructor check your work and sign this worksheet.   Completed ❑

    Instructor's signature: _____

Job 7

Wiring and Connector Service

Introduction

It is extremely important for an auto technician to be able to properly attach electrical components. Modern vehicle sensors operate on very low voltages, so connections must have low resistance. Technicians must know the proper methods for stripping wire insulation, cutting wire, crimping wire terminals, attaching wire connectors, and soldering. During this job, you will perform all of these basic tasks.

Objective

Using the listed tools and equipment, you will properly repair automotive wiring and make clean, low resistance wire connections to terminals.

Materials and Equipment

- Safety glasses.
- 6″ piece of primary wire.
- Used spark plug wire.
- Female slide connector without insulation.
- Spade connector with insulation.
- Rosin core solder.
- Soldering gun.
- Wire strippers.
- Needle-nose pliers.
- Small screwdriver.
- Burn-resistant work surface.

Instructions

Review textbook Chapter 12, *Wire and Wiring*. It discusses the terminology, components, and procedures that will be used in this job. Ask your instructor about any changes or safety precautions. As you read the job instructions, answer the questions and perform the tasks. Print your answers neatly and use complete sentences. Take your time and do not hesitate to ask for help.

Warning

Before performing this job, review all pertinent safety information in the text and discuss safety procedures with your instructor.

Name _____

Procedure

1. Locate a piece of primary wire. It should be about 6″ (152 mm) long. Using the wire strippers, as in **Figure 7-1,** cut and pull off about 1/4″ (6.35 mm) of insulation from each end of the primary wire. Select the correct size stripping jaws.

Completed ❑

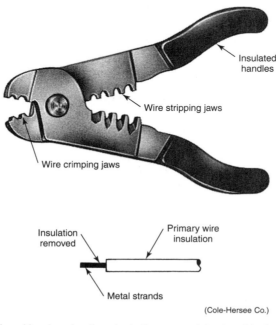

(Cole-Hersee Co.)

Figure 7-1: Primary wire insulation is stripped by clamping the wire in the appropriate size stripping jaws and pulling.

- If wire strippers are not available, use diagonal cutting pliers. Diagonals may be used to strip insulation by cutting part way through the insulation all the way around the wire. When the insulation has been cut, push on the side of the pliers with your thumb. The pressure will force the insulation free of the wire.

2. Fasten the terminal to the piece of wire as in **Figure 7-2.** Slide the stripped end of the wire into the connector. Then use the crimping tool to collapse the middle of the terminal tang. This will lock the terminal on the wire. Check the connection by pulling on the wire and terminal lightly.

Completed ❑

3. Next, crimp and attach the slide-type connector to the other end of the primary wire. If the slide connector has insulation on it, peel off the plastic insulation. You will be soldering the terminal shortly.

Completed ❑

Name _____

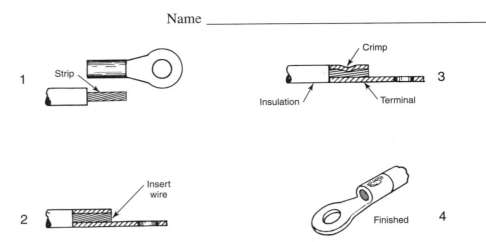

Figure 7-2: In Step 3, make sure to crimp the middle of the terminal lug or the connection may not hold.

4. Inspect the tip of the soldering gun or iron. The tip should be clean and tinned (coated with melted solder). Also, make sure you are using a work surface that will not burn easily.

Completed ❏

5. Touch the soldering gun tip on the connection, **Figure 7-3** and **Figure 7-4.** Preheat the wire and terminal. Next, touch the solder on the connection while still applying heat. Heat the connection until the solder melts and flows down into the terminal and wire. Avoid moving the wire for a moment after soldering. Movement will weaken the solder joint.

Completed ❏

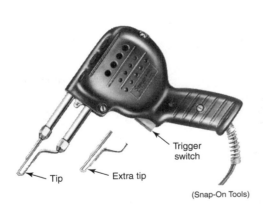

Figure 7-3: This type of soldering gun heats up in a matter of seconds, so be careful.

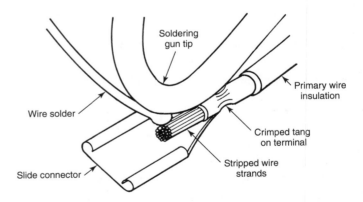

Figure 7-4: Heat both the solder and the wire until solder flows into the joint.

6. The jumper wire that you have just made should be extremely useful. It can be used during a wide range of electrical tests.

Completed ❏

Name _____

 Note: Steps 7–13 are optional. Most modern wires are replaced as assemblies with the terminals and boots installed.

7. Locate a piece of secondary wire (spark plug wire), **Figure 7-5.** It should be an old discarded wire. In the following directions, you will be required to remove and reinstall the terminals on the ends of the spark plug wire.

Completed ❑

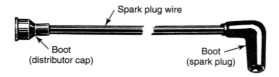

Figure 7-5: Use an old spark plug wire.

8. First, cut through the plug wire as close to the metal terminals as possible. Next, strip off about 3/8″ (9.53 mm) of insulation from each end of the wire, **Figure 7-6.** Be careful when stripping a carbon filled spark plug wire. Pulling too hard can break and ruin the carbon filled string in the wire.

Completed ❑

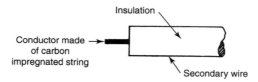

Figure 7-6: You must use a fairly small size jaw on your strippers since the diameter of the conductor is relatively small.

9. What does the wire conductor look like?

Completed ❑

_____

_____

10. Using needle-nose pliers and a small standard tip screwdriver, open the old terminals to remove the bits of spark plug wire. After you have removed the bits of wire, try to reshape the terminals into their original precrimped shape, **Figure 7-7.**

Completed ❑

Name _____

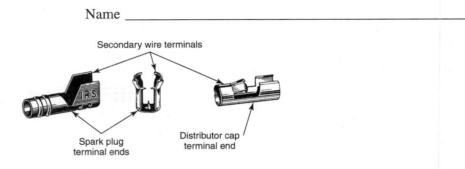

Figure 7-7: Try to return the used terminals to their original shapes for this exercise.

11. Next, make sure the rubber boots are in place on the spark plug wire.

Completed ❑

12. Fold the strands of the plug wire over the outside of the insulation and slip the terminal end into position. See **Figure 7-8.** The carbon strand should touch the metal portion of the terminal. If a staple is used, it should be inserted into the center of the carbon-impregnated strands before the wire is inserted into the terminal.

Completed ❑

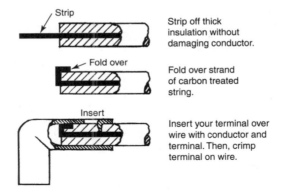

Strip off thick insulation without damaging conductor.

Fold over strand of carbon treated string.

Insert your terminal over wire with conductor and terminal. Then, crimp terminal on wire.

Figure 7-8: Installing a terminal connector on a secondary wire.

13. Finally, crimp the terminal to the spark plug wire. Be careful not to crush and distort the outer end of the terminal. If distorted, the terminal will not fit into the distributor cap or over the spark plug properly. Often, the proper set of jaws for crimping secondary wiring will be clearly marked on the crimping tool.

Completed ❑

14. As a quick review of this job, answer the following questions in your own words.

Completed ❑

Name _____

15. What is the difference between primary and   Completed ❑
    secondary type wire?

   _____

   _____

   _____

16. What type of solder should be used with   Completed ❑
    electrical wiring?

   _____

   _____

17. How is a terminal fastened to a wire   Completed ❑
    without soldering?

   _____

   _____

   _____

18. How can diagonal-cutting pliers be used to   Completed ❑
    strip wire?

   _____

   _____

   _____

19. What should be done to the strands of a   Completed ❑
    secondary wire before installing the
    terminal end?

   _____

   _____

20. What will happen if you pull too hard on a   Completed ❑
    secondary wire?

   _____

   _____

   _____

21. Show your primary and secondary wire to   Completed ❑
    your instructor.

22. Have your instructor check your work and   Completed ❑
    sign this worksheet.

   Instructor's signature: _____

Job 8

Enable and Disable SIR (Air Bag) System

Introduction

All vehicles built within the last few years have supplemental inflatable restraints (SIR), better known as air bags. SIR systems must be disconnected to service other vehicle components. Always disable the air bag system before attempting to service any component on, or near, the air bag system. These areas include the steering column and dashboard, and sometimes the engine compartment. The following job will show you how to disable and enable an air bag system.

Materials and Equipment

- Vehicle with an air bag system.
- Basic hand tools.
- Service information as needed.

Objective

Given the listed tools and equipment, you will disable and enable an air bag system.

Instructions

Review the air bag information in Chapter 13, *Chassis Electrical Service*, of the workbook and textbook. Have your instructor assign a vehicle and provide any additional information needed to complete the job. As you read the job instructions, answer the questions and perform the tasks. Print your answers neatly and use complete sentences. Take your time and do not hesitate to ask for help.

Warning

Before performing this job, review all pertinent safety information in the text and discuss safety procedures with your instructor.

Procedure

STOP Warning: Always obtain the proper service literature and follow manufacturer's procedures exactly before attempting to disable any air bag system.

1. Obtain a vehicle with an air bag system.

Completed ❑

Disable Air Bag System

2. Turn the steering wheel to the straight-ahead position.

Completed ❑

Name _____

3. Lock the ignition and remove the ignition key.

Completed ❑

4. Locate and remove the air bag fuses. **Figure 8-1** shows a typical air bag system fuse. Check the service information, as there may be more than one air bag system fuse.

Completed ❑

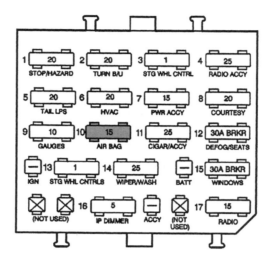

(General Motors)

Figure 8-1: The air bag fuse location will typically be marked inside the fuse box.

5. Locate the air bag connectors under the dashboard. Check the service manual for the exact connector locations. To access the connectors it may be necessary to remove the trim parts from the lower dashboard or center console.

Completed ❑

6. Disconnect the air bag connectors under the dashboard. Newer air bag connectors are yellow, but older systems may use another color for the connectors. **Figure 8-2** illustrates a common air bag connector location. Some air bag connectors, such as the one in **Figure 8-3,** have a locking pin to ensure that the connector does not come apart accidentally. This pin must be removed to separate the connector. Use the service literature to confirm that you are disconnecting the proper connectors.

Completed ❑

7. The air bags are now disconnected from all electrical power. Other repairs can be performed.

Completed ❑

Name _____

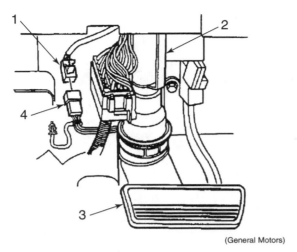

(General Motors)

Figure 8-2: The location of one typical air bag connector is shown here.

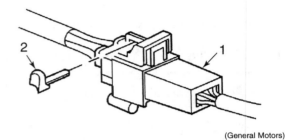

(General Motors)

Figure 8-3: The locking pin in this air bag connector ensures that the connector will not be disconnected accidentally.

> **STOP** **Warning: Although the air bags are disabled, they can still inflate without warning. Follow all of the following safety rules. For more information, consult the air bag section in Chapter 13 of your textbook.**

8. Review the following important air bag safety rules.   Completed ❑

   • Do not place the air bag module where temperatures will exceed 175°F (79.4°C).

   • If you drop any part of the air bag system, replace the part.

   • When carrying a live air bag module, point the bag and trim cover away from you.

   • Do not carry an air bag by the connecting wires.

   • Place a live air bag module on a bench with the bag and trim cover facing up.

   • Do not test an air bag with electrical test equipment unless the manufacturer's instructions clearly call for such testing.

Enabling Air Bag Systems

9. Make sure that the key is removed from the ignition switch.   Completed ❑

10. Reconnect the air bag connectors and reinstall any trim pieces that were removed.   Completed ❑

Name _____

11. Reinstall the air bag system fuses.   Completed ❑

12. Position yourself in the passenger compart-   Completed ❑
    ment so that you are away from the air bags.
    This will keep you from being injured if the
    system accidentally deploys.

13. Turn the ignition switch to the On position.   Completed ❑

14. Allow the air bag system self-diagnostic   Completed ❑
    program to run. This will ensure that the
    system is operating correctly.

15. Have your instructor check your work and   Completed ❑
    sign this worksheet.

    Instructor's signature: _____

Job 9

Test Starting and Charging Systems

Introduction

All vehicles have starting and charging systems. Even if you concentrate on another type of automotive work, you will come into contact with vehicles that will not start. You will then need to diagnose starting and charging system problems. Starting systems have remained the same for the last five decades. Charging systems have changed somewhat, but all vehicles built within the last 40 years use an alternator and voltage regulator to recharge the battery and provide other electrical needs. The technician can quickly determine the source of starting and charging problems by measuring actual voltage and current in the systems and comparing them to specifications.

Objective

Using the materials listed, you will correctly test a vehicle's starting and charging system.

Equipment and Materials

- Operating vehicle.
- Starting and charging system tester.

Instructions

Study Chapter 14, *Charging and Starting System Service*, before starting this job. Ask your instructor for the location of the starting and charging system tester to be used and the vehicle to be tested. The instructor should also inform you of safety precautions, equipment operating instructions, and any changes that are to be made to the procedure listed here.

As you read the job instructions, answer the questions and perform the tasks. Print your answers neatly and use complete sentences. Always follow instructions carefully when attaching the starting and charging system tester.

Warning

Before performing this job, review all pertinent safety information in the text and discuss safety procedures with your instructor.

Procedure

1. Put on safety glasses.   Completed ❑

2. Locate the proper starting and charging system tester.   Completed ❑

3. Determine the correct electrical specifications for the vehicle that you are testing.   Completed ❑

Name _____

List them here:

- Battery cranking voltage: _____
- Starter amperage draw: _____
- Charging system voltage at idle: _____
- Charging system voltage at 2500 rpm: _____
- Charging system amperage at idle: _____
- Charging system amperage at 2500 rpm: _____

4. Attach the starting and charging system tester to the vehicle according to the instructions. **Figure 9-1** shows the connections for a typical starting and charging system tester. Follow the specific procedure listed in your vehicle's service manual to disable the ignition system.

Completed ❑

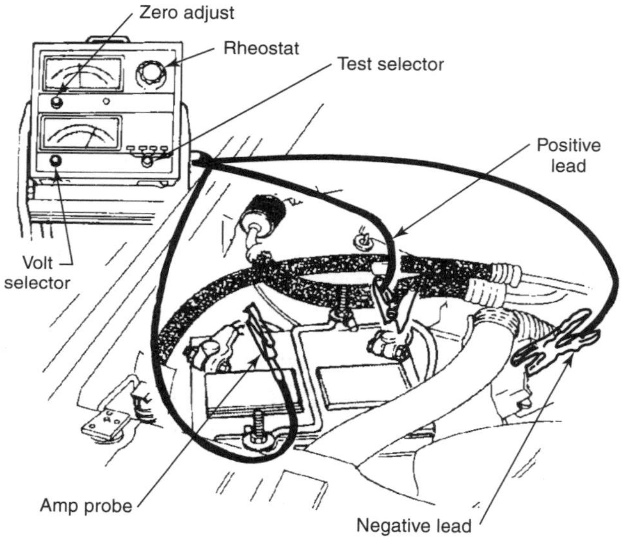

Figure 9-1: The connections for a typical starting and charging system tester.

5. Check the battery voltage and general battery condition with the ignition disabled.

Completed ❑

- What was the voltage reading? _____
- Were other problems found?

_____

_____

6. Have an assistant crank the engine with the ignition disconnected and check starter current draw and battery voltage as the starter is operated.

Completed ❑

Name _____

- How much current did the starter draw as it operated? _____
- What was the battery voltage reading as the starter operated?

_____

7. Reconnect the ignition and start the vehicle.   Completed ❑
   Check the charging system. Place a load on
   the system as needed to cause full alternator
   output.

   - What was the charging system voltage at idle? _____
     At 2500 rpm? _____
   - What was the maximum charging system amperage output?

   _____

   - If the alternator stator and diodes could be tested, were they
     in good condition? _____

8. Compare the starting and charging system   Completed ❑
   readings with system specifications.

   - Are all of the starting and charging system readings within
     specifications? _____
   - If not, what could the problem be? Consult your textbook for
     diagnosis information and ask your instructor for help if
     necessary.

   _____

   _____

9. Return the starting and charging system   Completed ❑
   tester and electrical specifications literature
   to the proper storage areas.

10. Have your instructor check your work and   Completed ❑
    sign this worksheet.

    Instructor's signature: _____

Job 10

Diagnose the Computer System

Introduction

Computers are used to monitor and control all major systems of a modern car, from the engine systems to anti-lock brakes. For this reason, today's technician must be well versed in analyzing and repairing computer problems. In order to analyze computer problems, the technician must have direct access to the computer. On OBD II vehicles, the only way to access the computer is to attach a scan tool to the diagnostic connector. Once the connection to the computer has been made, the technician can retrieve trouble codes and other computer control system information.

Objective

Given the needed tools and equipment, you will retrieve and interpret trouble codes on a computer controlled vehicle system and obtain other diagnosis information from the system.

Materials and Equipment

- Scan tool.
- Basic set of hand tools.
- Thermometer.
- Multimeter.
- Heat gun.
- Vehicle with a computer controlled system.

Instructions

Review Chapter 15, *Computer System Diagnosis and Repair*, for information on computer systems. After getting your instructor's approval, use the shop scan tool to analyze the operation of the vehicle's computerized engine control system. As you read the job instructions, answer the questions and perform the tasks. Print your answers neatly and use complete sentences.

Warning

Before performing this job, review all pertinent safety information in the text and discuss safety procedures with your instructor.

Procedure

Locate Service Manual Information

1. Select the type of service literature needed, read the service information and look up the computer trouble codes for the test vehicle.

Completed ❑

Name _____

2. If you were not going to use a scan tool for this job, how would you pull up trouble codes on this vehicle? If this is an OBD II vehicle, is there any way to retrieve trouble codes without using a scan tool?

Completed ❑

_____

_____

_____

3. On the test vehicle, how would you know when a computer trouble code existed?

Completed ❑

_____

_____

_____

4. Locations of the diagnostic connector on two vehicles from major manufacturers are shown in **Figure 10-1.** Where is the diagnostic connector on the test vehicle?

Completed ❑

_____

_____

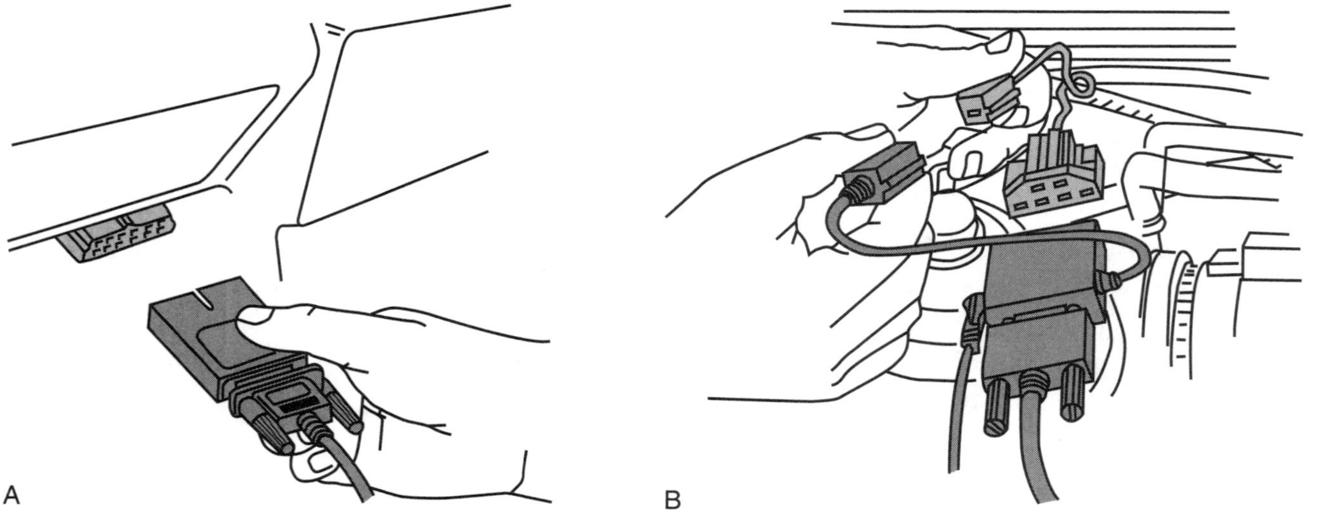

(Snap-On Tools)

Figure 10-1: The connectors for two major manufacturers are shown here. A—A GM ALDL connector is shown here. B—This Ford connector has a separate power lead.

5. Read the owner's manual for the scan tool being used. See **Figure 10-2.**

Completed ❑

6. Do you need to install a new scan tool cartridge for this particular vehicle? _____

Completed ❑

Name _____

If so, explain how this is done.

_____

_____

_____

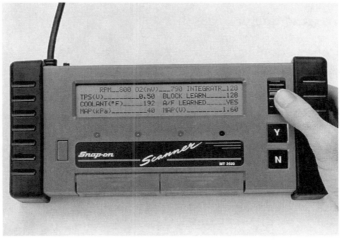

(Snap-On Tools)

Figure 10-2: Many manufacturers recommend the use of a scan tool to retrieve trouble codes.

Engine-off Scan

7. Connect the scan tool to the diagnostic connector on the vehicle. The ignition switch should be in the Off position.

   Completed ❑

8. To perform an engine-off scan test, turn the ignition to the On position and trigger the scan tool to retrieve trouble codes.

   Completed ❑

9. Does the scan tool show any trouble codes? _____ If so, what trouble code numbers are given?

   Completed ❑

_____

_____

10. Using a service manual, explain what each trouble code number indicates.

    Completed ❑

_____

_____

_____

Name _____

Engine-Running Scan

11. Next, do an engine-running scan. Start and warm the engine to full operating temperature to move the computer system into closed-loop operation.

Completed ❑

12. Trigger the scan tool to read trouble codes.

Completed ❑

13. Does the scan tool show trouble codes? _____ If so, explain what each code means.

Completed ❑

_____

_____

_____

Wiggle Test

14. To perform a wiggle test, wiggle the wires to the sensors and actuators while scanning for trouble codes. This test might help you uncover an intermittent problem.

Completed ❑

15. Did you find any problems during the wiggle test? _____ If so, explain.

Completed ❑

_____

_____

_____

Output Device Tests

16. To test the output devices, perform an output cycling test. During this test, the scan tool will signal the car's computer to operate injectors, idle solenoids, fan relays, and other actuators. This is sometimes called "forcing." Since most actuators convert electrical signals into physical movements, they can be seen or heard moving when cycled on and off. The scan tool will then tell you if the output devices are working properly.

Completed ❑

17. If the scan tool does not have the capability to perform an output cycling test, many actuators can be tested by measuring their internal resistance or by using jumper wires to apply an external voltage. Follow the manufacturer's instructions.

Completed ❑

Name _____

18. Explain the results of your actuator tests.   Completed ❑

_____

_____

_____

Pinpoint Tests

19. After you have retrieved trouble codes, you   Completed ❑
must do pinpoint tests to find the exact
source of the trouble. A trouble code simply
indicates which circuit might be at fault.

Sample Pinpoint Test

 **Caution: Depressurize the cooling system
before removing the sensor.**

20. Using the service manual for reference,   Completed ❑
remove the coolant temperature sensor
from the engine. See **Figure 10-3.**

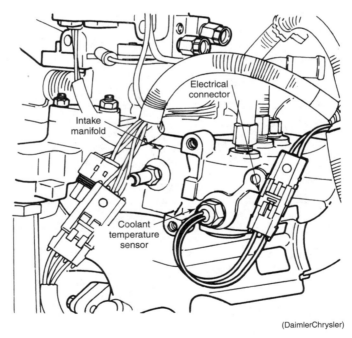

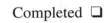

(DaimlerChrysler)

Figure 10-3: This coolant temperature sensor is mounted in the intake manifold coolant passage area. Disconnect the wires at the electrical connector before removing the sensor.

**Note: If an infrared temperature sensor is
available, it is possible to perform
Steps 21–23 without removing the sensor.**

Name _____

21. To simulate a test of the sensor, place the tip of the sensor in a beaker of heated water.   Completed ❑

22. Measure the sensors internal resistance with an ohmmeter. Take resistance readings at the following temperatures and record them in the chart.   Completed ❑

Water Temperature	Sensor Resistance
80°F	
100°F	
120°F	
140°F	
160°F	

23. Does the service manual give resistance specifications for the sensor? _____ If so, how do your readings compare to specifications? _____   Completed ❑

_____

_____

24. Next, disconnect the wires to a fuel injector. Using the ohmmeter, measure and record the injector's internal resistance. _____   Completed ❑

25. Does the injector resistance indicate a good or bad injector? _____ Explain.   Completed ❑

_____

_____

Erasure of Computer Trouble Codes

26. Reconnect all wires.   Completed ❑

27. Read and follow the service manual instructions for erasing stored trouble codes. How do you erase the stored trouble codes on the test vehicle?   Completed ❑

_____

_____

_____

Name _____

28. Erase any codes in the test vehicle fol-   Completed ❑
    lowing the instructions outlined in Step 27.

29. Return all equipment to its proper location.   Completed ❑

30. Have your instructor check your work and   Completed ❑
    sign this worksheet.

    Instructor's signature: _____

Job 11

Service the Ignition System

Introduction

Early electronic ignition systems used a conventional distributor, replacing the old contact points with electronic components. Newer electronic ignition systems eliminate the distributor and use crankshaft and camshaft position sensors and several coils to create and distribute high voltage. The latest direct ignition systems have a coil mounted directly on the spark plug. Electronic parts are dependable and do not require periodic maintenance. Some parts such as the coil(s), distributor cap, rotor, and wires (when used), may require periodic inspection. Spark plugs may be designed to last for 100,000 miles (160,000 km) or more. Instead of performing ignition system maintenance, the technician usually will be called on to diagnose an electronic ignition problem.

Objective

Given the needed tools and equipment, you will make visual and electrical checks to an electronic ignition system and disassemble and check an electronic ignition distributor.

Materials and Equipment

- Vehicle with a distributorless or direct electronic ignition system.
- Electronic ignition distributor.
- Service literature as needed.
- Basic hand tools.
- Multimeter.
- Dielectric grease.

Instructions

Review the Chapter 16, *Ignition System Service*, in your workbook and textbook. After getting your instructor's approval, obtain the needed tools and equipment. As you read the job instructions, answer the questions and perform the tasks. Print your answers neatly and use complete sentences. Do not be afraid to ask your instructor questions. As part of this job, you will rebuild and test an electronic ignition system. You will also disassemble and reassemble an electronic ignition distributor.

Warning

Before performing this job, review all pertinent safety information in the text and discuss safety procedures with your instructor.

Name _____

Procedure

Checking Distributorless or Direct Ignition System Components

1. Select a vehicle with a distributorless or direct ignition system, or a shop engine with distributorless or direct ignition.   Completed ❏

2. Make a visual inspection for the following: Completed ❏
   - Flashover or carbon tracking at the coil or coils. List any defect(s). _____
   - Damage to the plug wires and boots (if used). List any defect(s). _____
   - Corrosion on primary wiring or connections. List any defect(s). _____
   - Disconnected or damaged primary connectors. List any defect(s). _____

3. Based on the findings above, does the ignition system need any service? _____ List the service needed. _____   Completed ❏

⬦ **Note: If the system does not have spark plug wires, skip Steps 4, 5, and 6.**

4. Remove the terminals of one plug wire.   Completed ❏

5. Using an ohmmeter, check the wire resistance as shown in **Figure 11-1.**   Completed ❏

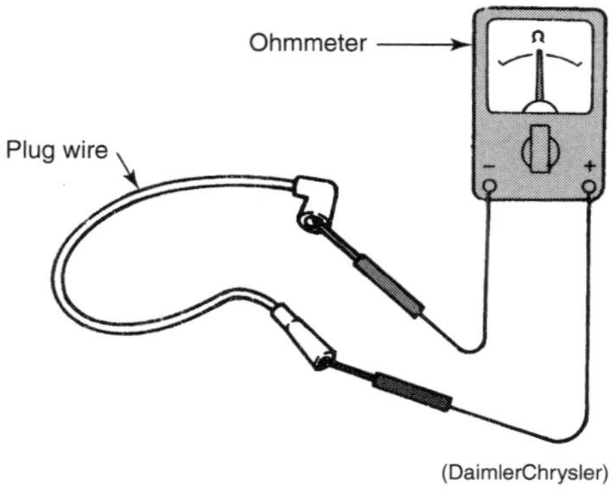

(DaimlerChrysler)

Figure 11-1: This spark plug wire is being tested with an ohmmeter.

6. Based on the ohmmeter readings, is the plug wire good? _____   Completed ❏

Name _____

7. Locate the pickup coil or crankshaft position sensor as applicable.   Completed ❑

8. Remove the electrical connector from the pickup coil or crankshaft position sensor.   Completed ❑

9. Test the coil or sensor with an ohmmeter.   Completed ❑

10. Compare the readings to specifications. Do the readings for the coil or sensor agree with the specifications? _____ If not, what could be the cause?   Completed ❑

_____

_____

Disassembling and Checking an Electronic Ignition Distributor

11. Have your instructor issue a distributor to you. Next, locate the proper service literature for that distributor.   Completed ❑

12. Remove the distributor cap from the distributor body. How is this done?   Completed ❑

_____

_____

13. Inspect the inside of the cap for signs of cracks, carbon traces, and burning. Can you find any signs of failure? _____   Completed ❑

14. Remove the ignition coil from the distributor cap. How do you remove the coil and the wire leads?   Completed ❑

_____

_____

15. Using an ohmmeter, check the internal resistance of the coil windings. See **Figure 11-2** and **Figure 11-3.**   Completed ❑
    • Coil Primary Winding Resistance: _____
    • Coil Secondary Winding Resistance: _____

16. Is the coil resistance within specifications?   Completed ❑

_____

Name _____

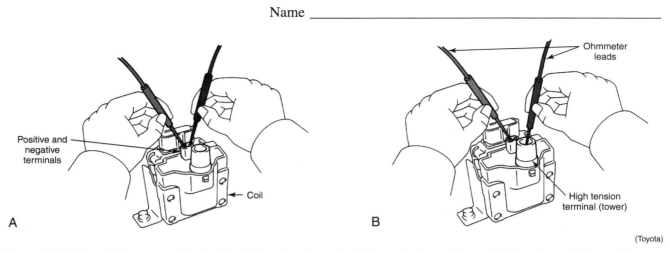

(Toyota)

Figure 11-2: Test points for a conventional coil are being shown here. A—Primary winding resistance. B—Secondary winding resistance.

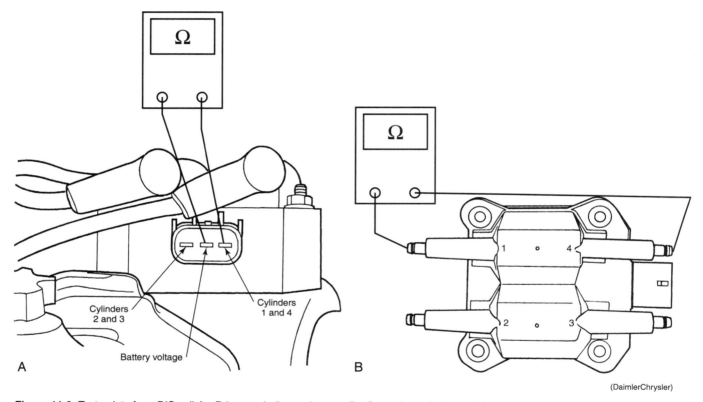

(DaimlerChrysler)

Figure 11-3: Test points for a DIS coil A—Primary winding resistance. B—Secondary winding resistance.

17. Check the distributor shaft bushings for wear. Try wiggling the top of the shaft sideways while you watch for movement in the distributor body. If the shaft wiggles sideways, the shaft bushings are worn.

                                                         Completed ❑

18. Describe the condition of the distributor shaft bushings.

                                                         Completed ❑

_____

_____

Name _____

19. Search to the service literature for informa-   Completed ❑
    tion about removing the distributor shaft.
    What do you have to do before removing
    the distributor shaft?

_____

_____

20. Remove the distributor shaft.   Completed ❑

21. Inspect the shaft itself for signs of wear at   Completed ❑
    the bushing bearing surfaces. Use a
    micrometer to measure shaft diameter at its
    most worn point and at an unworn point.
    • Shaft diameter (most worn): _____
    • Shaft diameter (unworn): _____

22. Is the distributor shaft worn beyond specifi-   Completed ❑
    cations? _____ Explain.

_____

_____

23. Use a telescoping gauge and micrometer to   Completed ❑
    measure distributor bushing inside diam-
    eter. Record the measurements.
    • Lower bushing inside diameter: _____
    • Upper bushing inside diameter: _____

24. Use the ohmmeter to check the condition of   Completed ❑
    the pickup coil. **Figure 11-4** shows a GM
    HEI pickup coil being checked. What is the
    pickup coil resistance? _____

Detach leads
from module

Module

A   B

(Champion)

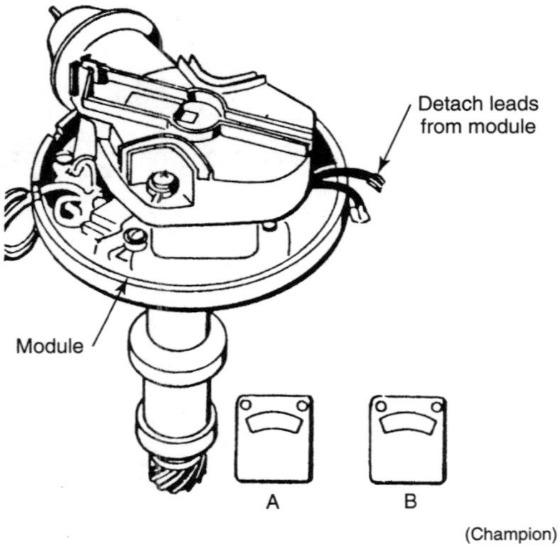

Figure 11-4: Check the pickup coil with an ohmmeter. A—Replace the pickup coil if this reading is less than infinite. B—The reading should be between 500 and 1500 ohms for this particular coil.

Name _____

25. Wiggle the pickup coil wires while measuring resistance. Sometimes the pickup coil wires can break internally and cause a resistance fluctuation. What did the wiggle test show you?   Completed ❑

_____

_____

26. Remove the ignition module. How many wires are connected to the module? _____   Completed ❑

27. Find a procedure for testing the module in the service manual. Briefly describe it here.   Completed ❑

_____

_____

28. Using a high impedance ohmmeter, measure the resistances across the terminals of the ignition module. Record your readings. _____   Completed ❑

29. To install the module, wipe a thin layer of dielectric grease across the bottom of the unit. This helps keep it from overheating. Install the module and tighten each fastener equally.   Completed ❑

30. Follow the service manual instructions to reassemble the distributor.   Completed ❑

31. Return all tools and parts to storage.   Completed ❑

32. Have your instructor check your work and sign this worksheet.   Completed ❑

Instructor's signature: _____

Job 12

Service the Fuel Injection System

Introduction

The engine will not run without fuel. Modern gasoline fuel injectors have solenoid operated flow control valves. The ECM uses inputs from various engine sensors to pulse the fuel injectors on and off. A pressure regulator controls overall fuel system pressure. Most modern systems are multiport systems with one injector per cylinder. Electric pumps are always used with fuel injection systems. Modern diesel fuel injection systems are also ECM controlled. Clearances in fuel injection parts are extremely small, and the slightest impurities can cause trouble. It is critical that you know how to troubleshoot and repair fuel injection systems.

Objective

Given the needed tools and equipment, you will troubleshoot and repair fuel injection system defects.

Materials and Equipment

- Vehicle with fuel injection.
- Basic hand tools.
- Service information as needed.
- Fuel pressure gauge.
- Hand vacuum pump.
- Fuel injection system tester.

Instructions

Review the fuel injection information in Chapter 18, *Fuel Injection System Service*, in your textbook and workbook. After getting your instructor's approval, use the following instructions and a service manual to analyze the operation of the vehicle's fuel injection system. As you read the job instructions, answer the questions and perform the tasks. Print your answers neatly and use complete sentences.

Warning

Before performing this job, review all pertinent safety information in the text and discuss safety procedures with your instructor.

Procedure

1. In the appropriate service literature, read the service information on the fuel injection system for the test vehicle.

Completed ❑

Name _____

- Vehicle make: _____
- Vehicle model: _____
- Vehicle year: _____
- Engine size: _____
- Transmission type: _____

Fuel System Inspection

2. Begin fuel injection service by inspecting the system for obvious problems. If possible, start the engine. Look for fuel leaks, disconnected wires, leaking vacuum hoses, and similar troubles.

Completed ❑

3. Could you find any visible problems with the fuel injection system? _____ If yes, explain what was found. _____

Completed ❑

4. Have someone turn the key to the Run position while you listen for the electric fuel pump. If the shop is noisy, you may have to get close to the pump to hear it. Can you hear the fuel pump running?

Completed ❑

5. Start the engine and use a stethoscope to listen to each of the injectors. Each injector should make a clicking sound, which indicates that it is open and closing. The sound from each injector should also be similar.

Completed ❑

6. Did all the injectors make the same sound? _____ If not, which injectors are defective?

Completed ❑

Checking Fuel Pressure

7. Locate the fuel pressure relief valve on the engine. It is usually on the fuel rail of multi-port injection systems. Central fuel injection systems usually have the relief valve on the throttle body.

Completed ❑

Note: Some vehicles do not have a service port. When servicing a system without a service port, follow the manufacturer's instructions carefully.

Name _____

8. Using the service manual instructions, relieve fuel pressure from the fuel system. See **Figure 12-1.**

Completed ❑

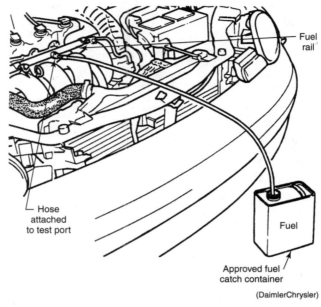

Figure 12-1: Use an approved fuel container to catch fuel during the pressure release procedure. Always relieve fuel pressure before working on the system.

> **Warning: Extreme care must be taken when relieving fuel system pressure. Always wear eye protection and keep a fire extinguisher handy. When loosening a fuel fitting or service port, cover the area with a shop cloth to contain any fuel that may spray out. Dispose of fuel-soaked cloths properly.**

9. Describe the procedure you used to relieve fuel pressure.

Completed ❑

_____

_____

_____

10. Connect a fuel pressure gauge in accordance with the service manual instructions.

Completed ❑

11. Turn the key to the On position but do not start the engine.

Completed ❑

12. Record the fuel pressure. _____

Completed ❑

13. Start the engine and record the fuel pressure at idle. _____

Completed ❑

Name _____

14. How does the fuel pressure reading compare to the specification when the key is on but the engine is not running? _____

   • How does the reading compare to the specification when the engine is running at idle? _____

Completed ❑

⬧ **Note: Some fuel pressure regulators do not have a vacuum line. If the system does not have a vacuum line, skip Step 15.**

---

15. With the engine running, remove the vacuum line to the pressure regulator. Fuel pressure should rise when the line is removed. Does the pressure increase? _____

Completed ❑

⬧ **Note: Your instructor may direct you to perform the following alternate method of fuel pressure regulator checking.**

Install a hand vacuum pump on the vacuum fitting to the fuel pressure regulator, which is usually on the fuel rail. See **Figure 12-2.** Start the engine and let it idle. Measure fuel pressure as you apply vacuum to the regulator. Explain what happens as you apply vacuum.

Completed ❑

_____

_____

_____

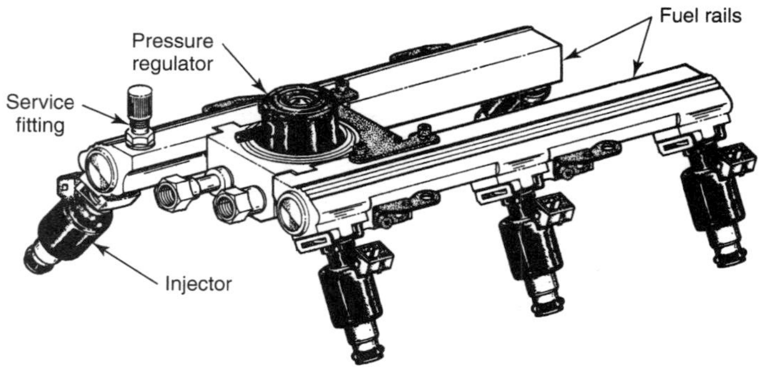

(AC Delco)

Figure 12-2: The placement of one manufacturer's pressure regulator is shown here.

Name _____

16. Find and remove the main fuel filter, not the   Completed ❏
    in-tank filter. If it is clogged, it can reduce
    fuel pressure. Where was the main fuel
    filter located? _____

17. Reinstall the fuel filter, Figure **12-3,** and   Completed ❏
    check for fuel leaks.

(DaimlerChrysler)

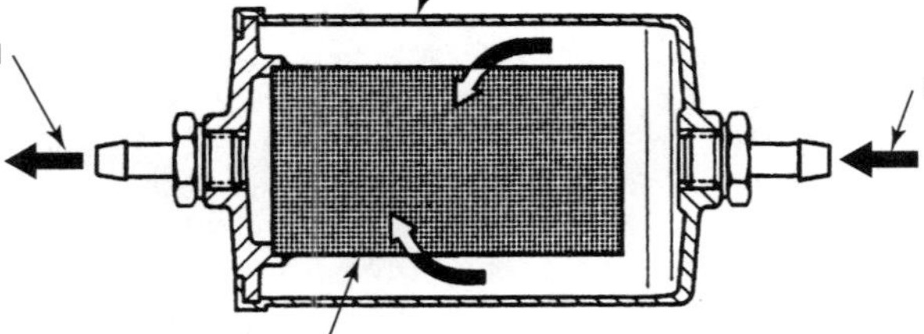

Figure 12-3: A cross section of a typical fuel filter is shown here. The entire assembly must be discarded when it becomes clogged. Be sure to install the new filter with the flow arrows pointing in the right direction.

Testing Fuel Injectors

18. What type of fuel injector tester are you   Completed ❏
    going to use? _____

19. Read the owner's manual for the injector   Completed ❏
    tester.

20. Install the injector tester on the engine   Completed ❏
    according to the manufacturer's instructions.

21. Start the engine and use the injector tester   Completed ❏
    to check each injector. What did the tester
    tell you about the condition of each
    injector?

    _____

    _____

22. If possible, use the injector tester to shut off   Completed ❏
    fuel to each cylinder in turn. Compare rpm
    drops on each cylinder. Did the rpm drop
    for each cylinder match the others within 50
    rpm? _____ If not, what could the problem
    be? Remember that a weak cylinder may
    not be caused by a fuel injector problem.

    _____

Name _____

23. Remove the tester from the vehicle. Return   Completed ❏
all equipment to its proper storage location.

24. Clean up any spilled fuel and properly   Completed ❏
dispose of fuel soaked rags.

25. Have your instructor check your work and   Completed ❏
sign this worksheet.

Instructor's signature: _____

Job 13

Test Exhaust System and Replace Exhaust System Components

Introduction

Except for the addition of catalytic converters and oxygen sensors, modern exhaust systems are similar to those used decades ago. Actual replacement procedures remain relatively simple. While exhaust systems usually last much longer than they used to, the technician will eventually have to locate and correct an exhaust system problem. Most exhaust system problems consist of leaks and restrictions. You must know how to troubleshoot and repair exhaust systems.

Objective

Given the needed tools and equipment, you will test for exhaust system leaks and back pressure, and replace defective exhaust system components.

Materials and Equipment

- Vehicle in need of exhaust system service.
- Service literature as needed.
- Basic hand tools.
- Vacuum gauge or pressure gauge as specified by your instructor.
- Air chisel with adapters.
- Pipe-straightening cone.
- Pipe expander.

Instructions

Review the exhaust system service material in Chapter 19, *Exhaust System Service*, in your textbook and workbook. After getting your instructor's approval, use the following instructions and a service manual to test and replace exhaust system components. As you read the job instructions, answer the questions and perform the tasks. Print your answers neatly and use complete sentences. Ask your instructor for help as needed.

Warning

Before performing this job, review all pertinent safety information in the text and discuss safety procedures with your instructor.

Name _____

Procedure

Caution: When performing any procedure that requires running the vehicle, make sure that exhaust fumes are being properly evacuated from the shop.

Exhaust Leak Check

1. Start the engine and set speed to fast idle.   Completed ❑

2. Raise the vehicle on a lift.   Completed ❑

3. Look and listen for leaks. Leaks can be heard, and rusted out components are usually visible. Were any leaks found? _____ List the defects found. _____   Completed ❑

_____

Caution: Exhaust system parts become hot quickly. Do not touch any exhaust parts with your bare hands.

4. Lower the vehicle, return the engine to idle and stop.   Completed ❑

Exhaust Restriction Check

Note: Use one or both of the following methods to check for exhaust restriction.

5. Attach a vacuum gauge to the intake manifold and check for low vacuum at high engine speed. Vacuum at 2500 rpm should not be lower than vacuum at idle.   Completed ❑

6. To use the other method of checking for an exhaust restriction, attach a pressure gauge to the exhaust manifold and check for high pressure at high engine speed. At 2000 rpm, the pressure should not be more than 3 psi (21 kPa).   Completed ❑

   • Method used: _____

   • Reading: _____

   • Specification: _____

Name _____

7. Does the exhaust appear to be restricted? _____ If yes, what could be the cause of the restriction?

Completed ❑

_____

_____

Changing Exhaust System Parts

8. Raise the vehicle. Use a lift if possible. It is far more convenient to lift the vehicle high enough to maneuver the pipe and mufflers.

Completed ❑

9. Apply penetrating oil to the pipe bracket fasteners and to the muffler outlet joint clamp.

Completed ❑

10. Remove the clamps and bracket fasteners.

Completed ❑

11. Loosen the outlet joint and pull the pipe free with a chain wrench. If a pipe will be reused, remove it carefully. If the parts are welded together, cut off the pipe where necessary to engage the new parts.

Completed ❑

12. If any parts are held together by bolted flanges, remove the bolts and separate the flanges.

Completed ❑

13. After removal, use a pipe-end straightening cone or a pipe expander to straighten the ends of all pipes that will be reused.

Completed ❑

14. Clean the inside or outside of the nipples that will be reused. If a nipple is distorted, use a straightening cone to restore it to a perfectly round state.

Completed ❑

15. Compare the old and new exhaust system parts. Do they match? _____

Completed ❑

16. Scrape old gasket material off of any flanged connectors that will be reused.

Completed ❑

17. Apply a coating of exhaust system sealant to the section of the pipe that will be installed inside of another pipe.

Completed ❑

18. Slide the pipes and nipples together. Make certain that the depth is correct.

Completed ❑

Name _____

19. Slide the clamps into position and tighten them lightly.   Completed ❏

20. Make sure that all parts are installed in the correct position and in the proper direction. Make sure that the exhaust system parts are aligned and will not contact any other part of the vehicle.   Completed ❏

21. Tighten all clamps and brackets. Do not over tighten clamps.   Completed ❏

22. Start the engine and check for leaks.   Completed ❏

23. Return all equipment to its proper storage location and dispose of old parts.   Completed ❏

24. Have your instructor check your work and sign this worksheet.   Completed ❏

   Instructor's signature: _____

Job **14**

Diagnose Driveability Problems

Introduction

Modern vehicles no longer receive periodic tune-ups to correct performance, smoothness, or fuel efficiency problems. Instead, vehicle systems are carefully checked to locate defective components. You must know how to use the seven-step troubleshooting process to find and correct driveability problems.

Objective

Given the proper tools and equipment, you will follow the seven-step troubleshooting process to determine the cause of a driveability problem.

Materials and Equipment

- Basic hand tools.
- Driveability worksheet.
- Service literature as needed.
- Scan tool.
- Multimeter.
- Other test equipment as needed.

Instructions

Review Chapter 20, *Driveability Diagnosis*, in your textbook and workbook. After getting your instructor's approval, use the following instructions and a service manual to troubleshoot and correct a driveability problem. As you read the job instructions, answer the questions and perform the tasks. Print your answers neatly and use complete sentences. Refer to the seven-step troubleshooting process in **Figure 14-1,** and ask your instructor for help as needed.

Warning

Before performing this job, review all pertinent safety information in the text and discuss safety procedures with your instructor.

Procedure

1. Determine the exact problem by performing the following:
   - Question the vehicle's driver.
   - Consult previous service records.
   - Road test the vehicle.

Completed ❑

Name _____

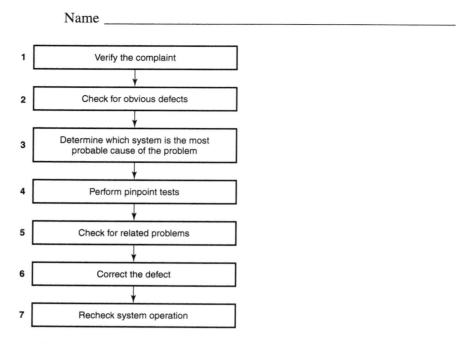

Figure 14-1: The seven-step diagnostic process is the quickest way to isolate the cause of a driveability problem.

2. Does your description of the problem agree with the driver's original description of the problem? _____ If the answer is no, explain.

Completed ❑

_____

_____

3. Use the Driveability Worksheet at the end of the job to record the information you've gathered.

Completed ❑

4. Check for obvious problems by performing the following:

Completed ❑

- Make a visual under-hood inspection.
- Retrieve trouble codes.
- Make basic electrical checks.

5. Describe any obvious problems found.

Completed ❑

_____

_____

6. Do you think that this could be the cause of the problem determined in Step 1? _____ Explain your answer.

Completed ❑

_____

_____

Name _____

> **Note: In cases where an obvious problem is located before all steps are completed, the instructor can approve the elimination of some of the following steps.**

7. Determine which component or system is causing the driveability problem by performing the following:

   • Obtain the proper service manual and/or other service information.

   • Use the service information to compare the type of problem with the trouble codes retrieved.

   • Determine whether the indicated problems could be the cause of the problem.

   • Use the scan tool to monitor fuel trim.

   • Use the scan tool to access the misfire monitor.

   • Make other checks as needed to determine which system or part is causing the problem.

   Completed ❑

8. List any readings that are not within speci-fications, or other abnormal conditions.

   Completed ❑

   _____

   _____

9. Do you think this could be the cause of the problem? _____ Explain your answer.

   Completed ❑

   _____

   _____

10. Eliminate potential causes by checking the components that could cause the problems found in Step 8.

    • Make physical checks.

    • Make electrical checks.

    • Make pressure, vacuum, temperature, or other checks as necessary.

    • List any defective components or systems found. _____

    Completed ❑

    _____

11. Do you think these defects could be the cause of the original problem? _____ Explain your answer.

    Completed ❑

    _____

    _____

Name _____

12. Isolate and recheck possible causes of the problem.   Completed ❑
    - Recheck damaged components.
    - Repeat electrical checks.

13. Did the rechecking procedure reveal any new problems or establish that suspected components were in fact good? _____ If there were any differences with the conclusions in Step 11, what was the reason?   Completed ❑

    _____

    _____

14. Correct the defect by making necessary repairs or adjustments. Briefly describe the services performed. _____   Completed ❑

    _____

15. Recheck system operation by performing either or both of the following: Completed ❑
    - Make checks using test equipment.
    - Road test the vehicle.

16. Did the service operations in Step 14 correct the problem? _____ If not, what steps should you take now? _____   Completed ❑

    _____

    _____

16. Clean the work area and dispose of old parts.   Completed ❑

17. Return all service manuals and test equipment to storage.   Completed ❑

18. Have your instructor check your work and sign this worksheet.   Completed ❑

    Instructor's signature: _____

Name _____

Driveability Worksheet

Name: _____   Date: _____

VIN: _____   Year: _____   Make: _____   Model: _____

Style: _____   Color: _____   Engine: _____   Trans: _____   A/C: _____   P/S: _____

Describe the vehicle's problem as accurately as you can: _____

_____

_____

_____

Check the boxes that best describe when the vehicle's problem occurs.

☐ Hot   ☐ Cold   ☐ Constant   ☐ Intermittent   ☐ Recurring

Check the boxes that best describe the type of symptom(s) that occur.

☐ Stalling	☐ Lack of power	☐ Overheating
☐ Hard start	☐ Surge/Chuggle	☐ Noise
☐ Does not start	☐ Pinging	☐ Vibration/Harshness
☐ Rough idle	☐ Miss/Cuts out	☐ Fluid leak
☐ Incorrect idle	☐ Poor fuel economy	☐ Unusual odors
☐ Hesitation	☐ Indicator light(s) on	☐ Smoke

Other: _____

_____

Background information (any service or repair work performed recently, other problems, etc.): _____

_____

_____

Job 15

Test Emission Systems

Introduction

Emission control systems are an integral part of all modern engines. Most emission controls are tied into the vehicle computer. Unlike emission controls on older vehicles, they cannot simply be bypassed when they cause trouble. You must know how to troubleshoot and repair emission control system problems.

Objective

Given the needed tools and equipment, you will check vehicle exhaust emissions and determine whether they are within legal requirements. You will also check the operation of emission control devices and determine whether they are performing properly.

Materials and Equipment

- Basic hand tools.
- Service literature as needed.
- Exhaust gas analyzer.
- Scan tool.
- Multimeter.
- Other test equipment as needed.

Instructions

Review the emissions control systems material in Chapter 21, *Emission System Testing and Service*, in your textbook and workbook. After getting your instructor's approval, use the following instructions and a service manual to test for emissions problems. As you read the job instructions, answer the questions and perform the tasks. Print your answers neatly and use complete sentences. Ask your instructor for help as needed.

Warning

Before performing this job, review all pertinent safety information in the text and discuss safety procedures with your instructor.

Procedure

STOP **Warning: Carbon monoxide will build up quickly during the following procedure. Do not perform the procedure in a closed area. Provide adequate ventilation.**

Name _____

1. Have your instructor assign a vehicle for use in performing the following tests.   Completed ❏

Checking Emission Controls and Engine Condition with an Exhaust Gas Analyzer

2. Place the exhaust gas analyzer hose in the exhaust pipe.   Completed ❏

3. Make other engine connections as needed.   Completed ❏

4. Start the engine.   Completed ❏

5. Observe exhaust readings at different engine speeds. Write the readings in the following chart. Indicate whether each reading is a percentage or a grams-per-mile figure.   Completed ❏

Exhaust Gas	Idle	1000 rpm*	2500 rpm*
HC			
CO			
NO$_x$			
CO$_2$			
O$_2$			

*The manufacturer may require testing at different engine speeds. Follow the manufacturer's instructions.

6. Turn off the engine.   Completed ❏

7. Do the analyzer readings indicate that the emission controls and engine are operating properly? _____ If not, what could be the cause? Refer to the reading and cause chart, **Figure 15-1.**   Completed ❏

Checking Individual Emission Control Components

8. Check EGR valve operation by one of the following procedures: Completed ❏

   • Make sure the valve moves when the engine speed is increased.
   • Apply vacuum to the diaphragm and note valve movement, **Figure 15-2.**
   • Use a test light or multimeter to check EGR solenoid operation.
   • Use a scan tool to check electronic solenoid operation.

Name _____

Excessive Hydrocarbon (HC) Reading: Excessive HC is usually caused by a problem that results in an incomplete burning of fuel. Sometimes accompanied by a "rotten egg" smell.

Poor cylinder compression
Leaking head gasket
Ignition misfire
Poor ignition timing
Defective input sensor
Defective output device
Defective ECM

Open EGR valve
Sticking or leaking injector
Improper fuel pressure
Leaking fuel pressure regulator
Oxygen sensor contaminated or
  responding to artificial lean or rich condition
Fuel filler cap improperly installed

Excessive Carbon Monoxide Reading: Excessive CO is caused by a problem that results in a rich air-fuel mixture. However, excessive CO is often created by an insufficient amount of air or too much fuel reaching the cylinder. Will sometimes coincide with a high HC and/or low O_2 reading.

Plugged air filter
Engine carbon loaded
Defective input sensor
Defective ECM
Sticking or leaking injector

Higher than normal fuel pressure
Leaking fuel pressure regulator
Oxygen sensor contaminated or
  responding to artificial lean condition

Excessive Hydrocarbon (HC) and Carbon Monoxide (CO) Readings: When both HC and CO are excessive, this often indicates a problem with the emissions control system or an on-going problem, usually indicated by a rich air-fuel mixture, that has damaged an emissions control component. You should check all of the systems previously mentioned along with these listed below.

Plugged PVC valve or hose
Fuel contaminated oil
Heat riser stuck open
AIR pump disconnected or defective

Evaporative emissions canister saturated
Evaporative emissions purge valve stuck open
Defective throttle position sensor

Excessive Oxides of Nitrogen (NO$_x$): Excessive oxides of nitrogen (NO$_x$) are created when combustion chamber temperatures become too hot or by an excessively lean air-fuel mixture.

Vacuum leak
Leaking head gasket
Engine carbon loading
EGR valve not opening
Injector not opening
Low fuel pressure

Low coolant level
Defective cooling fan or fan circuit
Oxygen sensor grounded or
  responding to an artificial rich condition
Fuel contaminated with excess water

Excessively Low Carbon Dioxide (CO$_2$) Reading: A low CO_2 reading is usually caused by a rich air-fuel mixture or a dilution of the exhaust gas sample. You should check all of the possibilities mentioned earlier, starting with the ones listed below.

Exhaust system leak
Defective input sensor
Defective ECM

Sticking or leaking injector
Higher than normal fuel pressure
Leaking fuel pressure regulator

Low Oxygen (O$_2$) Reading: Low oxygen readings are usually caused by a lack of air or rich air-fuel mixture, the same factors that can create an excessive CO reading. Note: Not all analyzers check the exhaust gas for O_2 content.

Plugged air filter
Engine carbon loaded
Defective input sensor
Defective ECM
Sticking or leaking injector

Higher than normal fuel pressure
Leaking fuel pressure regulator
Oxygen sensor contaminated or
  responding to artificial lean conditioning
Evaporative emissions system valve defective

High Oxygen (O$_2$) Reading: A high oxygen level is an indication of a lean air-fuel mixture, dilution of the air-fuel mixture, or dilution of the exhaust gas sample by outside air. When a high oxygen reading is present, the CO reading is usually very low or does not register.

Vacuum leak
Low fuel pressure

Defective input sensor
Exhaust system leak near the tailpipe

Figure 15-1: These are the most common causes of high emissions readings. Each section explains how the excessive reading is created.

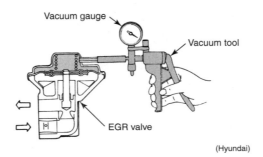

(Hyundai)

Figure 15-2: Apply vacuum to the EGR valve diaphragm.

Name _____

9. Is the EGR system operating properly?   Completed ❑
   _____ If not, what could the problem be?

   _____

10. Check air pump and injection system valve   Completed ❑
    operation by the following procedure:
    • Observe pump output with the engine running.
    • Observe whether the diverter valve operates when the engine
      is accelerated and decelerated.
    • Observe the operation of the switching valve as the engine
      warms up.
    • Observe the condition of the check valve(s).

11. Does the air injection system appear to be   Completed ❑
    working properly? _____ If not, what could
    be the problem? _____

12. Use a scan tool to check the operation of the   Completed ❑
    catalytic converter.

    ┌───┐
    │ **Note: This procedure can only be performed** │
    │ **on an OBD II-equipped vehicle.**   │
    └───┘

    • Attach a scan tool to the vehicle diagnostic connector.
    • Retrieve trouble codes as necessary.
    • Operate the engine and observe the readings from the oxygen
      sensors.

13. Do the oxygen sensor inputs indicate that   Completed ❑
    the converter is operating properly? _____
    If not, what could be the cause? _____

    _____

14. Check the PCV system condition by the   Completed ❑
    following procedure:
    • Remove the PCV valve from its grommet and shake it.
    • Use a tester to ensure that air is flowing into the crankcase
      when the engine is running.

15. Is the PCV system operating properly?   Completed ❑
    _____ If not, what could the problem be?

    _____

Name _____

16. Check evaporative emissions system opera-   Completed ❑
    tion by the following procedure:

    • Observe purge airflow with engine running, **Figure 15-3.**

    • Check the fuel tank, canister, and lines for damage or discon-
      nected components.

    • Pressurize the system with nitrogen and test for leaks

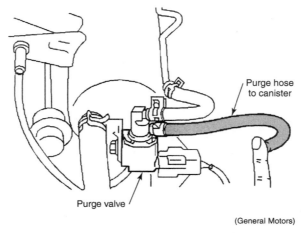

Purge hose
to canister

Purge valve

(General Motors)

Figure 15-3: Check for vacuum at the hose. There should be no vacuum with the engine cold, and full vacuum with the engine warm.

17. Does the evaporative emissions system   Completed ❑
    appear to be working properly? _____
    If not, what could be the problem?

    _____

18. Return all tools and equipment to storage.   Completed ❑

19. Clean the work area and dispose of old parts.   Completed ❑

20. Have your instructor check your work and   Completed ❑
    sign this worksheet.

    Instructor's signature: _____

Job 16

Perform a Maintenance Tune-up

Introduction

Tune-ups are no longer performed to correct drivability problems. The technician may, however, be called on to perform what is known as a maintenance tune-up. A maintenance tune-up consists of replacing the spark plugs and filters, and checking other vehicle components. This task is preventive maintenance and is not expected to solve any drivability problems.

Objective

After studying related textbook material and satisfactorily performing this task, you will perform a maintenance tune-up on a late-model vehicle.

Materials and Equipment

- Vehicle in need of a maintenance tune-up.
- Basic hand tools.
- Applicable service literature.
- Test equipment as needed.

Instructions

Review maintenance tune-up information in Chapter 20, *Driveability Diagnosis*, in your textbook and workbook. Check with your instructor before starting. The instructor may want you to perform other tasks on the vehicle, or to combine this job with another job. As you read the job instructions, answer the questions and perform the tasks. Print your answers neatly and use complete sentences. Ask your instructor for help as needed.

Warning

Before performing this job, review all pertinent safety information in the text and discuss safety procedures with your instructor.

Procedure

Performing a Preliminary Check for Defective Parts and Other Problems

1. Before beginning the tune-up check for problems that will not be cured by a tune-up.

   Completed ❑

2. Visually inspect all vacuum hoses, vacuum diaphragms and vacuum motors. If you suspect a problem, check vacuum units with a hand vacuum pump.

   Completed ❑

Name _____

3. Visually inspect electrical devices, wiring, and connections for obvious signs of corrosion or overheating. If necessary, follow up the visual inspection with electrical checks.   Completed ❑

4. Check the ignition secondary components visually, or with test equipment.   Completed ❑

5. Check the cylinder power balance.   Completed ❑

6. If any defects are found, consult your instructor before proceeding.   Completed ❑

Replacing Tune-up Related Parts

7. If the engine has aluminum cylinder heads, allow the engine to cool for one hour or more. If the engine has cast iron heads, skip this step.   Completed ❑

8. Remove a spark plug boot. Use a special boot removal tool if available to reduce the chance of damaging the boot.   Completed ❑

9. Remove the spark plug.   Completed ❑

> ◈ **Note: Observe the plug to determine if there are any fuel system or internal engine problems. Refer to textbook Chapter 16, Ignition System Service, if necessary.**

---

10. Gap the replacement plugs. **Figure 16-1** shows a typical plug gapping procedure. Bend the side electrode only.   Completed ❑

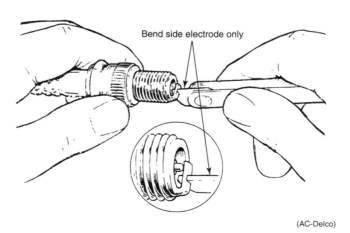

(AC-Delco)

Figure 16-1: When adjusting spark gap, bend only the side electrode.

Name _____

11. If the spark plug uses a gasket, make sure that it is in place on the new plugs. Tapered seat plugs do not use gaskets.   Completed ❑

12. Install the new plug and tighten it to the proper torque.   Completed ❑

13. Repeat steps 8–12 to replace all of the vehicle's spark plugs.   Completed ❑

14. Replace the distributor rotor if used.   Completed ❑

15. Replace the PCV valve and PCV filter if used.   Completed ❑

16. Replace the air and fuel filters, and the carbon canister filter if used.   Completed ❑

Cleaning Components

17. Clean the following components as recommended by the manufacturer, or as needed.   Completed ❑
    • Throttle body.
    • Fuel injectors.
    • Air cleaner housing.
    • PCV hoses.

Making Checks and Adjustments

18. Check the coolant level and condition.   Completed ❑

19. Check the air-fuel ratio and emissions with an emissions tester.   Completed ❑

20. Check the charging system voltage and amperage. If any problems are found, consult your instructor.   Completed ❑

21. Check and adjust the following when applicable: Completed ❑
    • Thermostatic air cleaner operation.
    • Turbocharger boost pressure.
    • Hot and cold idle speed.
    • Ignition timing.
    • Ignition advance devices.
    • Carburetor idle mixture.
    • Carburetor choke operation.
    • Head bolt torque (only when recommended).

Name _____

22. Road test the vehicle to confirm that service has been performed properly.   Completed ❑

23. Return all equipment to storage.   Completed ❑

24. Clean the work area and dispose of old parts.   Completed ❑

25. Have your instructor check your work and sign this worksheet.   Completed ❑

    Instructor's signature: _____

Name _____ Date _____

Score _____ Instructor _____

Job 17

Service the Cooling System

Introduction

An automotive cooling system must be serviced periodically. The cooling system removes around 30 percent of all of the heat energy produced by combustion. If ignored, the cooling system can cause serious mechanical breakdowns in a short period of time. Deteriorated hoses and belts, system leaks, bad coolant pumps, faulty radiator caps, and stuck thermostats can all cause sudden and complete failure of the system. After extended use, antifreeze can break down, allowing the system to corrode. It is important that you know how to service a cooling system.

Objective

Given the proper tools and equipment, you will learn to service an automotive cooling system.

Materials and Equipment

- Vehicle with cooling system in need of service.
- Basic hand tools.
- Cooling system pressure tester.
- Safety glasses.
- Antifreeze tester.
- Replacement coolant.

Materials for the Thermostat Test

- Thermostat.
- Heat-resistant container.
- Mechanic's wire.
- Heat source (oven or hot plate).

Instructions

Before starting this job, review Chapter 23, *Cooling System Service*, in your textbook and workbook. Since this job contains two sections, your instructor may want you to perform the thermostat test at home on a stove as a homework assignment. Ask about these details. The cooling system testing and inspecting portion of the job will be performed in the shop on a school-owned assembly. As you read the job instructions, answer the questions and perform the tasks. Print your answers neatly and use complete sentences.

Warning

Before performing this job, review all pertinent safety information in the text and discuss safety procedures with your instructor.

Name _____

Procedure

Inspecting the Cooling System

1. Obtain a vehicle in need of cooling system service.   Completed ❑

2. Visually inspect the cooling system. Squeeze the hoses to check for hardness, cracks, or softness. Look at **Figure 17-1.**   Completed ❑

Figure 17-1: Squeeze the hoses to check for hardness or for swelling and softness.

3. Allow the system to cool for at least 30 minutes.   Completed ❑

4. Remove the radiator cap and check the condition of the coolant. There should be no signs of rust and the system should be filled properly.   Completed ❑

5. Measure the freeze point of the coolant with an antifreeze tester. Since testers vary, follow manufacturer's operating instructions. The freezing point of the coolant should be appropriate for your area. The freezing point should be at least −35°F (−37°C). This indicates a 50-50 antifreeze and water mix, which is necessary in all vehicles for corrosion protection.   Completed ❑

> **Note: Freezing points will vary for different types of antifreeze. Textbook Chapter 23, Cooling System Service, contains the needed information concerning freeze points of various antifreeze mixtures.**

Name _____

6. Check that the radiator fins are free of leaves and bugs. The rubber seal on the radiator cap should be soft and unbroken.

Completed ❑

7. Explain the condition (good, fair, replace) of each of the system's check points in the spaces provided in the following chart.

Completed ❑

Check Point	Condition
Radiator Hoses	
Heater Hoses	
Water Pump Bearings	
Coolant Level	
Coolant Condition	
Antifreeze Protection	
Radiator Cap	
Radiator Fins	
Belt Condition	
Belt Tension	

8. Why did any of the cooling system check points fail in Step 7?

Completed ❑

_____

_____

Servicing Belts

9. Inspect the drive belts. Look for signs of slippage or cracking. See **Figure 17-2.**

Completed ❑

Figure 17-2: Check belts for cracks, splits, frayed edges, glazing, and other damage. Replace the belt if wear is evident.

Name _____

 Note: If the belts are in good condition, skip Steps 10 and 11.

10. If any belt is damaged, replace it by loosening the bracket bolts on the alternator, air conditioning compressor, power steering pump, or air pump. You may need a droplight to find all of the bolts.

Completed ❑

11. Install the new belt(s).

Completed ❑

12. Adjust the belts as necessary. Make the belts as loose as possible without allowing them to slip, squeal, or flop in operation. This will lengthen the life of the bearings in the alternator, a/c compressor, coolant pump, and power steering pump. Special belt tension gauges are available that will measure exact tension. To adjust belts by hand, refer to **Figure 17-3.** In general, a belt, depending on length, should deflect about 1/2″–5/8″ (13 to 9.5 mm) with 25 pounds (11 Kg) of push applied.

Completed ❑

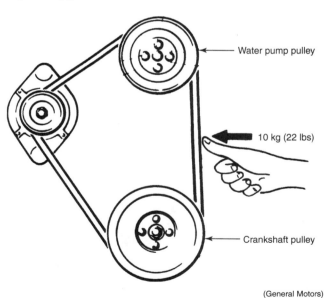

Water pump pulley

10 kg (22 lbs)

Crankshaft pulley

(General Motors)

Figure 17-3: Check the belt tension. A loose belt can cause slippage and overheating. An overtightened belt can cause premature bearing wear in the water pump and alternator.

Note: The alternator bearings are not submerged in lubricant, as with the a/c compressor, coolant pump, and power steering pump. The alternator belt should not be tightened as much as the other belts.

Name _____

13. List the types of belts (alternator, power steering) used on the engine.

Completed ❑

_____

_____

Draining Coolant

14. Locate the drain cock on the bottom of the radiator. Some vehicles have drains installed on the engine block. Open the engine block drains also.

Completed ❑

15. Open the drain cock and allow the coolant to drain into a container.

Completed ❑

Thermostat Service

16. Remove the thermostat from under the top radiator hose fitting. This fitting is also called a thermostat housing. Scrape off all of the old gasket material from both the engine and the thermostat housing.

Completed ❑

Note: The following step is optional. Ask your instructor if you should create your own thermostat gasket or use a preformed thermostat gasket.

---

17. Make a new gasket by the following procedure. Obtain a piece of gasket material, a pair of scissors, and a small ball peen hammer. Lay the thermostat housing over the gasket material and trace the outside shape of the housing with a pencil or pen. Cut out this shape with scissors. Hold this pattern over the thermostat opening in the engine, **Figure 17-4.** Then, to cut out the inside holes for the thermostat and bolts, tap lightly around the edge of the openings with the peen end of the hammer. Make sure that you do not shift the gasket while tapping or the holes will be misaligned. After tapping and perforation, the inside of the gasket holes will easily tear out, **Figure 17-5.**

Completed ❑

18. Reinstall the thermostat, making sure that it is right side up (pellet toward engine or pin pointing toward radiator), as shown in

Completed ❑

Name _____

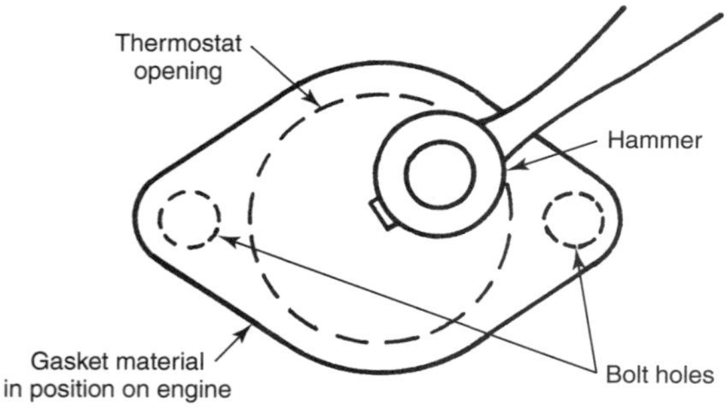

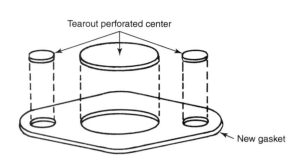

Figure 17-4: Lay a piece of gasket material over the thermostat opening on the engine. Tap along the hole edges (dotted lines) to cut out the gasket.

Figure 17-5: After tapping with the hammer, tear out the center of the holes.

Figure 17-6. Coat the sealing surfaces with nonhardening sealer and fit the gasket and housing into place.

- When tightening the thermostat housing bolts, be extremely careful not to overtighten the bolts. Snug them a little at a time to pull the housing down straight and prevent distortion. Overtightening the housing can cause breakage.

19. How do you know the thermostat is *not* installed backwards?

Completed ❏

_____

_____

Refilling the Coolant System

20. Close the drain cocks on the bottom of the radiator and engine block.

Completed ❏

21. Refill the system with a 50-50 mixture of the proper kind of antifreeze and water. Many modern vehicles take special long-life coolant. Never mix coolant types.

Completed ❏

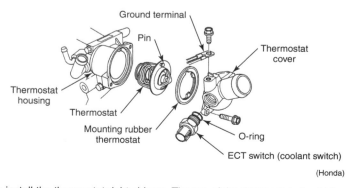

(Honda)

Figure 17-6: Make sure that you install the thermostat right side up. The top of the thermostat should face away from the engine.

Name _____

22. When the radiator is filled, lightly replace the radiator cap and start the engine.

Completed ☐

23. Allow the engine to reach operating temperature to open the thermostat. When the thermostat opens, the coolant level in the radiator will drop.

Completed ☐

24. Continue adding 50-50 mixture of antifreeze and water until the level stabilizes.

Completed ☐

25. Open any bleed valves on the thermostat or other coolant passage to bleed air from the engine block.

Completed ☐

26. Fill the coolant reservoir as necessary.

Completed ☐

Pressure Testing the System

27. Using a system pressure tester, check the condition of the radiator cap. Mount it on the tester as in **Figure 17-7.** Pump up pressure until the pressure needle levels off at a specific pressure.

Completed ☐

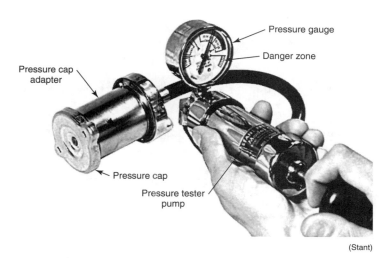

(Stant)

Figure 17-7: A good radiator cap will hold the pressure specified on the top of the radiator cap.

- A good radiator cap will hold the pressure listed on top of the cap and will not leak. A bad cap, besides leaking, may not hold pressure within specified limits (pressure high or low).

28. Explain the results of the pressure cap test.

Completed ☐

_____

_____

Name _____

29. Again using the pressure tester, test for cooling system leaks by mounting the tester on the radiator. See **Figure 17-8.** Pressurize the system to the pressure rating given on the top of the radiator cap (usually 8 psi–16 psi). If the pressure holds for two minutes without bleeding off, the system is not leaking. While waiting, you should watch for leaks.

Completed ❑

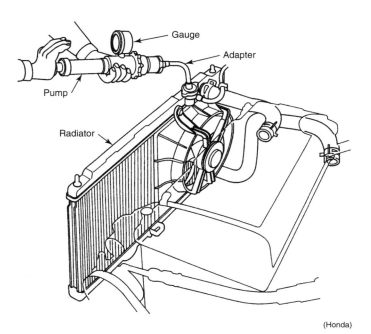

(Honda)

Figure 17-8: When pressure testing the system, pump the system to the pressure listed on the radiator cap and no more.

30. Explain the holding pressure and condition of the cooling system.

Completed ❑

_____

_____

31. Return all of tools and equipment.

Completed ❑

32. Have your instructor check this portion of the job. At this time ask your instructor about the details of the second portion of the job.

Completed ❑

Thermostat Testing

33. First, visually inspect the thermostat. Hold it up to the light and check to see if there is

Completed ❑

Name _____

a gap or opening around the edge of the sealing valve. If there is, the thermostat may be defective and may require replacement. Regardless of the condition, it will be tested.

34. How is the thermostat's appearance? Completed ❏

_____

35. Arrange the water-filled container, thermostat, thermometer, and heating source as shown in **Figure 17-9.** Stir the water gently while heating and watching the thermostat. Completed ❏

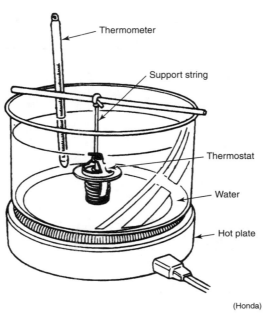

(Honda)

Figure 17-9: This is a standard setup for testing a thermostat's opening temperature. Both thermometer and thermostat should be kept away from the sides and bottom of the container.

36. Record whether the thermostat valve is closed, starting to open, or fully opened in the following chart. Write the answer next to the appropriate temperature. For example, the thermostat valve may be closed at 180°F (82°C) and open at 210°F (99°C). Watch very closely for the opening temperature when the valve first starts to open. It is important. Completed ❏

Name _____

Water Temperature	Valve Position
100°F (38°C)	
150°F (66°C)	
170°F (77°C)	
175°F (79°C)	
180°F (82°C)	
185°F (85°C)	
190°F (88°C)	
195°F (91°C)	
200°F (93°C)	
205°F (96°C)	
210°F (99°C)	
212°F (100°C)	

37. At what temperature did the thermostat start to open? _____   Completed ❑

38. At what temperature was it fully open? _____   Completed ❑

39. A good thermostat should start to open at 5°F–10°F (2.7°C–5.6°C) above or below the operating temperature stamped on the thermostat. It should be fully open at a temperature 20°F–25°F (11.2°C–13.9°C) above the opening temperature.   Completed ❑

40. Can the thermostat be reused? _____ Explain.   Completed ❑

_____

_____

41. Have your instructor check your work and sign this worksheet.   Completed ❑

Instructor's signature: _____

Job 18

Service Cylinder Heads

Introduction

During the reconditioning of a cylinder head, it is very important to closely inspect all of the components for wear or damage. If a cylinder head is cracked, warped, or burned between the combustion chambers, it must be repaired or replaced. Other parts should be checked and compared to specifications. A good technician knows that overlooking any problem can cause engine failure. A poorly reconditioned cylinder head will have to be redone at the shop's expense.

Objective

Using the tools and equipment listed below, you will disassemble and reassemble a cylinder head, and inspect, test, and measure cylinder head components.

Materials and Equipment

- Valve spring compressor.
- Brass hammer.
- Safety glasses.
- Straightedge.
- Flat feeler gauge.
- Ruler.
- Sliding caliper.
- Caliper or micrometer.
- Cylinder head assembly.

Instructions

Review Chapters 24 and 25 in your textbook and workbook. Ask your instructor for any added details for the job. As you read the job instructions, answer the questions and perform the tasks. Print your answers neatly and use complete sentences.

Remember to keep all of your parts in containers. If you drop any of the small keepers, make sure that you find them. Inform your instructor if you lose anything or run into problems.

Warning

Before performing this job, review all pertinent safety information in the text and discuss safety procedures with your instructor.

Name _____

Procedure

Checking the Head for Warpage

1. Identify the cylinder head used in the job by listing the following information.

   Completed ❑

   • Engine make: _____

   • Engine model: _____

2. Before disassembling the cylinder head, check it for warpage. Lay a straightedge across the surface of the head at various angles. Try to slide different size feeler gauge blades between the straightedge and the head. See **Figure 18-1.** The largest blade that will fit between the straightedge and head indicates the amount of cylinder head warpage.

   Completed ❑

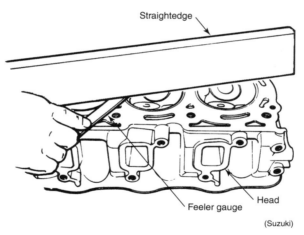

Figure 18-1: Use a straightedge and feeler gauge to test the cylinder head for warpage.

3. When a head is warped beyond specifications, it must be milled or machined to straighten its deck surface. Engine overheating often causes the cylinder head to warp. A warped cylinder head may cause the head gasket to blow (begin leaking), which may then lead to water and oil leakage and more overheating. If antifreeze finds its way into the combustion chamber, the piston can be destroyed.

   Completed ❑

   • When an engine has driven with a leaking head gasket, serious head and block damage can occur. The head and block surfaces can be burned away by the hot, high-pressure combustion leak. If the burn depth is not too deep, it can be machined away by milling.

Name _____

4. A small amount of head distortion is acceptable, around 0.003″ (0.076 mm) on any 6″ (152 mm) surface. When machining, always check with specs for maximum amount of milling permissible.

Completed ❑

5. What was the largest size of feeler gauge that would fit between the straightedge and the head? _____

Completed ❑

6. Describe the condition (straightness) of the head and discuss the action that should be taken.

Completed ❑

_____

_____

Head Disassembly

7. Begin cylinder head disassembly by striking the retainers to free them from their keepers. Use a brass hammer to keep from damaging the valve stems.

Completed ❑

8. Using a spring compressor, as shown in **Figure 18-2,** squeeze the valve springs and remove the keepers. One end of the compressor fits on the head of the valve and the other over the spring retainer. Hold on to the compressor firmly and keep it square. If it starts to slip, open and reposition the compressor.

Completed ❑

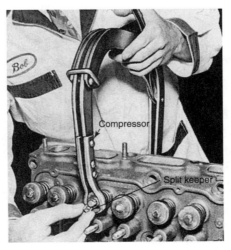

(General Motors)

Figure 18-2: Compress the valve springs and remove the keepers.

Name _____

9. Open the compressor and remove the retainer, spring and the seal.   Completed ❑

10. Place all of the organized parts in a container. Do *not* remove the valves from the head at this time.   Completed ❑

11. Repeat steps 8–10 on the rest of the valves. Don't lose any of the parts.   Completed ❑

Checking Valve Guides

12. With the springs removed, check the condition of all of the valve guides. Look at **Figure 18-3.** Pull each valve open about 1/4″ (6.35 mm) and wiggle it sideways and up and down. The valves should not be excessively loose in their guides. If a valve wiggles excessively, wear is present in the guide or on the valve stem.   Completed ❑

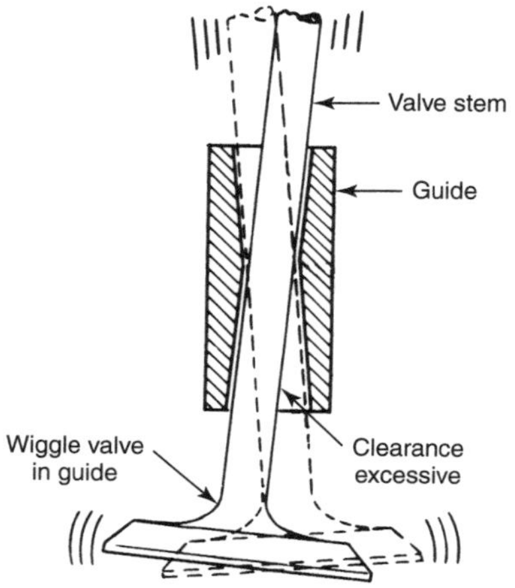

(General Motors)

Figure 18-3: Stem-to-guide clearance near the end of the guide must be within limits. Wiggle the valve to check for stem-to-guide wear.

13. When a valve guide is worn beyond specifications, it must be repaired or replaced. When a valve stem is worn, the valve must be replaced. Mark any of the worn guides so that they can be serviced. Valves with worn stems require replacement.   Completed ❑

14. List the condition of each of the valve guides in the following chart. Label the guides as being in "good" (no wiggle), "fair" (slight wiggle under 1/32″), or "bad"   Completed ❑

Name _____

(over 1/32″ wiggle) condition. Start at the punch-marked end of the cylinder head and work across. The instructor's punch marks will be on the end of the head, not on the machined face of the head.

Guide number	1	2	3	4	5	6	7	8 4 or 8 cyl.	9	10	11	12 6-cyl.
Guide condition												

Inspecting Valves

15. Remove one valve at a time, keeping the valves in order. Inspect each valve for wear or damage by comparing them to **Figure 18-4.** The surfaces of the stems should not be worn or rough. The faces of the valves should not be excessively burned. The heads of the valves should have a margin.

Completed ❑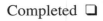

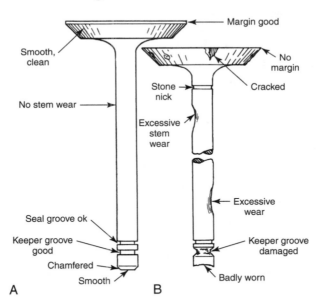

Figure 18-4: Valve A is acceptable. Valve B is not. Study the problems.

16. A margin is a flat lip between the face and head of the valve. See Figure 18-4. If the margin is gone and the valve is relatively sharp, the valve must be replaced. Also, look on the end of the valve stem and check for wear.

Completed ❑

Name _____

17. List the condition of each of the valves in the following chart. Check the condition of the stems, stem tips, margins, and faces. Categorize them as being in "good" (no detectable wear), "fair" (normal amount of correctable wear), or "bad" (unrepairable wear or damage) condition.

Completed ❑

Valve number	1	2	3	4	5	6	7	8 4 cyl	9	10	11	12 6-cyl
Stem condition												
Margin condition												
Face condition												

Grinding Valves

18. If this is an actual reconditioning of a cylinder head (valve job), grind the valves and seats at this time. If you are using a shop-owned cylinder head, go on to the next step of the job. If this is a real repair, check with your instructor for additional information. You may be told to perform Job 19, Service Valves, at this point.

Completed ❑

 Note: Do not attempt to use valve-grinding equipment without your instructor's permission.

19. To assure that you can identify the parts of a cylinder head correctly, use a ruler, sliding caliper, or micrometer to measure the following dimensions.

Completed ❑

- Intake Valve Head Diameter
- Exhaust Valve Head Diameter
- Valve Length
- Valve Stem Diameter
- Valve Margin Width (Maximum)
- Valve Margin Width (Minimum)
- Intake Port Height
- Valve Spring Free Height

Name _____

Reassembling the Head

20. Begin reassembly of the cylinder head by oiling the valve stems and then installing the valves into their original guides.

Completed ❑

21. Basically, there are two types of valve seals presently in use, umbrella and O-ring types. Look at **Figure 18-5.** Naturally, the umbrella type is installed before the valve spring. However, the O-ring type seal must be installed after the spring and retainer.

Completed ❑

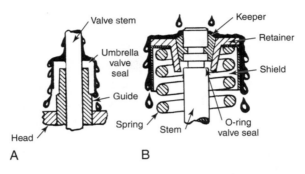

A   B

Figure 18-5: The two types of valve seals are shown here. A—An umbrella seal is installed before the spring. B—The spring and retainer must be compressed before installing an O-ring seal.

 Note: If you install an O-ring type seal before compressing the spring over the valve, the seal may be cut and damaged. The engine may smoke and consume an excessive amount of oil.

22. Explain how you installed the oil seals.

Completed ❑

_____

_____

23. Reassemble the rest of the parts of the cylinder head. To seat the keepers in their retainers, tap the ends of the valve stems with a brass hammer. A steel hammer can damage the stems.

Completed ❑

Checking for Valve Leakage

Note: The following step is optional. Ask your instructor if you should complete it.

Name _____

24. The heads can be tested for leaks by pouring   Completed ❑
    water or parts solvent into each of the ports
    in the head. Place the head with the ports
    facing up and pour the water or parts cleaner
    into the ports. If fluid leaks out around the
    head of any of the valves, the valve or seat
    must be reground or lapped. This operation
    can save you the unpleasant task of having
    to repeat the valve job. After you've
    completed the test, blow the head dry.

25. Have your instructor check your work and   Completed ❑
    sign this worksheet.

    Instructor's signature: _____

Job 19

Service Valves

Introduction

This exercise, along with Job 18, *Service Cylinder Heads*, should give you enough experience and knowledge to perform an actual engine valve job. Remember that the fit between a valve face and its seat is critical. The slightest mistake can cause compression leakage, rough idle, poor gas mileage, and increased emissions. In such a case, the head would have to be removed from the engine and the valve job done over.

Objective

Given a cylinder head and the tools listed, you will recondition the intake and exhaust valves and the cylinder head valve seats.

Materials and Equipment

- Safety glasses.
- Valve spring compressor.
- Brass hammer.
- Steel rule.
- Valve grinding machine.
- Valve seat grinder.
- Air or electric drill.
- Rotary carbon brush.
- Cylinder head assembly.

Instructions

Review the information in Chapter 25, *Cylinder Head and Valve Service*, in your textbook and workbook. Ask your instructor for the location of the cylinder head to be used in this job. This job may be performed in conjunction with Job 18, *Service Cylinder Heads*. You should also have seen demonstrations on the use of valve- and seat-grinding equipment and be familiar with its use.

As you read the job instructions, answer the job questions and perform the job tasks. Print your answers neatly and use complete sentences. Ask your instructor for help as needed.

Warning

Before performing this job, review all pertinent safety information in the text and discuss safety procedures with your instructor.

Name _____

Procedure

Disassembling the Cylinder Head

 Note: Some of the steps in this section may have been performed as part of Job 18, Service Cylinder Heads.

1. Clean the carbon from the combustion chamber, using the drill and wire brush, **Figure 19-1.** Wear eye protection as carbon and wire bristles can fly. Even if the head is already clean, practice using a wire brush.

Completed ❑

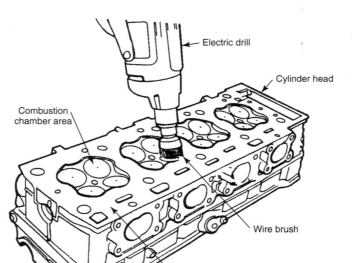

(General Motors)

Figure 19-1: Use an electric drill and a wire brush to remove carbon from a cylinder head combustion chamber.

2. Using a valve spring compressor, remove the valve springs from the cylinder head. If the retainers are stuck to the keepers, tap the retainers with a brass hammer.

Completed ❑

Checking Guide Wear

3. Pull each valve open about 1/4″ (6.4 mm) and check for valve guide and stem wear. See **Figure 19-2**. There should be no detectable movement when the valve is wiggled from front to back and from side to side. A loose valve can result in valve

Completed ❑

Name _____

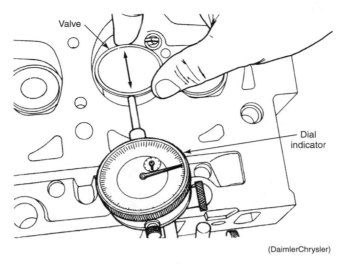

(DaimlerChrysler)

Figure 19-2: A dial indicator can be used to measure actual guide wear.

breakage, excessive oil consumption, clattering noises that sound like loose valve train parts, and compression loss.

4. What is the condition of the valve guides?   Completed ❑

Checking Valve Condition

5. Inspect valve condition. Slide the valves out of the cylinder head and inspect them for rough and worn stems, burned faces, worn stem tips, or other problems as in **Figure 19-3.**   Completed ❑

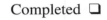

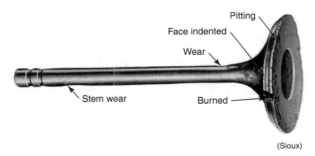

(Sioux)

Figure 19-3: Always check for valve defects.

⬥ **Note:** If the valve will not pull out of the head freely, check to see if the stem is mushroomed. If it is, you will need to file off and chamfer the end of the valve stem with a hand file. Never hammer a valve out of a head or the guide can be broken.

Name _____

6. Now, clean any carbon from the valve heads. Use an electric wire wheel, **Figure 19-4.** Wear leather welding gloves and eye protection. Also, check that the tool rest is close to the wire wheel.

Completed ❑

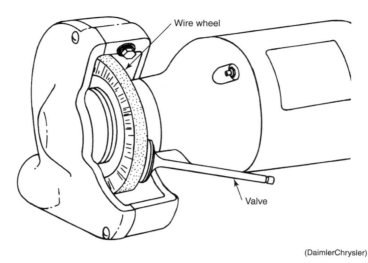

(DaimlerChrysler)

Figure 19-4: A wire wheel will quickly remove carbon from a valve. Always wear eye protection when cleaning the valves.

Grinding Valves

7. Go to the valve-grinding machine and prepare it for operation by checking the fluid levels and plugging it in. Also, familiarize yourself with its controls. Move the lever and turn the control wheel while watching the resulting movement of the machine.

Completed ❑

8. Dress (smooth) the grinding stone according to the manufacturer's instructions. If the valves are being ground for practice on a shop cylinder head, ask your instructor whether you should dress the stone. Stone dressing is normally done during an actual valve job.

Completed ❑

9. Insert the valve into the valve grinder. Check that the chuck jaws grasp the valve stem on the shiny, machined portion of the stem nearest the valve head. This operation is illustrated in **Figure 19-5.** Do not let the valve protrude out or slide in too far.

Completed ❑

Name _____

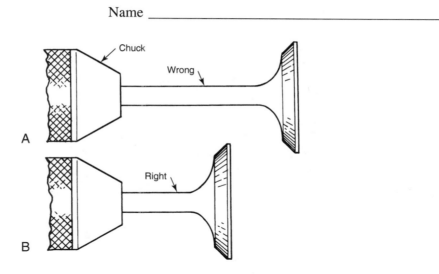

Figure 19-5: Insert the valve to the proper depth or wobble may result. A—Valve protruding too far out of the chuck. B—Valve depth is correct.

10. Turn the grinder on and see if the valve is wobbling in the chuck. If it is, check for dirt on the stem or chuck. Remount the valve and try again. When wobble cannot be corrected, the valve is bent and must be replaced.

Completed ❑

11. Now, determine the valve face angle by looking it up in a service manual. Adjust the grinder to the appropriate angle by loosening the chuck hold-down nut and swiveling the chuck mechanism until its degree marks line up accordingly. Next, observe the valve face angle relative to the grinding stone. The two should be perfectly parallel, as in **Figure 19-6.** Most valves use a 45° or 30° angle.

Completed ❑

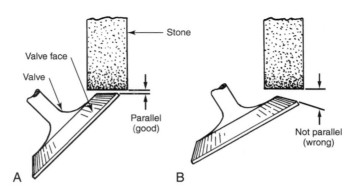

Figure 19-6: To adjust the chuck angle, loosen the hold-down nut and swivel the chuck until the proper angle marks are lined up. The stone should be parallel to the valve face.

Name _____

12. If an interference angle is desired, set the grinder to produce a cut that is 1° smaller than the valve seat angle. For instance, if the valve seat is to be cut at 45°, you should set the chuck on the valve grinder at 44°. If the valve seat is to be cut at 30°, you should set the chuck on the valve grinder at 29°. This will provide quick valve sealing when the engine is started. If in doubt, refer to service manual.

Completed ❑

13. When beginning to grind the valve, feed the valve into the stone very slowly. Make sure the coolant is flowing over the valve head. The valve should be positioned in front of the stone, while turning the depth wheel a little at a time.

Completed ❑

14. As soon as the stone touches the valve, start moving the valve back and forth, **Figure 19-7.** Valves and stones can be easily ruined during this step of the repair, so take your time.

Completed ❑

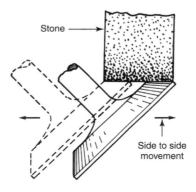

Figure 19-7: Keep the valve directly in front of the grinding wheel and feed it in slowly.

15. What is the valve face angle? _____

Completed ❑

16. While continuing to move the valve back and forth, slowly feed the valve into the turning stone, **Figure 19-8.** Feed very slowly while noting the depth of cut. Use the markings on feed wheel to monitor the depth of cut. Watch the face of the valve carefully. As soon as the face is cleaned up (black pits and marks removed), stop. Back off the depth wheel. You only want to grind off enough metal to true the face of the

Completed ❑

Name _____

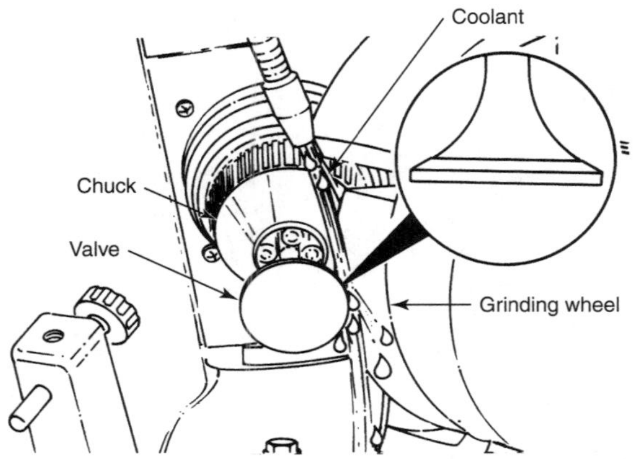

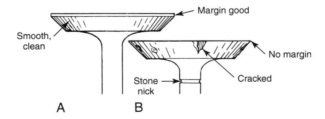

(Cummins Engine Co.)

Figure 19-8: When grinding a valve face, proceed slowly and grind a little at a time.

valve—the less material ground off, the better.

17. How much material did you have to grind off the valve face? _____   Completed ❑

18. Stop the machine and inspect the valve margin, as in **Figure 19-9.** Measure it with the steel rule. In general, if the margin is less than 1/32″ (0.79 mm), the valve should be replaced. A thin valve margin will quickly overheat and burn when operating in an engine. There is not enough metal left to dissipate combustion heat, especially on the exhaust valves.   Completed ❑

Figure 19-9: Your valve should look like A, not B. Notice that valve B is sharp and cannot be used.

19. How wide is the valve margin? _____ Is this satisfactory? _____   Completed ❑

20. Next, dress, or true up, the end of the valve stem. Mount the valve in the V-block on the opposite end of the grinder chuck, **Figure 19-10.** Grind off as little metal as possible, 0.01″ (0.25 mm) or less. Excessive grinding will remove the thin, hardened outer surface of the stem, causing rapid stem wear.   Completed ❑

Name _____

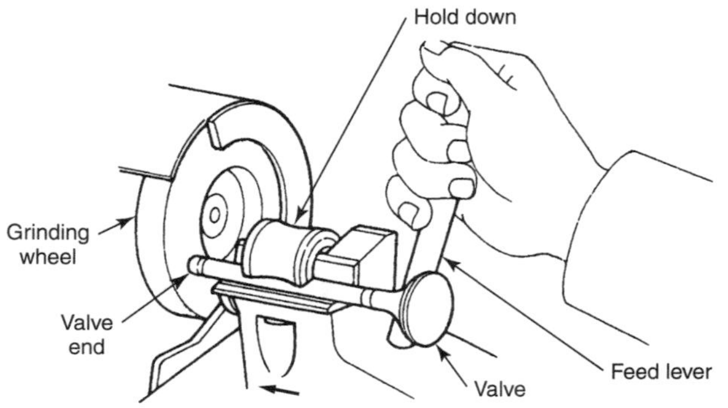

Figure 19-10: Remove the minimum amount of metal necessary when dressing the end of the valve stem.

21. How much material did you grind off the   Completed ❏
    stem? _____

22. Grind the other valves by repeating   Completed ❏
    Steps 9–21.

Grinding or Cutting Valve Seats

> **Note: Many shops use valve seat cutters rather than grinding stones. Unlike stones, cutters do not have to be trued to maintain the accuracy of their cutting face. If your shop uses cutters rather than stones, some of the steps below will not apply. When valve seat cutters are used, always read and follow the cutter manufacturer's directions.**

23. To begin the grinding operation of the valve   Completed ❏
    seat, find the correct size stone or cutter.
    The cutter should be slightly larger than the
    valve seat and head of the valve. Never use
    a stone that is too small or it will rapidly cut
    into the port area of the head. Refer to
    Figure 19-11.

24. Clean the guide. Next, install the pilot,   Completed ❏
    stone, and sleeve, **Figure 19-11**. Inspect the
    fit of the stone closely, **Figure 19-12.** Turn
    the stone with your hand to check its oper-
    ation. Also, check that it has the correct
    angle.

25. Usually, the stone-dressing tool is adjusted   Completed ❏
    to the valve face angle (30° or 45°). What
    diameter and angle stone did you select?

    _____ _____

Name _____

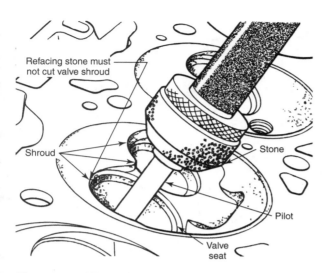

Figure 19-11: Wear safety glasses when refacing (grinding) a valve seat.

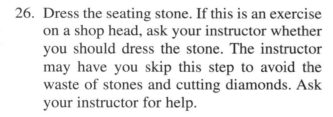

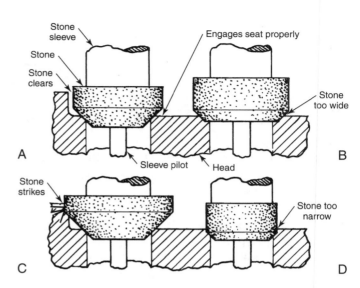

Figure 19-12: The stone must be the correct width. A—This stone is acceptable. B—This stone is too wide and will produce a horizontal step at the bottom of the seat. C—This stone is too wide. D—This stone is too narrow and will make a vertical step at the top of the step.

26. Dress the seating stone. If this is an exercise on a shop head, ask your instructor whether you should dress the stone. The instructor may have you skip this step to avoid the waste of stones and cutting diamonds. Ask your instructor for help.

Completed ❑

27. What angle should be ground on the stone?

_____

Completed ❑

28. Place a drop of oil on the pilot shaft. Insert the stone assembly over the pilot. Next, using the drive motor, spin the stone in the seat for a second or less. Do not push down on the drive motor. Let the weight of the motor do the cutting. Grind off as little metal as needed to clean up the seat. The finished seat should be free of pits (black spots). Remember, grinding the seats sinks the valve down into the head. This will change valve height, leading to valve adjustment problems.

Completed ❑

Checking Valve Contact

29. To check your work (face-to-seat contact), place pencil marks on the valve face, as shown in **Figure 19-13.** With the valve in

Completed ❑

Name _____

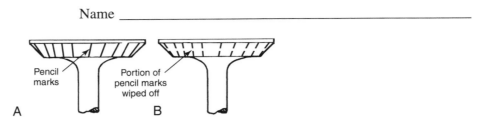

Figure 19-13: Pencil marks on the valve face will reveal the seating surface between the valve face and valve seat. A—Marks applied. B—A portion of the marks has been wiped away by turning the valve 1/4 turn in the seat.

place in the cylinder head, give the valve about one-quarter of a turn while holding down on the valve.

30. Remove the valve and inspect the pencil marks on the face of the valve. The portion of the pencil marks rubbed off indicates the contact point between the valve face and the seat. The contact point should be in the middle of the valve face. It should be about 1/16″ (1.59 mm) wide and should extend all the way around the face, **Figure 19-14.**

Completed ❑

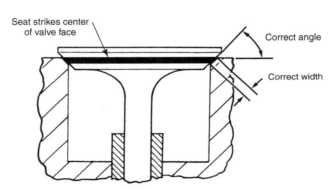

Figure 19-14: The correct contact width should be about 1/16″ (1.59 mm). The top and bottom portions of the seat can be removed to change the seat's width and location.

31. Was the pencil mark wiped off all the way around the valve?

Completed ❑

32. Describe the contact surface between the valve seat and the valve face.

Completed ❑

_____

_____

34. Repeat Steps 24–30 on the other valves. Note that the intake and exhaust valves may use different angles, and that the stone will probably need redressing during the grinding operation.

Completed ❑

Name _____

35. As a quick review, answer the following   Completed ❏
    questions about valve service.

    • What is the function of a valve margin? _____

    _____

    • What is an interference fit between a valve and seat? _____

    _____

    • What are the characteristics of a good contact surface between a valve and seat? _____

    _____

36. Before reassembling the head, have your   Completed ❏
    instructor check your work and sign this
    worksheet.

    Instructor's signature: _____

Job 20

Service Pistons
and Cylinders

Introduction

Piston and cylinder problems can result in loss of compression, oil burning, and noises. Check the troubleshooting charts in Chapter 22, *Engine Mechanical Troubleshooting*, for methods of diagnosing piston, piston ring, and cylinder problems. Remember that oil burning can be caused by other problems, such as leaking valve stem seals, clogged cylinder head drain holes, or a leaking intake manifold gasket on a V-type engine. Before deciding that the pistons, rings and cylinders are the problem, perform all possible diagnostic tests to pinpoint the exact cause of oil consumption. Your tests will let you determine if the engine needs a piston and cylinder service.

Objective

Given an engine short block and set of tools, you will correctly recondition a piston assembly and cylinder and replace piston rings.

Materials and Equipment

- Engine assembly.
- Basic hand tools.
- Ridge reamer.
- Safety glasses.
- Ring expander.
- Inside micrometer.
- Outside micrometer.
- Set of flat feeler gauges.
- Ring compressor.
- Engine oil.
- Cylinder hone.
- Air or electric drill.
- Two short pieces of fuel line hose.
- Torque wrench.

Instructions

Before beginning this job, review the information in Chapter 26, *Engine Block, Crankshaft, and Lubrication System Service*, of your textbook and workbook. Ask your instructor for any added directions and for the location of the special equipment and engine to be used. As you read the job instructions, answer the questions and perform the tasks. Print your answers neatly and use complete sentences.

Name _____

This is a rather involved and time-consuming exercise. If you see that time is running out, inform your instructor so that the engine parts may be stored.

> **Note: This procedure details the complete removal, repair, and reinstallation procedure for one engine piston and cylinder. Ask your** instructor whether you should service one piston and cylinder in the engine assembly or remove all pistons and service all cylinders. If all pistons and cylinders will be serviced, it is usually better to perform a sub-task (such as piston removal or cylinder honing) to all cylinders at once, rather than performing the complete procedure to one cylinder at a time.

Warning

Before performing this job, review all pertinent safety information in the text and discuss safety procedures with your instructor.

Procedure

Part Disassembly

1. Position the first piston to be removed at BDC (bottom dead center). The crankshaft may be turned with a breaker bar and socket which fits the large bolt on the front of the crankshaft snout. If needed, a large pry bar can be wedged between the flywheel bolts to turn the crank. A special flywheel-turning tool may also be used to engage the flywheel teeth and turn the crankshaft.

Completed ❑

2. How did you position the piston at BDC?

Completed ❑

_____

_____

Inspecting the Cylinders

3. Inspect the engine cylinder for wear or damage. Typical defects are scratches, grooves, and signs of overheating. To check for excessive cylinder wear, rub your fingernail across the top of the cylinder. This will reveal any ridge formed in the top of the cylinder. See **Figure 20-1.**

Completed ❑

> **Warning: A ring ridge is formed by the wearing action of the piston rings. If a ring ridge is present, it must be removed before** piston removal. Forcing the piston out of the cylinder and over a ring ridge can break the rings and damage the piston grooves and lands.

Name _____

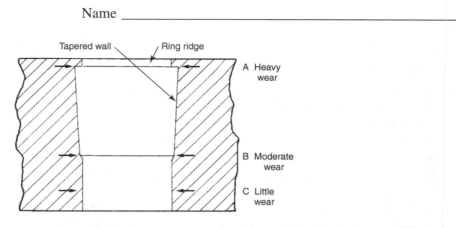

Figure 20-1: This is a typical cylinder wear pattern. The cylinder diameter at point A (top of ring travel) minus the cylinder diameter at point B (bottom of ring travel) indicates the amount of taper. Note the sharp edge formed at the top of ring travel. The ridge at the bottom of ring travel is less pronounced.

4. Does the cylinder have a ring ridge? _____   Completed ❑

Removing the Ring Ridge

5. To remove the ring ridge, begin by stuffing rags into the bottom of the cylinder bore.   Completed ❑

6. Insert the ridge-reaming tool into the cylinder, **Figure 20-2.**   Completed ❑

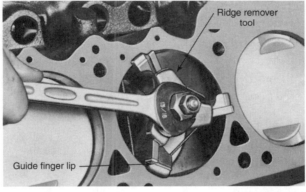

(DaimlerChrysler)

Figure 20-2: Removing a ring ridge. This reamer is supported by lips on the top of the guide fingers.

7. Adjust the cutters out against the ridge. Next, turn the reamer with the ratchet and socket until the ridge is cut flush with the worn part of the cylinder wall. The new reamed surface must blend smoothly with the existing cylinder.   Completed ❑

8. Remove the rags and blow out the cylinder to remove metal shavings that might scratch the cylinder during piston removal.   Completed ❑

Name _____

Removing the Pistons

9. Unscrew the nuts holding the rod cap on the connecting rod. Keep them in order and right side up. It is desirable to replace the connecting rod nuts exactly as removed.

Completed ❑

10. Check that the rod cap and rod are numbered. If the rods and caps are not numbered, mark them with a punch or number punch. When all of the pistons are removed, it is essential that all of the caps, connecting rods, pistons, and related fasteners be reinstalled in the same location. What is the number of the rod you are removing? _____

Completed ❑

11. Check that the piston head has a mark to indicate which direction it should face when it is reinstalled. If there is no mark, make a mark on the connecting rod to indicate the front of the engine. This step is important for reassembly.

Completed ❑

12. Once the rod cap is removed, push the piston out of the cylinder by pushing the connecting rod towards the top of the block. If the piston is difficult to remove, lightly tap on the connecting rod with a plastic-faced hammer or block of wood.

Completed ❑

Note: If directed by your instructor, remove the other pistons by repeating Steps 1 through 12 before proceeding with the other steps of this procedure.

Remove Crankshaft (Optional)

Note: If directed by your instructor, perform the following steps to remove the crankshaft. If the crankshaft will not be removed, skip Steps 13 through 20.

13. Mark and remove all connecting rod caps.

Completed ❑

14. Push the connecting rods away from the crankshaft to avoid damage to the journals.

Completed ❑

Name _____

15. Mark the crankshaft main bearing caps as shown in **Figure 20-3.** Be sure to mark all main bearing caps so that they can be reinstalled in the same position and direction.

Completed ❑

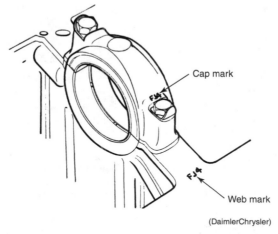

Cap mark

Web mark

(DaimlerChrysler)

Figure 20-3: Mark the bearing caps before they are removed. They must be reassembled in the proper position and direction.

16. Loosen and remove the main bearing cap fasteners.

Completed ❑

17. Remove the main bearing caps from the block.

Completed ❑

18. Lift the crankshaft from the engine block.

Completed ❑

19. Check the condition of the bearings and crankshaft journals. Consult your instructor if a problem is found.

Completed ❑

20. Proceed with piston removal and cylinder inspection.

Completed ❑

Measuring Cylinder Wear

21. Using the service literature, find the standard bore size and write it down. _____

Completed ❑

22. Measure the diameter of the cylinder with an inside micrometer or telescoping gauge and outside mike, **Figure 20-4.** You must take measurements in the cylinder locations described in **Figure 20-5.**

Completed ❑

23. Make measurements at right angles to and along the engine centerline, near the top of ring travel. This will let you calculate how much the cylinder is out of round.

Completed ❑

Name _____

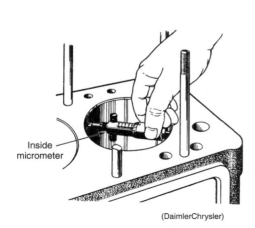

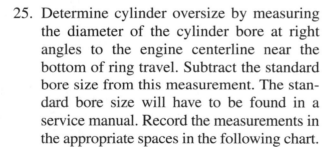

(DaimlerChrysler)

Figure 20-4: Use your inside micrometer or outside micrometer and telescoping gauge to measure the bore diameter.

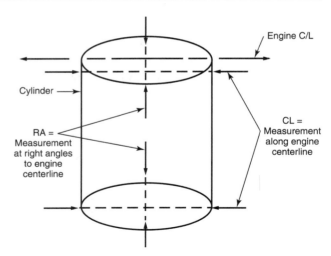

Figure 20-5: To determine cylinder taper, out-of-round, and oversize, measure the cylinder diameter at the locations indicated by the four dashed lines.

24. When checking cylinder taper, check the top and bottom of ring travel at right angles to the engine centerline.   Completed ❑

25. Determine cylinder oversize by measuring the diameter of the cylinder bore at right angles to the engine centerline near the bottom of ring travel. Subtract the standard bore size from this measurement. The standard bore size will have to be found in a service manual. Record the measurements in the appropriate spaces in the following chart.   Completed ❑

Measurement	Result
Bottom of Cylinder R/A	
Standard Bore Size	
Cylinder Oversize	

26. Next, calculate the out-of-round. Measure the cylinder at right angles to the engine centerline near the top of ring travel. Then, measure the cylinder along the engine centerline at the same location. Subtract the smaller value from the larger.   Completed ❑

Measurement	Result
Top of the Cylinder R/A	
Top of the Cylinder C/L	
Out-of-Round	

Name _____

27. Finally, the cylinder's taper needs to be determined. Begin by measuring the cylinder at right angles to the engine center-line near the top of ring travel. Next, measure the cylinder at right angles to the engine centerline near the bottom of ring travel. Subtract the second measurement from the first to determine cylinder taper.

Completed ❑

Measurement	Result
Top of Cylinder R/A	
Bottom of Cylinder R/A	
Taper in Cylinder	

Honing the Cylinders

28. To hone or deglaze the cylinder, clamp the hone into a low-speed electric drill.

Completed ❑

29. Insert the hone into the cylinder bore. Squirt a moderate amount of hone oil in the cylinder.

Completed ❑

30. Turn on the drill. At the same time, move the drill up and down the full length of the cylinder. Be careful not to pull the hone too far out of the bore, or hone damage may result.

Completed ❑

31. Move the hone up and down at a rate that will produce a 50° crosshatch pattern as shown in **Figure 20-6.** This will help the rings seat and seal during engine start up.

Completed ❑

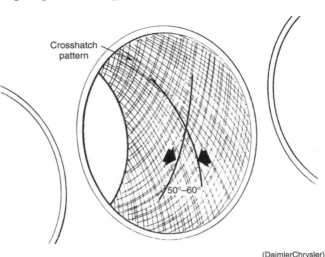

Crosshatch pattern

50°–60°

(DaimlerChrysler)

Figure 20-6: A spring-loaded hone was used to create this desirable crosshatch pattern.

Name _____

32. Before withdrawing the hone, hand-squeeze or adjust the stones together to prevent vertical scratches.

Completed ❑

33. What is the approximate angle of the hone marks in the cylinder? _____

Completed ❑

Cleaning the Cylinders

34. After honing, always clean the cylinder thoroughly. Any grit left in the engine will act as grinding compound that can wear moving parts of the engine. Scrub the cylinder with soap and hot water and then rinse it with clean hot water. Wipe the cylinder dry with a clean rag.

Completed ❑

35. Wipe the cylinder with a clean oil-soaked rag. Wipe until all of the grit is removed. The oil will help pick up heavy particles from inside the hone scratches.

Completed ❑

Finding Piston Clearances

36. Next, determine the piston clearance. Measure the diameter of the piston across the skirts, even with the piston pin. Use a large outside micrometer.

Piston diameter: _____

Completed ❑

37. Go back to Step 25 of the job and look up the cylinder diameter (bottom of cylinder R/A). Cylinder diameter: _____

Completed ❑

38. Finally, subtract cylinder bore diameter by piston diameter to determine piston clearance.

Piston clearance (cylinder diameter minus piston diameter): _____

Completed ❑

⬥ **Note: If directed by your instructor, service the other cylinders by repeating Steps 22 through 38 before proceeding with the other steps of this procedure.**

Checking Piston Pins

39. To check the general wrist-pin fit in the piston, clamp the rod I-beam lightly in a vise, as in **Figure 20-7**. Try to rock the piston sideways (opposite normal swivel).

Completed ❑

Name _____

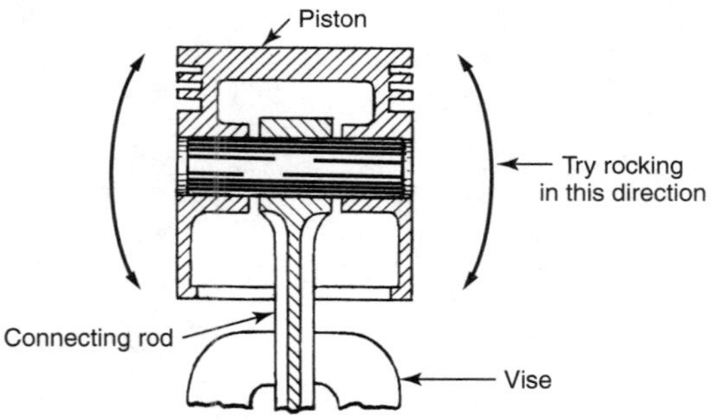

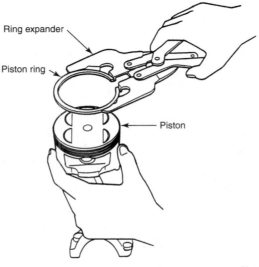

Figure 20-7: To check for pin looseness, rock the piston as indicated here.

Any detectable movement indicates piston pin looseness. During an actual repair this would require service (pin or bushing replacement).

40. Describe the condition of the piston pin.   Completed ❑

_____

_____

Removing the Piston Rings and Cleaning the Ring Grooves

41. Position the piston and rod assembly in the vise so that the piston skirt is resting on top of the vise jaws and cannot swivel. Clamp the jaw around the connecting rod. Using the ring expander, **Figure 20-8,** remove the piston rings from the piston. To prevent ring breakage, open the rings only enough to clear the piston lands. Keep them in order and right side up.

Completed ❑

(Honda)

Figure 20-8: During removal, be careful not to overexpand the rings.

Name _____

42. Clean the carbon from the inside of the ring grooves with a ring groove cleaner, **Figure 20-9.** Select the correct-width scraper for each groove and be careful not to cut too much. Ideally, you do not want to cut any of the metal in the groove, just the carbon.

Completed ❑

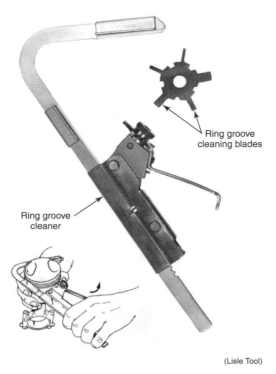

(Lisle Tool)

Figure 20-9: Try to remove all of the carbon but none of the metal from the grooves.

43. If a groove cleaner is not available, an old, broken ring can be used to scrape carbon from inside the grooves.

Completed ❑

Checking Ring Groove Wear

44. Measure the ring side clearance as demonstrated in **Figure 20-10.** Fit a ring into the top and middle ring grooves. Determine the largest size feeler gauge that will fit between the side of each groove and ring. The size of that gauge is the ring side clearance for that ring. The top piston groove will usually be the most worn.

Completed ❑

45. What is the top ring side clearance? _____

Completed ❑

46. What is the second ring side clearance? _____

Completed ❑

Name _____

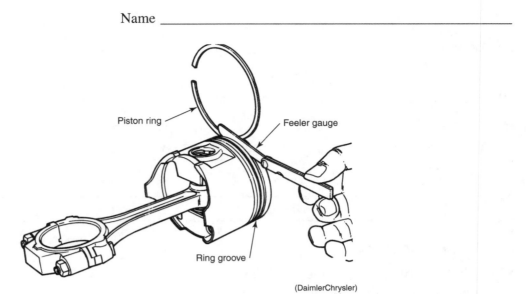

(DaimlerChrysler)

Figure 20-10: Ring side clearance is checked as shown.

Checking Ring Gap

47. Next, determine the piston ring gap. Install the compression ring into the cylinder squarely. Push it to the bottom of ring travel with the head of the piston.

Completed ❑

48. Determine the largest size feeler gauge that will fit in the gap between the ends of the ring, **Figure 20-11.** When a ring gap is too small, the heat of operation will cause expansion. This expansion could crush the ring outward against the cylinder wall with tremendous force, damaging or scoring the ring and cylinder.

Completed ❑

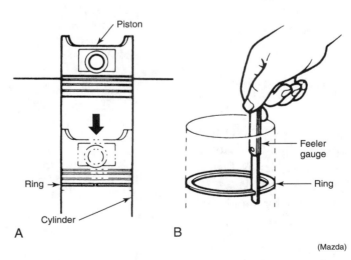

(Mazda)

Figure 20-11: Measuring ring gap. A—Using a piston to push the ring to the bottom of ring travel in the cylinder. B—Measuring the ring gap with a feeler gauge.

Name _____

49. What is the upper compression ring gap?   Completed ❑

_____

50. What is the lower compression ring gap?   Completed ❑

_____

51. What can happen when a ring gap is   Completed ❑
    too small?

_____

_____

Installing Piston Rings

52. Install the rings on the piston. Take your   Completed ❑
    time and follow the directions given in the
    Figure 20-12. Start with the oil rings, making
    sure that you butt the ends of the expander.

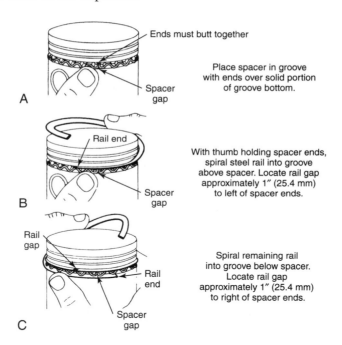

Figure 20-12: The basic steps for oil ring installation are shown here.

53. With the oil ring in place, you may either   Completed ❑
    install compression rings by hand or with a
    ring expander. The ring expander is faster
    and reduces the chance of ring breakage. If
    you must spiral the rings on by hand, be
    careful not to extend the rings too much.

54. Double-check that you have the rings right   Completed ❑
    side up. They are sometimes marked with a

Name _____

dot or small circle on the top. Space the ring gaps away from each other as described in Figure 20-12. Ask your teacher for help if you are in doubt about any task.

Note: If directed by your instructor, service the other pistons by repeating Steps 39 through 54 before proceeding with the other steps of this procedure.

Install Crankshaft (Optional)

Note: If directed by your instructor, perform the following steps to reinstall the crankshaft.

---

55. Install the upper halves of the main bearings in the engine block. Be sure to check for dirt and metal particles between the back of the bearing caps and the block before installation.

Completed ❑

56. Lightly lubricate the bearing halves.

Completed ❑

57. Carefully install the crankshaft into the engine block.

Completed ❑

58. Install the main bearing caps, being sure to install them according to the marks made during removal.

Completed ❑

Note: If your instructor directs, check the bearing clearance with Plastigauge.

---

59. Install and tighten the main bearing cap fasteners.

Completed ❑

60. Check that the crankshaft turns easily with the main bearing caps properly torqued.

Completed ❑

61. Proceed with piston installation.

Completed ❑

Installing Pistons

62. Oil the piston, pin, and rings generously, **Figure 20-13.**

Completed ❑

Name _____

(Federal Mogul)

Figure 20-13: The cylinder, piston, rings, pin, and rod bearings must be heavily oiled.

63. Tighten the ring compressor around the piston rings. Check that the small indentations on the ring compressor are down near the bottom of the piston. These small lips prevent the compressor from sliding into the cylinder with the piston.

Completed ❑

64. Remove the rod cap. Install the bearing. Then, oil the outside surface of the bearing.

Completed ❑

65. Slide two pieces of rubber gas line (or other suitable protector) over the rod bolts to protect the crank, **Figure 20-14.**

Completed ❑

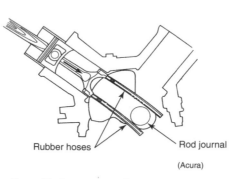

Rubber hoses   Rod journal

(Acura)

Figure 20-14: Slide pieces of rubber hose over the rod bolts to protect the crankshaft.

66. Ensure that the piston marking faces the front of the engine.

Completed ❑

67. Install the piston in the cylinder. Tap lightly on the piston head with a hammer handle, **Figure 20-15.** Keep the compressor flat on the block. If the oil ring pops out between the compressor and cylinder, reinstall the compressor and try again.

Completed ❑

Name _____

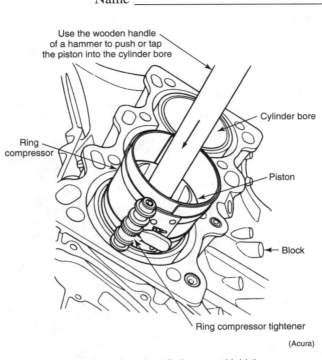

Use the wooden handle
of a hammer to push or tap
the piston into the cylinder bore

Cylinder bore

Ring
compressor

Piston

Block

Ring compressor tightener

(Acura)

Figure 20-15: A properly adjusted ring compressor makes piston installation easy. Hold the compressor tightly against the block surface.

STOP **Warning: Do *not* hammer the piston to the bottom of the bore until you can reach under the engine to guide the rod over the crank. If a rod bolt is hammered into the crank, the crank can be damaged.**

68. While carefully guiding the connecting rod over the crank journal with the bearing insert in place, tap the piston fully down into the cylinder.

Completed ❑

69. Double-check that the piston and rod assembly is facing in the proper direction. Reversing the rod can cause bearing and crankshaft damage.

Completed ❑

70. How do you know that the piston and rod assembly is facing in the right direction?

Completed ❑

_____

_____

Check Bearing Clearance

71. Next, use Plastigage to determine the rod bearing clearance. With the rod cap bearing insert clean and dry, lay a strip of Plastigage across the bearing.

Completed ❑

Name _____

72. Install and torque the rod cap to specifications.

Completed ❑

73. Remove the cap and compare the flattened Plastigage to the paper scale provided with the Plastigage, as shown in **Figure 20-16.**

Completed ❑

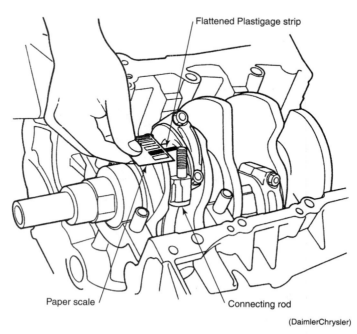

(DaimlerChrysler)

Figure 20-16: Check the width of the flattened Plastigage with the paper scale.

74. As the bearing clearance gets smaller, the width of the crushed Plastigage gets wider and vice versa. Also note, a 2 on the Plastigage scale would equal a bearing clearance of 0.002″ (1=0.001″, 1.5=0.0015″, etc.).

Completed ❑

75. What is the rod bearing clearance? _____

Torquing the Rod Cap Nuts

76. Carefully remove the flattened Plastigage from the rod insert.

Completed ❑

77. Oil the bearing and crank journal and install the rod cap. The rod identification numbers should be lined up.

Completed ❑

78. Torque the rod cap nuts to the service manual's specifications. Next, make sure the crank will still rotate. What is the rod nut torque specification? _____

Completed ❑

Name _____

79. Repeat Steps 52 through 78 for all pistons that will be reinstalled.

Completed ❑

80. Clean the tools and work area.

Completed ❑

81. Return all tools to storage, and clean the work area.

Completed ❑

82. Have your instructor check your work and sign this worksheet.

Completed ❑

Instructor's signature: _____

Job 21

Assemble Engine

Introduction

Once all of the repair jobs have been performed, it is time to finish the engine assembly. Final assembly usually includes installing the cylinder heads, intake and exhaust manifolds, valve train components, valve timing mechanism, oil pan, oil pump, coolant pump, and other engine parts. Proper assembly methods must be used to ensure efficient, dependable engine operation. For example, the cylinder heads, manifolds, timing gears, and valve components must be torqued to exact specifications and in the correct sequence. If final engine assembly is not done properly, all of your previous work will be wasted.

Objective

Given the listed tools and equipment, you will correctly assemble an engine.

Materials and Equipment

- Engine in need of reassembly.
- Service literature as necessary.
- Torque wrench.
- Basic hand tools.
- Engine oil.
- Plastigauge.
- Gaskets and other parts as needed.

Instructions

Review the information in textbook Chapters 26, 27, and 28, and then check with your instructor for additional information. As you read the job instructions, answer the questions and perform the tasks. Print your answers neatly and use complete sentences. Take your time and follow each step carefully.

Refer to the appropriate service literature for added instructions if needed. Also, feel free to ask your instructor for help if you have difficulty. Note that this procedure assumes that the block and heads have been overhauled, and that the engine is installed on an engine stand.

Warning

Before performing this job, review all pertinent safety information in the text and discuss safety procedures with your instructor.

Procedure

Torque Specifications

1. Record the following engine information: Completed ❑

   - Engine make: _____
   - Engine year: _____
   - Engine size: _____

Name _____

2. Look up the following torque specifications in the shop manual. Record your findings in the appropriate spaces.

Completed ❑

- Cylinder head bolts: _____
- Intake manifold bolts: _____
- Rocker arm bolts (if applicable): _____
- Oil pump bolts: _____
- Coolant pump bolts: _____
- Timing mechanism bolts: _____
- Front cover bolts: _____

Installing the Oil Pump and Oil Pan

Note: Some oil pumps are installed in the engine front cover and driven directly by the crankshaft. Refer to the manufacturer's service literature for information on how to assemble and install these types of pumps.

3. Rotate the engine block until the bottom of the engine faces up.

Completed ❑

4. Place the oil pump in position on the engine block, and install and tighten the oil pump bolts.

Completed ❑

Note: Some oil pumps are driven by a shaft from the distributor or distributor adapter gear. If the pump shaft can only be installed from the bottom, install it as part of the oil pump assembly.

5. Scrape the oil pan and block mating surfaces to remove old gasket material.

Completed ❑

6. Place the oil pan gaskets and seals on the engine block. Use the proper type of sealer if required.

Completed ❑

7. Place the oil pan on the block.

Completed ❑

8. Install and tighten the oil pan fasteners. Use the correct sequence and tighten the fasteners to the proper torque.

Completed ❑

Installing the Cylinder Heads on the Block

9. Rotate the engine until the top of the engine faces up.

Completed ❑

Name _____

10. Locate the picture in the service literature that illustrates the head bolt tightening sequence for the engine. Compare the illustration with the head bolt-tightening sequence in **Figure 21-1.** Notice the basic crisscross pattern.

Completed ❏

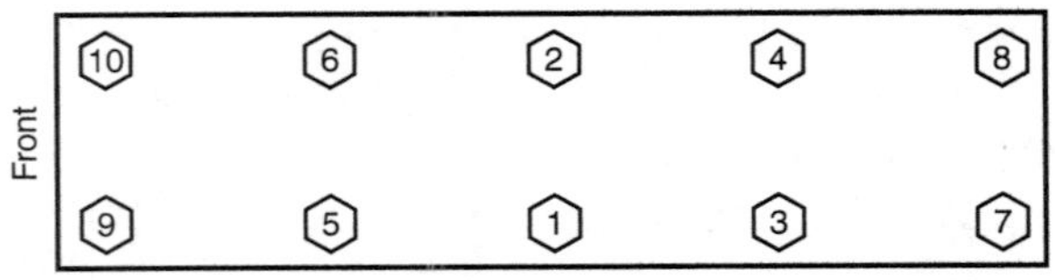

Figure 21-1: The numbers indicate the tightening order of the head bolts.

11. In **Figure 20-2,** draw the head bolt tightening pattern for the engine that you are working on. Draw circles for the bolts and number them in the tightening sequence.

Completed ❏

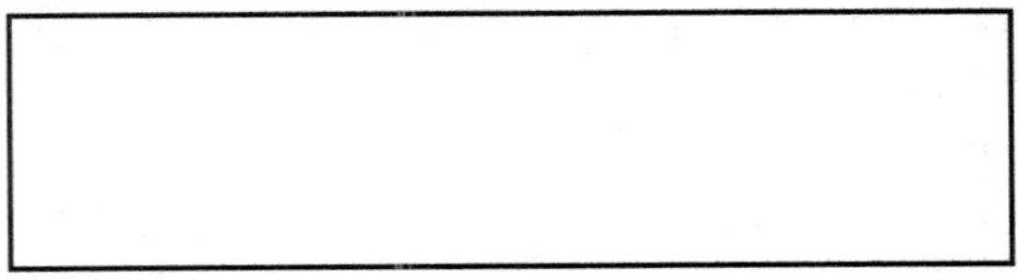

Figure 21-2: Draw the tightening sequence for your cylinder head as it appears in the service manual.

12. Scrape the block and head mating surfaces clean. Remove all traces of old gasket material from the engine.

Completed ❏

13. Place the head gasket on the block. There will usually be writing on the gasket to let you know the direction it should face. Use gasket sealer only if recommended by manufacturer.

Completed ❏

14. Carefully position the cylinder head on the block. Aligning pins will help in installation and reduce the chance of damaging the head gasket.

Completed ❏

15. Squirt a small amount of oil on the head bolt threads and thread them into the block by hand. Use a speed handle or ratchet (not an impact wrench) to turn them until they seat in the block. Do not tighten the bolts at this time

Completed ❏

Name _____

Torquing the Cylinder Heads

16. Using a torque wrench, tighten the head bolts to one-half of their torque specification. Follow the order or sequence given in the service literature.

    Completed ❑

    • What is one-half of the head bolt torque specification? _____
    • Where is the location of the first bolt to be tightened? _____

17. Torque the head bolts to three-fourths of the torque specification. Follow the correct sequence.

    Completed ❑

18. Tighten the head bolts to their full torque value. Finally, using the tightening sequence, make sure each bolt is at full torque. This completes the head installation. Some manufacturers recommend that the head bolts be retorqued after the engine has been started and has reached full operating temperature. The instructions with the new head gaskets will normally give this information. What is the full tightening specification of the head bolts? _____

    Completed ❑

19. Did any of the bolts turn when you tightened them for the final time? _____ On Figure 21-2, indicate the bolts that turned.

    Completed ❑

    Can you think of a reason that some of the bolts could be turned more? _____

Installing the Intake Manifold

Note: On some inline engines, the intake and exhaust manifolds are installed at the same time.

20. Find a view illustrating the torque sequence for the intake manifold. **Figure 21-3** shows the tightening sequences for a typical V-type intake manifold and a typical inline intake manifold.

    Completed ❑

21. In **Figure 21-4,** draw the intake manifold bolt tightening pattern for the engine that you are working on. Draw circles for the bolts and number them in the tightening sequence.

    Completed ❑

Name _____

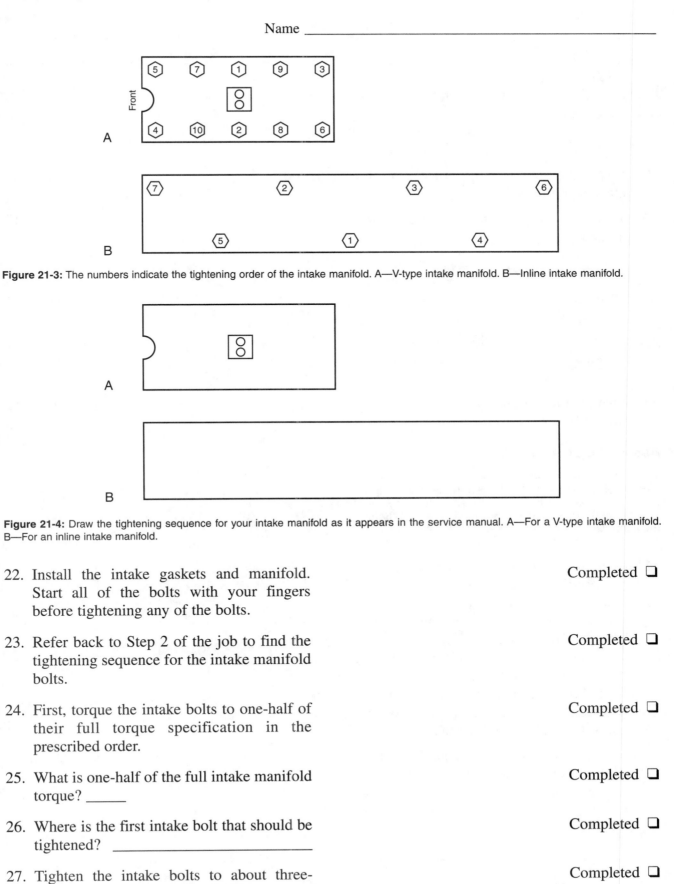

Figure 21-3: The numbers indicate the tightening order of the intake manifold. A—V-type intake manifold. B—Inline intake manifold.

Figure 21-4: Draw the tightening sequence for your intake manifold as it appears in the service manual. A—For a V-type intake manifold. B—For an inline intake manifold.

22. Install the intake gaskets and manifold. Start all of the bolts with your fingers before tightening any of the bolts.   Completed ❑

23. Refer back to Step 2 of the job to find the tightening sequence for the intake manifold bolts.   Completed ❑

24. First, torque the intake bolts to one-half of their full torque specification in the prescribed order.   Completed ❑

25. What is one-half of the full intake manifold torque? _____   Completed ❑

26. Where is the first intake bolt that should be tightened? _____   Completed ❑

27. Tighten the intake bolts to about three-fourths of the full torque specification.   Completed ❑

Name _____

28. Torque the intake bolts to their full specification following the correct sequence. Go over all of the bolts one or two times to finish the installation. As with the cylinder head, some manufacturers recommend retightening the bolts after engine operation.

          Completed ❑

29. What is the full torque specification for the intake manifold bolts? _____

          Completed ❑

30. Did any of the bolts turn during the final torque sequence? _____

          Completed ❑

Installing the Exhaust Manifold

31. Install the exhaust gaskets and the manifold on the head.

          Completed ❑

32. Start all of the bolts with your fingers. Next, tighten the bolts beginning at the center and alternating outward.

          Completed ❑

33. What is the torque specification for the exhaust bolts? _____

          Completed ❑

Assembling the Valve Train

> **Note: Not all of the following steps will apply to every engine. This procedure assumes that, if the engine has an in-block camshaft, the camshaft has been installed previously.**

34. Install the push rods and rocker arms in their original locations. If a rocker shaft is used, start the rocker bolts with your fingers. Next, use the proper sequence to tighten them a little at a time. Start in the middle and tighten each bolt one-half turn at a time. This will prevent a strain on any one bolt. Remember, the pressure of the compressed valve springs will be pushing against the rocker shaft assembly.

          Completed ❑

35. On some engines, you may have to adjust the valves at this time. Look up the valve-clearance adjusting procedure for your engine. In your own words, how do you adjust the valves?

          Completed ❑

_____

_____

_____

Name _____

Installing the Timing Mechanism

36. The timing mechanism times the crankshaft and camshaft to each other. This mechanism can be two meshing gears, two sprockets and a timing chain, or several sprockets and a timing belt. Find the installation procedure in the service manual for your engine. **Figure 21-5** illustrates the three types of timing mechanisms.

Completed ❑

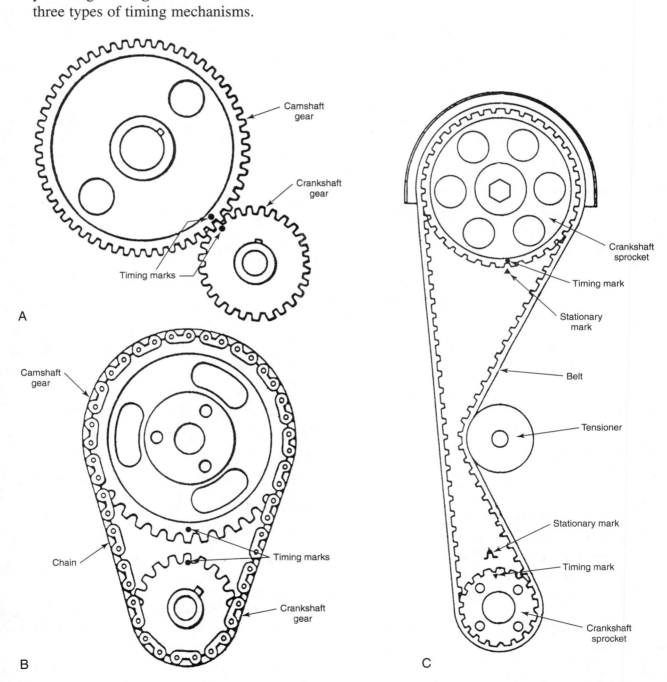

Figure 21-5: Three types of timing mechanisms are shown here. A—A gear-driven timing mechanism. B—A chain-driven timing mechanism. C—A belt-driven timing mechanism.

Name _____

37. Following the service instructions, loosely place the timing gears or sprockets on the crankshaft and camshaft. Rotate the crankshaft and camshaft until the timing marks are aligned.   Completed ❑

38. Once the timing marks are aligned, complete the installation of the timing mechanism by installing the timing chain or belt, and installing the fasteners. It may be necessary to remove one of the gears to install the timing chain. Make sure that all washers are in position, and that fasteners are tightened to their proper torque.   Completed ❑

39. After all parts are installed, recheck the timing marks to ensure that they are in position.   Completed ❑

Install the Timing Cover and Coolant Pump

40. Scrape the timing cover, coolant pump, and block mating surfaces to remove old gasket material.   Completed ❑

41. Place the timing cover gaskets and seals on the engine block. Use the proper type of sealant if required.   Completed ❑

42. Place the timing cover on the block.   Completed ❑

43. Install and tighten the timing cover fasteners. Use the correct sequence and tighten the fasteners to the proper torque.   Completed ❑

44. Place the coolant pump on the block.   Completed ❑

45. Install and tighten the pump fasteners.   Completed ❑

Installing Other Engine Parts

46. Install the vibration damper. If the vibration damper has a rotor for use with a crankshaft position sensor, align the rotor and sensor with the necessary special tool.   Completed ❑

47. Place the flywheel on the crankshaft flange.   Completed ❑

Note: Flywheel installation may have to wait until the engine is removed from the engine stand.

Name _____

48. Install and tighten the flywheel bolts. These bolts are critical, and must be torqued correctly.   Completed ❑

49. If the valves do not have to be readjusted after the engine is started, install the valve covers.   Completed ❑

50. Install any remaining engine parts. If the engine has a distributor, time it to the engine during installation.   Completed ❑

51. Clean and return all of the tools, and clean the work area.   Completed ❑

52. Have your instructor check your work and sign this worksheet.   Completed ❑

Instructor's signature: _____

Job 22

Adjust Clutch and Transmission Linkage

Introduction

Both manual and automatic transmissions and transaxles have linkage that can go out of adjustment. Manual clutch linkage usually requires adjustment as the clutch disc wears. Transmission shift linkage may not need adjustment until many miles have been accumulated. The technician must learn how to adjust clutch and transmission linkage. Misadjusted linkage can cause severe clutch wear and gear clash on a manual transmission. Improper linkage adjustments on an automatic transmission can cause improper shift points and internal transmission damage.

Materials and Equipment

- Vehicle with a manual transmission or transaxle in need of adjustment.
- Vehicle with an automatic transmission or transaxle in need of adjustment.
- Basic hand tools.
- Measuring tool or drill bit as needed.
- Service information as needed.

Objective

Given the listed tools and equipment, you will adjust a manual clutch, manual transmission/transaxle linkage, and automatic transmission/transaxle throttle linkage.

Instructions

Review the information in Chapters 29, 30, and 31 of your workbook and textbook. Ask your instructor for any additional instructions and to assign vehicles for your use. As you read the job instructions, answer the questions and perform the tasks. Print your answers neatly and use complete sentences.

Warning

Before performing this job, review all pertinent safety information in the text and discuss safety procedures with your instructor.

Procedure

Adjusting a Manual Clutch

1. Locate a vehicle with a manual transmission or transaxle.   Completed ❑

2. Check clutch adjustment by measuring the amount of free play at the clutch pedal.   Completed ❑
   - Specification: _____
   - Measured free play: _____

Name _____

3. If the clutch linkage is hydraulically operated, check the fluid level in the master cylinder before going to Step 4.

Completed ❏

4. If free play is not within specifications, adjust the play by loosening the locknut(s) and turning threaded adjuster according to manufacturer's instructions. **Figure 22-1** shows one method of adjustment.

Completed ❏

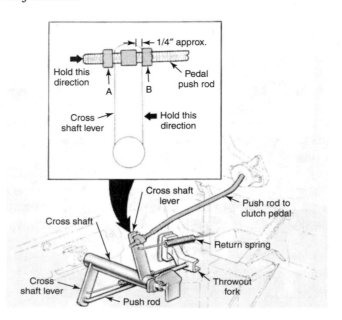

Figure 22-1: One clutch adjustment arrangement is shown here.

 Note: Some clutches have self-adjusting linkage and no periodic service is needed.

---

5. Recheck clutch operation after adjustment has been made.

Completed ❏

Adjusting Manual Transmission or Transaxle Linkage

6. Road test the vehicle and note the shift operation. Check for a loose feel, failure to engage one or more gears, gear clash, binding, and vibration.

Completed ❏

 Note: Be sure that the above problems are not the result of incorrect clutch adjustment.

Name _____

7. If shift linkage operation is not acceptable, inspect all shift linkage and fasteners in the passenger compartment. If necessary, raise the vehicle and inspect all shift linkage at the transmission or transaxle. The linkage on many front-wheel drive vehicles can be accessed without raising the vehicle. If the linkage and related fasteners are in good condition, proceed to adjust the linkage as explained in Steps 8 and 9 as applicable.

Completed ❑

8. Most cable linkage will have an adjuster at the shifter side of the cable, **Figure 22-2.** This can be adjusted as needed. A drill bit or special alignment tool is usually placed in an alignment hole, either between two shifter rods, or between a shifter rod and a stationary part. The stationary alignment holes can be located on the vehicle body or the transmission or transaxle case. See **Figure 22-3.** Adjust the shifter until the drill bit or tool passes through the holes easily. Then, tighten the lock nuts.

Completed ❑

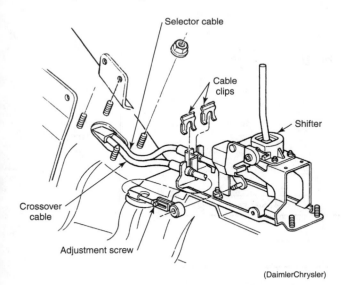

(DaimlerChrysler)

Figure 22-2: A typical adjustment setup is shown here.

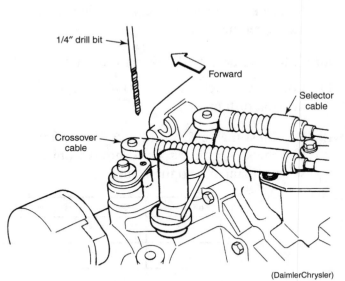

(DaimlerChrysler)

Figure 22-3: Adjust the shifter until the drill bit or tool passes through the holes easily.

9. Rod and lever linkage will have adjustable swivels under the shifter or where the shifter mechanism enters the gear case. It may be necessary to place a drill bit or special alignment tool through two or more

Completed ❑

Name _____

alignment holes, **Figure 22-4.** Adjust the shifter until the drill bit or tool passes through the holes easily. Tighten the lock nuts or swivel nuts as necessary and recheck.

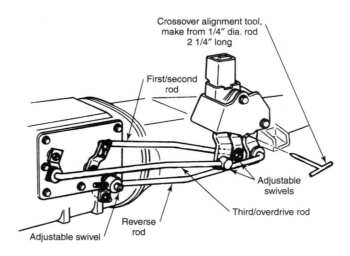

Figure 22-4: Adjusting shift linkage on a floor shift transmission. Note the use of an alignment pin to hold the gearshift in the correct position during adjustment.

10. Recheck shifter operation after adjustments have been made.   Completed ❑

Adjusting Automatic Transmission or Transaxle Throttle Linkage

11. Locate a vehicle with an automatic transmission or transaxle.   Completed ❑

12. Road test the vehicle and note shift points. List the shift points specifications from the service manual, and the actual shift points: Completed ❑

Shift Range	Service Manual Specifications	Actual Shift Points
1–2		
2–3		
3–4		
4–5		
5–6		
6–5		
5–4		
4–3		
3–2		

Name _____

13. Is the transmission shifting within the proper ranges? _____   Completed ❑

14. Do the shifts feel: firm? _____ Harsh? _____ Soft and slipping? _____.   Completed ❑

    If any problems are noted in Steps 13 and 14, what could be the cause? _____

    _____

15. If the shift points are incorrect, open the hood and locate the throttle cable. It should look something like **Figure 22-5.**   Completed ❑

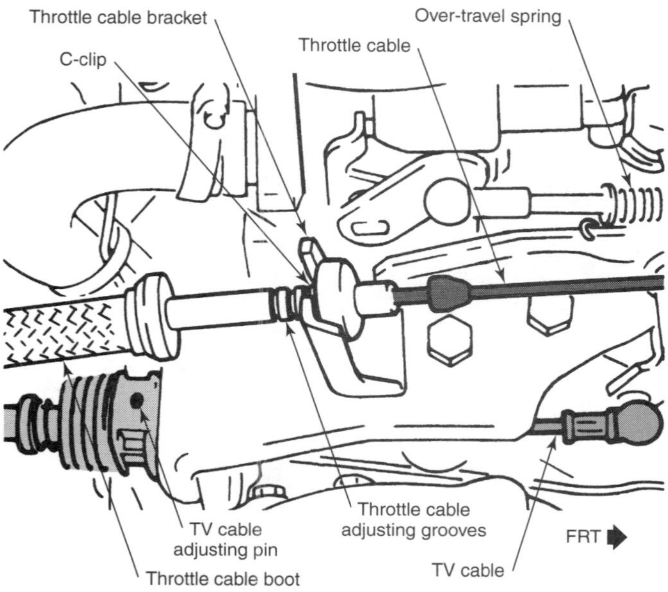

Figure 22-5: A typical throttle cable.

16. With the engine off, have an assistant press the accelerator pedal to the floor.   Completed ❑

> Note: Some throttle cables are self-adjusting. Pressing the accelerator pedal to the floor completes the adjustment.

17. Check the throttle plate or plates and ensure that they are completely open.   Completed ❑

18. If the throttle plates are not completely open, adjust the linkage between the accelerator pedal and the throttle plates. The adjustment device is usually located at the throttle body.   Completed ❑

Name _____

19. Once the throttle plates are opening completely when the accelerator pedal is pressed to the floor, unlatch the throttle-to-transmission cable adjuster clip or loosen the throttle-to-transmission linkage adjustment locknut as applicable.

Completed ❑

20. With the throttle plates open completely, pull the throttle-to-transmission linkage to the wide-open throttle position if necessary.

Completed ❑

21. Latch the adjuster clip to tighten the lock nut as applicable.

Completed ❑

22. Road test the vehicle and check transmission shift points and shift quality.

Completed ❑

23. Return all equipment to its proper storage location and clean shop area.

Completed ❑

24. Have your instructor check your work and sign this worksheet.

Completed ❑

Instructor's signature: _____

Job 23

Service CV Axle and Drive Shaft Flexible Joints

Introduction

A faulty CV joint or drive shaft universal joint can cause a wide range of problems. A dry drive shaft U-joint can make a high-pitched chirping sound, especially when operated in reverse. If highly worn, a bad U-joint can also cause a metallic crunching or grinding sound, which resembles the sound of popcorn popping. Bad inner CV joints usually make noise, while bad outer CV joints cause vibration. A badly worn universal joint or CV joint may break and separate from the remainder of the drive train. The drive shaft or axle could then pound large dents in the floorboards or break other components under the vehicle. Proper diagnosis, inspection, and repair of flexible joints are very important.

Objective

Using the listed tools and equipment, you will properly remove, disassemble, inspect, and assemble the parts of a universal joint and a CV joint.

Materials and Equipment

- Vehicle in need of universal joint service.
- Vehicle in need of CV joint axle service.
- Basic hand tools.
- Correct service literature.
- Lift or floor jack and jack stands.
- Large brass or ball-peen hammer.
- Safety glasses.
- Large screwdriver.
- Pin punch.
- Needle-nose pliers.
- Snap ring pliers.
- One small and one large socket.
- Large driving punch.
- CV boot tool.
- Vise.
- Replacement parts.

Name _____

Instructions

Ask your instructor for the location of the drive shaft and universal joint assembly and the CV joint assembly. They may be located in or out of a vehicle. As you read the job instructions, answer the questions and perform the tasks. Print your answers neatly and use complete sentences. Ask your instructor for help as needed.

Warning

Before performing this job, review all pertinent safety information in the text and discuss safety procedures with your instructor.

Procedure

> **Caution: If you must clamp the drive shaft in the vise, make sure that you do not dent or bend the drive shaft. Shaft unbalance and vibration may occur when the shaft is returned to service.**

Replacing a Universal Joint

1. Raise the vehicle on a lift or raise it with a floor jack and support it with jack stands.   Completed ❑

2. Mark the mating surfaces of the differential yoke and drive shaft yoke. This will allow you to reassemble the driveline in exact alignment and prevent a possible imbalance and vibration.   Completed ❑

3. Remove the fasteners holding the rear of the drive shaft to the pinion flange.   Completed ❑
   - What type of fastener is used to hold the rear of the drive shaft to the pinion? _____
   - Is this a one piece or two-piece drive shaft? _____

4. If necessary, remove the center support bolts.   Completed ❑

5. Push the drive shaft forward to clear the pinion flange.   Completed ❑

6. Loosen the fasteners and remove the drive shaft from the transmission. Be careful not to scrape the slip yoke on the ground or drop the U-joint bearing caps.   Completed ❑

7. Clamp the drive shaft lightly in a vise. Be careful not to dent or bend the drive shaft.   Completed ❑

8. If U-joint snap rings are used, remove them from the drive shaft.   Completed ❑

Name _____

9. Use the vise and the proper size sockets to remove the bearing caps from the drive shaft yoke(s). This operation is shown in **Figure 23-1.** Make sure that the small socket is smaller than the trunnion and that the larger socket is large enough to accept the bearing cap.

Completed ❑

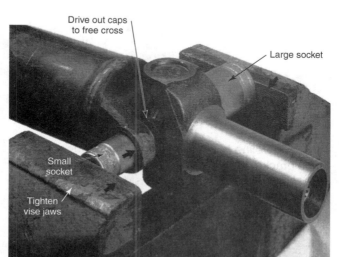

(DaimlerChrysler)

Figure 23-1: Check that the sockets are the correct sizes, and then press out the bearings.

10. Remove the cross from the yoke.

Completed ❑

11. Install the new U-joint in the drive shaft. **Figure 23-2.** If you find it difficult to drive the bearing caps completely into position (clearing snap ring grooves), one of your needle bearings has probably fallen sideways in the cap. If so, remove the caps and start over. Do not try to force the cap in with excessive force or damage will occur.

Completed ❑

Figure 23-2: Use a vise to force the bearing caps inward until they are flush with the yoke lug surface. Do not tighten beyond this point.

Name _____

12. Install the U-joint snap rings, **Figure 23-3.** Did you have difficulty getting the snap rings to seat? _____ If so, what was the cause?

Completed ❑

_____

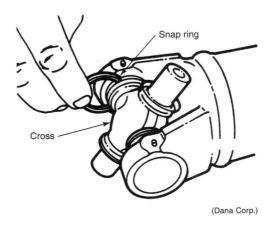

(Dana Corp.)

Figure 23-3: Reinstall the snap rings.

13. Install the drive shaft to the rear of the transmission and tighten the fasteners.

Completed ❑

14. Install the drive shaft on the pinion flange.

Completed ❑

15. Install the center support if applicable.

Completed ❑

16. Install and tighten the center support and other drive shaft fasteners.

Completed ❑

17. Lower the vehicle and check the operation of the drive shaft.

Completed ❑

18. Return all tools and equipment to storage.

Completed ❑

19. Clean the work area.

Completed ❑

Replacing a CV Joint

20. Raise the vehicle on a lift or raise it with a floor jack and support it with jack stands.

Completed ❑

21. Remove the wheel and tire on the side of the CV axle to be removed.

Completed ❑

   • Side to be serviced (right or left): _____

   • Does this axle have an intermediate shaft? _____

22. Remove the center nut holding the CV axle to the hub.

Completed ❑

Name _____

23. Remove the necessary suspension parts to gain axle clearance.   Completed ❏

24. Remove the CV axle from the steering knuckle. Was a special tool needed to remove the shaft from the hub? _____   Completed ❏

25. Remove the CV axle from the transaxle.   Completed ❏

26. Clamp the CV axle lightly in a vise.   Completed ❏

27. Remove the clamps holding CV joint boot and remove the boot.   Completed ❏

28. Remove the C clips or snap rings that hold the CV joint together, **Figure 23-4.**   Completed ❏

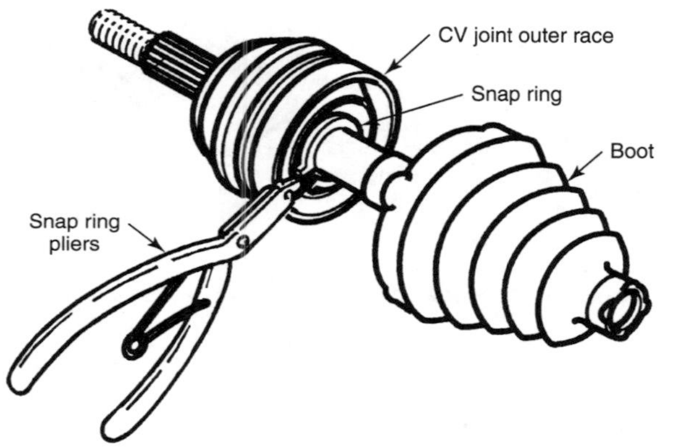

CV joint outer race

Snap ring

Boot

Snap ring pliers

(General Motors)

Figure 23-4: Remove the snap ring so the joint may be disassembled.

29. Separate the joint. Refer to the service manual instructions, if needed. Light taps from a brass hammer may be needed. Is the CV joint a Rzeppa or tripod type? Refer to **Figure 23-5,** which shows the two types of joints. _____   Completed ❏

30. If this is a Rzeppa joint, remove the balls from the cage. The cage must be tilted inside the housing to free each ball. If this is a tripod joint, remove the tripod caps from the cross.   Completed ❏

31. While wearing rubber gloves and eye protection, clean the CV joint parts in a cold solvent tank or parts washer.   Completed ❏

Name _____

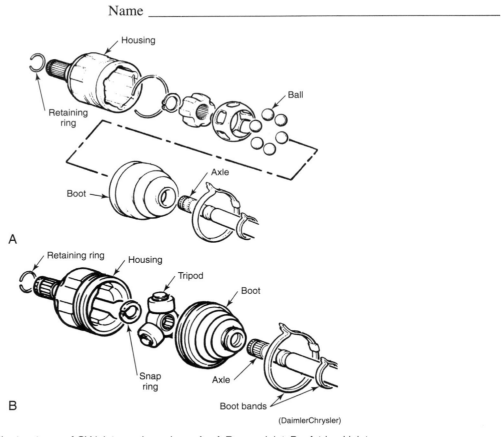

Figure 23-5: Exploded views of the two types of CV joint are shown here. A—A Rzeppa joint. B—A tripod joint.

32. Inspect the parts of the joint (bearings, housings, snap rings, ring grooves) for signs of wear. Look for pitting, marring, and other surface imperfections. Describe the condition of the CV joint.   Completed ❑

_____

_____

33. Obtain new parts as needed.   Completed ❑

34. Reassemble the CV joint.   Completed ❑

35. Add sufficient lubricant of the proper type.   Completed ❑

> ⚠ **Caution: Do not use chassis grease in a CV joint.**

36. Install the CV boot using new clamps. A special tool may be necessary, **Figure 23-6.**   Completed ❑

37. Install the drive shaft in the transaxle.   Completed ❑

Name _____

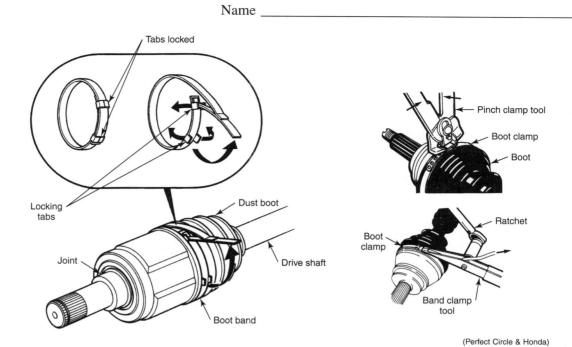

(Perfect Circle & Honda)

Figure 23-6: Special straps and a special pinch tool may be required to install the restraining boot straps.

38. Install the drive shaft in the steering knuckle.   Completed ❑

39. Reassemble the front suspension as necessary.   Completed ❑

40. Install and tighten the CV axle nut.   Completed ❑

41. Install the wheel and tire.   Completed ❑

42. Lower the vehicle and check CV axle operation.   Completed ❑

43. Return all tools and equipment to storage.   Completed ❑

44. Clean the work area.   Completed ❑

45. Have your instructor check your work and sign this worksheet.   Completed ❑

Instructor's signature: _____

Job 24

Service Disc Brakes

Introduction

A vehicle equipped with disc brakes has more stopping power than an equivalent drum brake vehicle. Disc brakes have fewer parts than drum brakes. This makes them easy to service. They must be repaired properly, however, to avoid pulling or vibrating brakes. During this job, you will perform various kinds of service on a disc brake assembly.

Objective

Using the tools and equipment listed for this job, you will service a disc brake assembly.

Materials and Equipment

- Vehicle in need of brake service.
- Basic hand tools.
- Ruler or sliding caliper.
- Large C-clamp.
- Two large screwdrivers.
- Safety glasses.

Materials and Equipment for Rebuilding the Caliper

- Air nozzle.
- Brake fluid.
- Electric drill.
- Cylinder hone.
- Piston installation tool (if necessary).

Instructions

Review Chapter 34, *Brake Service*, in your workbook and textbook before beginning this job. Ask your instructor whether or not you will rebuild the caliper and for the location of the vehicle or the shop disc brake unit to be serviced. As you read the job instructions, answer the questions and perform the tasks. Print your answers neatly and use complete sentences. If you run into any problems, feel free to ask for help.

Warning

Before performing this job, review all pertinent safety information in the text and discuss safety procedures with your instructor.

Name _____

Procedure

Brake Disassembly

 Note: Skip Steps 2–7 if the caliper assembly is a shop unit.

1. Park the vehicle on a lift and draw some of the fluid out of the master cylinder or reservoir. This will allow the calipers to be pushed back without causing the master cylinder to overflow.

   Completed ❑

2. Raise the vehicle and remove the wheels and tires as necessary.

   Completed ❑

3. Remove the caliper locating pins, stabilizer bolts, or cap screws so that the caliper can be removed. Since bolt locations and procedures vary, you will need to inspect the construction of your particular unit and decide exactly which fasteners must be removed to free the caliper. How many bolts held the caliper in place? _____

   Completed ❑

4. Slide the caliper from the rotor. If the caliper will not slide off the rotor, you may have to use a large screwdriver or C-clamp to push the piston back into its bore. Look at **Figure 24-1.**

   Completed ❑

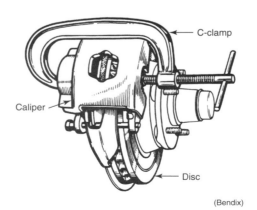

(Bendix)

Figure 24-1: To free the brake pads, use a C-clamp to force the piston back into the bore.

Name 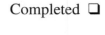 _____

5. Lay the caliper on the upper control arm or a tie rod, or hang it on a piece of mechanic's wire. The weight of the caliper should not be hung on the rubber brake hose or hose damage may occur.

Completed ❑

6. Now, remove the brake pads, retaining clips, shims, and antirattle clips when used. If antirattle clips are used on the pads, note how the clips fit into place. One example is given in **Figure 24-2.**

Completed ❑

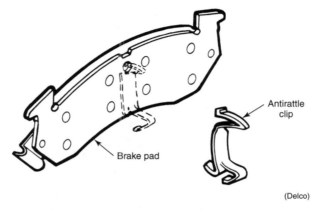

(Delco)

Figure 24-2: This antirattle clip keeps the pad from rattling. Note how the clip fits on the pad.

7. Does the unit use antirattle clips? _____ How do they fit into place?

Completed ❑

_____

_____

Part Inspection

8. Inspect the condition of all of the parts to be reused. They should be checked for wear, breakage, distortion, and heat damage. Also, look around the edge of the caliper piston. If leakage is found, the caliper must be rebuilt or replaced.

Completed ❑

9. Describe the condition of the brake parts.

Completed ❑

_____

10. What is the maximum and minimum thickness of the brake pad linings?

Completed ❑

_____

_____

Name _____

Rebuilding a Caliper

 Note: Steps 11–18 should only be completed if you have your instructor's approval to disassemble and rebuild the caliper.

11. Disconnect the rubber brake hose at the caliper, not at the steel brake line. Be careful not to lose the special sealing flat washer on the end of the rubber brake hose fitting. Drain the fluid from the caliper.

    Completed ❏

12. As in **Figure 24-3,** position rags or a small block of wood inside the caliper. Then, keeping the hands out of the way, slowly apply air pressure to the inside of the caliper cylinder.

    Completed ❏

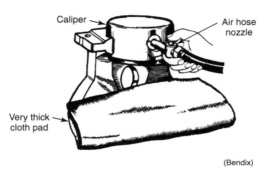

Caliper

Air hose nozzle

Very thick cloth pad

(Bendix)

Figure 24-3: Apply only enough air pressure to slowly push the piston out of its bore. Keep your hands clear and wear eye protection.

 Warning: Too much air pressure can cause the piston to shoot out of the caliper with tremendous force.

13. If the piston is frozen and will not come out with moderate air pressure, tap on the piston lightly with a soft hammer or mallet and try again. In extreme cases, you can reinstall the caliper on the vehicle and force the piston out with brake system pressure.

    Completed ❏

14. Now, remove the dust boot and piston seal from the caliper. During a real repair, these parts would always be replaced.

    Completed ❏

15. Inspect the condition of the caliper cylinder bore. Check it for scratches, pits, and scoring. Crocus cloth may be used to clean up minor imperfections. You may need to hone the cylinder. See **Figure 24-4.**

    Completed ❏

Name _____

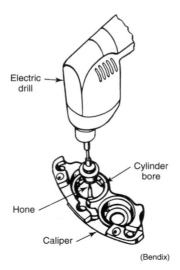

Electric drill
Cylinder bore
Hone
Caliper

(Bendix)

Figure 24-4: If the cylinder is in good condition, crocus cloth can be used instead of a hone.

> **Caution: If excessive honing is required, the caliper must be replaced!**

16. After honing or sanding, clean the cylinder and use a ruler, sliding caliper, or inside micrometer to estimate the diameter of the caliper cylinder.

    Caliper cylinder bore diameter: _____   Completed ☐

17. Lubricate the caliper bore, piston, and seal with clean brake fluid. Next, fit the seal into the groove in the cylinder.   Completed ☐

18. Now, slide the dust boot over the piston, **Figure 24-5.** Next, hold the piston and boot over the bore and use your fingers to work the boot bead into place. A driving tool may be needed. Depending on the type of brake, you may need to change your procedures slightly. If you are having problems, ask your instructor for help.   Completed ☐

19. After fitting the dust boot into position, push the piston down into the bore. You may need to use a C-clamp to push the piston fully into place. See **Figure 24-6.** Be careful not to position the piston sideways during installation or you may damage the piston or cut the seal.   Completed ☐

Name _____

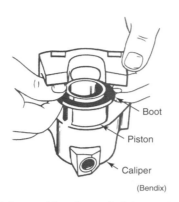

(Bendix)

Figure 24-5: Before pushing piston all of the way into the bore, fit the boot into its grooves.

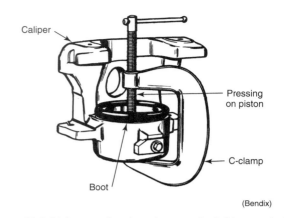

(Bendix)

Figure 24-6: Make sure the piston is not cocked sideways during the installation.

20. Inspect the condition of the rotor, **Figure 24-7.** It should be free of heavy scoring and runout. Normally, unless the rotor has grooves deeper than 0.015″ (0.38 mm) or runout in excess of 0.004″ (0.102 mm), manufacturers recommend that you avoid turning the rotor. Cutting a rotor on a lathe, when it is not needed, will only increase the chances of the rotor developing runout or warpage. However, it will probably be necessary to turn the rotors if the brakes vibrated or pulsed during normal stops.

Completed ❑

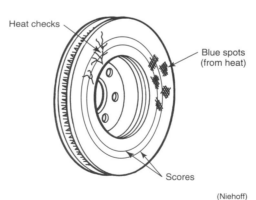

(Niehoff)

Figure 24-7: Check the condition of the rotor. Blue spots and minor scoring can be removed. The presence of any cracks will necessitate disc replacement.

21. Measure the thickness of the rotor. Look up and compare its thickness to specifications. Quite often, the minimum thickness is marked on the side of the rotor or disc.

Completed ❑

Name _____

22. As an optional activity, measure rotor runout with a dial indicator as described in the textbook.   Completed ❑

23. Record the following measurements.   Completed ❑
    - Actual rotor thickness: _____
    - Minimum thickness specification: _____
    - Rotor runout: _____

Bleeding Brakes

24. If applicable, bleed the system of air, as described in the textbook. Also, look at **Figure 24-8.** With the system pressurized or with someone pumping the brake pedal, open the bleeder screw until all of the air bubbles are removed from the system. The jar and rubber hose will help prevent a mess on the floor and the reentry of air.   Completed ❑

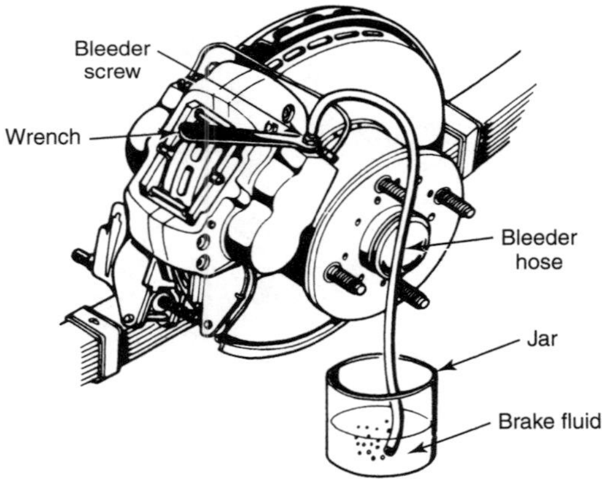

Bleeder screw

Wrench

Bleeder hose

Jar

Brake fluid

(Maserati)

Figure 24-8: When bleeding brakes, attach one end of the bleeder hose to the bleeder screw and submerge the other end in a jar partially filled with brake fluid.

25. Fill the master cylinder with fluid and check that you have a solid brake pedal.   Completed ❑

26. Clean the work area and return all tools and equipment to storage.   Completed ❑

27. Have your instructor check your work and sign this worksheet.   Completed ❑

    Instructor's signature: _____

Job 25

Service Drum Brakes

Introduction

Servicing a drum brake assembly is a fairly simple task, once you have mastered a few basic procedures and tools. Naturally, the first time you try to install brake shoes on a vehicle, it may feel like one of your hands is tied behind your back. However, as with any automotive operation requiring excellent eye-hand coordination, practice is necessary. For this job, you may simulate a brake repair on a shop-owned teaching unit.

Objective

Given the following list of tools and equipment, you will properly service a drum brake assembly.

Materials and Equipment

- Basic hand tools.
- Safety glasses.
- Brake retracting spring tool.
- Brake hold-down spring tool.
- Wheel bearing grease.
- Brake adjuster.
- Ruler.
- Small screwdriver.
- Brake adjusting gauge.
- Lug wrench.
- Electric drill.
- Brake fluid.
- Wheel cylinder hone.

Instructions

Review Chapter 34, *Brake Service*, in your workbook and textbook before beginning this job. Ask your instructor whether or not you will perform the wheel-cylinder rebuilding portion of the job. Also, inquire about the location of the brake assembly to be used.

As you read the job instructions, answer the questions and perform the tasks. Print your answers neatly and use complete sentences. If you run into any problems, feel free to ask for help.

Warning

Before performing this job, review all pertinent safety information in the text and discuss safety procedures with your instructor.

Name _____

Procedure

 Note: If you see that class time is running out, place all of the parts in a can or parts tray so that they will not be lost.

Disassembly

 Note: If the drum and brake assembly being used is a shop unit, skip Step 1 and Step 2.

1. Remove the wheel, tire and brake drum. If the drum will not come off, you may need to back up the star wheel, **Figure 25-1.** The adjustment may be too tight or the drum may be highly worn and grooved. A stuck rear brake drum could also be caused by rust between the axle flange and drum. To free a rusted drum, use light taps with a hammer. Only strike the drum on the edge nearest you or the drum could be broken. The inner edge is not supported and will break off easily.

Completed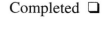

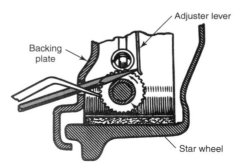

Backing off on adjusting screw
(access slot in backing plate)

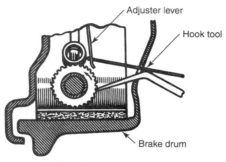

Backing off on adjusting screw
(access slot in brake drum)

Figure 25-1: The adjuster lever must be held out of the way before turning the adjuster star wheel.

Name _____

2. Did you have to back off the adjustment to remove the drum? _____ Explain.
Completed ❏

_____

_____

3. Carefully inspect the way the brake parts fit together, especially the automatic adjuster mechanism and the springs. A few minutes time here can save a lot of time later.
Completed ❏

4. If needed (no backing plate stops), install a spring-type clamp into the ends of the wheel cylinder, as in **Figure 25-2.** This will prevent the pistons and cups from popping out of the cylinder.
Completed ❏

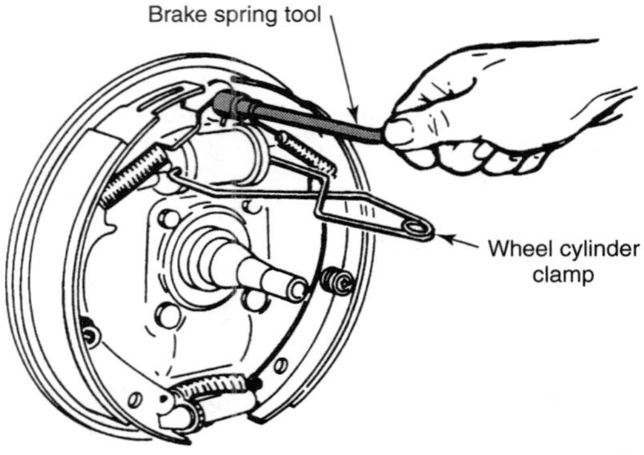

Brake spring tool

Wheel cylinder clamp

(FMC)

Figure 25-2: A cylinder clamp is placed over the ends of the cylinder. A brake spring tool is used to remove the shoe return springs.

5. Measure the dimensions of the following components.
Completed ❏

Primary lining length: _____

Secondary lining length: _____

Lining width: _____

Lining thickness: _____

6. Using the brake spring tool, Figure 25-2, remove the upper (primary and secondary) return springs, and the adjusting cable or lever. Keep the front and rear springs and other parts separate to simplify reassembly. Quite often, the springs may have different tension even though their physical appearance is similar. Always replace brake springs in the same location.
Completed ❏

Name _____

7. Next, remove the hold-down springs with the special hold-down spring tool or with pliers. See **Figure 25-3.**

Completed ❏

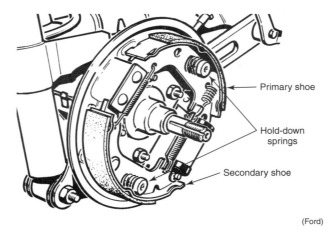

Primary shoe

Hold-down springs

Secondary shoe

(Ford)

Figure 25-3: Pliers or a special hold-down spring tool is used to remove the hold-down springs.

 Note: To remove the springs, you must use one finger to hold the pin tight against the inside or rear of the backing plate.

8. The brake shoes, star wheel, and lower return spring can now be removed as one unit.

Completed ❏

 Warning: Do not blow off brake parts with compressed air. Brake lining dust may contain asbestos, a known carcinogen. Use a special brake vacuum unit or washer to clean the brake assembly.

Inspecting Parts

9. Check the springs and all related links, cables, and other components for damage.

Completed ❏

10. Inspect the brake drum carefully. Look for gouges, heat checking, hot spots, and heavy glazing. Also, use a brake micrometer to check the drum for an out-of-round condition or excessive wear. If the drum shows any of these defects, it should be turned as explained in your textbook. If turning the drum will cause it to exceed its maximum diameter, the drum should be replaced.

Completed ❏

Name _____

11. Are any of the parts damaged? _____ Completed ❑
Explain.

_____

_____

Rebuilding Wheel Cylinders

> **Note: Steps 12–17 should only be performed if your instructor has given you permission to remove and rebuild a wheel cylinder**

12. If your instructor has given permission, remove and disassemble the wheel cylinder. Completed ❑

13. Observe the internal parts of the wheel cylinder. What is the wheel cylinder cup size? _____ Completed ❑

14. After disassembly, check the inside of the cylinder surface for wear, pits, scratches, and scoring. The slightest scratch or score may require cylinder replacement. Completed ❑

15. Hone the cylinder with the electric drill and a small hone. This procedure is shown in **Figure 25-4,** and described in detail in the textbook. Be extremely careful not to pull the hone too far out of the cylinder. The hone could break. Completed ❑

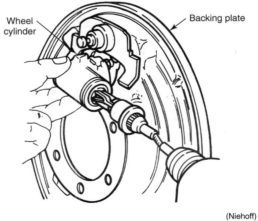

Wheel cylinder   Backing plate

(Niehoff)

Figure 25-4: Recondition the cylinder with light honing. The stones must be fine. Place crocus cloth over the stones to create the smooth, final finish.

16. When honing is complete, clean the cylinder with denatured alcohol or brake fluid and wipe dry. Completed ❑

Name _____

17. Reassemble the wheel cylinder. Then, rein-
    stall the cylinder on the backing plate. If the
    unit is installed on a vehicle always use new
    wheel cylinder cups and boots.

Completed ❑

Reassembling Parts

18. Apply a light coating of high-temperature
    wheel-bearing grease on the star wheel
    threads and raised pads on the backing
    plate, **Figure 25-5.** This will help avoid
    excessive friction, squeaks, and possibly an
    inoperable automatic brake-adjusting
    mechanism.

Completed ❑

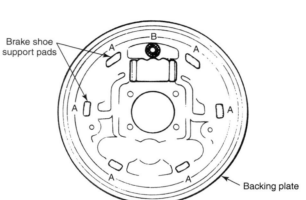

(DaimlerChrysler)

Figure 25-5: Clean and lubricate the brake shoe support pads.

19. Place the shoes into position on the backing
    plate and secure them with their hold-down
    springs. Check that you have the smaller
    primary lining towards the front of the
    vehicle. Also, check that the shoes are
    completely into position.

Completed ❑

20. Fit the anchor pin plate, adjusting cable or
    link, and primary and secondary springs
    into place. See **Figure 25-6.** You will need
    to stretch the springs with a special brake
    spring tool. With a cable type adjuster, the
    cable guide must be installed under the rear
    or secondary return spring.

Completed ❑

21. Is the brake mechanism a link or cable
    type? _____

Completed ❑

22. After screwing it together most of the way,
    slip the adjusting screw or star wheel
    between the bottom of the shoes.

Completed ❑

Name _____

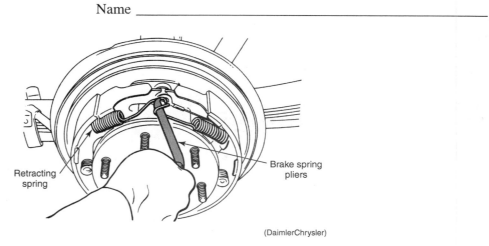

(DaimlerChrysler)

Figure 25-6: This tool is used to pry and stretch springs over the anchor pin.

23. Fit the short hook end of the lower return spring into its hole in the primary shoe.

Completed ☐

> **STOP** **Warning: The half hook on the end of the spring must fit (lock) as shown in the lower portion of Figure 25-7. If the hook is installed wrong, the spring can easily come off and jam against the rotating brake drum.**

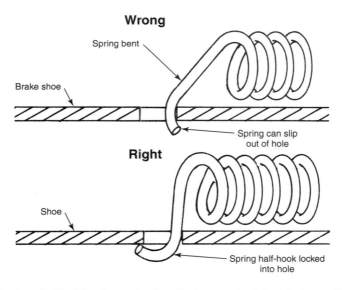

Figure 25-7: In cutaway views of an installed brake spring, notice how the lower spring is installed correctly (locked) into the hole in the shoe.

24. Attach the other end of the lower spring to the adjusting lever (cable-type) or to the other brake shoe (lever-type).

Completed ☐

25. Complete the installation and assembly of the remaining parts.

Completed ☐

26. Do the springs have half hooks? _____

Completed ☐

Name _____

Adjusting Brakes

27. Adjust the brakes by setting the lining-to-drum clearance. Install the special gauge into the brake drum. Refer to **Figure 25-8.**

Completed ❑

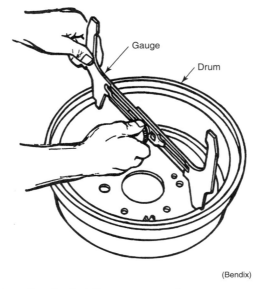

(Bendix)

Figure 25-8: When measuring the brake drum diameter, lock the gauge securely.

28. Spread the gauge to the maximum inside diameter of the drum and lock it.

Completed ❑

29. Tap the brake shoes from side to side to assure that they are completely seated on the anchor pin and centered on the backing plate.

Completed ❑

30. Fit the adjusted brake gauge over the brake linings, **Figure 25-9.**

Completed ❑

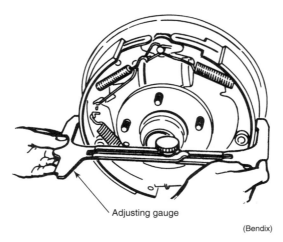

Adjusting gauge

(Bendix)

Figure 25-9: Check the lining diameter at several points across the linings.

Name _____

31. Adjust the star wheel until the linings touch the gauge.

Completed ❑

32. Fit the brake drum into place. The linings should almost touch the drum if adjusted properly.

Completed ❑

Bleeding Brakes

 Note: These steps apply to drum assemblies installed on a vehicle. If the wheel cylinders have been rebuilt, or if any hydraulic component was removed from the vehicle, you will need to remove air from the hydraulic system by bleeding the brakes.

33. Install the brake drum(s).

Completed ❑

34. Have an assistant pump the brake pedal.

Completed ❑

35. Open the bleeding screw at the fitting farthest away from the master cylinder while your assistant is holding the brake pedal down. Close the bleeding screw and then have your assistant quickly release the brake pedal. Repeat this procedure as needed. See **Figure 25-10.** A pressure tank or brake bleeder may also be used.

Completed ❑

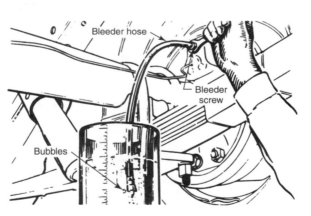

Figure 25-10: Bleed the brakes until there are no more bubbles in the fluid flowing from the hose. Note bubbles are still being forced from this system.

 Caution: Do not press the brake pedal with the drum removed or the wheel cylinder will pop apart.

36. After air bubbles stop coming out of the bleeder fitting, snug the fitting and check the feel of the brake pedal.

Completed ❑

Name _____

37. Refill the master cylinder with brake fluid.   Completed ❑

⬥ **Note: Do not allow the master cylinder to
become empty during this procedure. If the
master cylinder runs dry, air will be
reintroduced into the system.**

---

38. Repeat Steps 34–37 until all brakes are bled.   Completed ❑

39. Turn in your tools and clean the work area.   Completed ❑

40. Have your instructor check your work and   Completed ❑
    sign this worksheet.

    Instructor's signature: _____

Job 26

Diagnose Anti-Lock Brakes/Traction Control System

Introduction

Most modern vehicles have anti-lock brake systems (ABS) and traction control systems. Both systems may be controlled by a single control module and use some of the same sensors and output devices. The most common problem is an illuminated ABS or traction control system warning light. The technician must be able to diagnose problems in these systems.

Materials and Equipment

- Vehicle with an anti-lock brake system.
- Scan tool.
- Multimeter.
- Hydraulic pressure gauge.
- Basic hand tools.
- Service information as needed.

Objective

Given the listed tools and equipment, you will diagnose a problem in an anti-lock brake system.

Instructions

Review the information in Chapter 35, *Anti-Lock Brake and Traction Control System Service*, of your workbook and textbook. Ask your instructor to assign a vehicle and provide any additional information about this job. As you read the job instructions, answer the questions and perform the tasks. Print your answers neatly and use complete sentences. Take your time and do not hesitate to ask for help.

Warning

Before performing this job, review all pertinent safety information in the text and discuss safety procedures with your instructor.

Procedure

1. Determine the exact ABS or traction control complaint by questioning the vehicle driver and consulting previous service records.

Completed ❑

Name _____

2. Check the fluid level in the master cylinder. If the level is low, add fluid and look for leaks. Be sure to check fluid level before going on to Step 3.

Completed ❑

3. Road test the vehicle as necessary. Be sure to follow all traffic regulations and pick a secluded spot for emergency stops.

Completed ❑

4. Does your description of the ABS/traction control problem agree with the driver's description of the problem? _____ If the answer is no, explain your answer.

Completed ❑

_____

_____

5. Is there a problem, or is the system operating normally? For instance, if the ABS system only operates at 10 mph or above, the system is operating normally and you should instruct the customer that this is normal operation.

Completed ❑

6. If a problem is detected, check for obvious problems by visually inspecting all ABS and traction control components. A major part of this step is to check ABS and traction control system wiring and connections. Also, check the foundation brakes and power assist units. Many ABS problems are rooted in other parts of the brake system. Did you spot any obvious problems? _____ Explain your answer.

Completed ❑

_____

_____

Note: If an obvious problem is located before all steps are completed, your instructor can approve the elimination of some of the following steps.

7. Using the scan tool, retrieve the trouble codes. In **Figure 26-1,** the technician is using a Tech II scan tool to retrieve trouble codes. Your shop may have a different type of scan tool. Follow the manufacturer's

Completed ❑

Name _____

directions exactly. Were any codes present?
_____ List any trouble codes retrieved.

_____

_____

(DLC) (General Motors)

Figure 26-1: A scan tool is invaluable when checking the anti-lock brake system for trouble codes. This tool attaches to the data link connector .

Note: Do not remove any system fuses until the trouble codes have been recorded.

---

8. Using the service information, compare the type of problem with the trouble codes retrieved. Could the problem indicated by the trouble codes be the cause of the customer complaint? _____ Explain your answer.

Completed ❑

_____

_____

9. Check the ABS/traction control system fuses. If any fuse is blown, locate the cause before replacing the fuse.

Completed ❑

10. Check any components identified by the trouble codes. This may involve making visual, electrical, or pressure checks. If the wheel speed sensors and rotors are adjustable, check the adjustment. A typical adjustment procedure is shown in **Figure 26-2.**

Completed ❑

Name _____

Check the sensors and rotors for road tar or other debris that could interfere with signal generation.

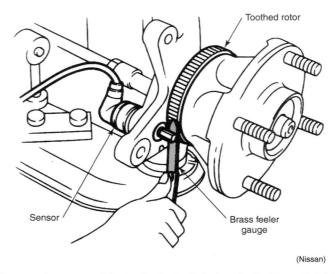

(Nissan)

Figure 26-2: The gap between the speed sensor and the toothed rotor is being checked here. Set this gap to the vehicle manufacturer's specification.

11. Test the G-force sensors, control module or ECM, and the hydraulic pump, actuators, solenoids, and lines as necessary. Refer to the manufacturer's service literature for exact instructions.   Completed ❑

12. Describe any problems found, including readings that are not within specifications or other abnormal conditions.   Completed ❑

13. Do you think that the ABS/traction control problem located in the previous step could be the cause of the problem determined in Steps 1 through 5? _____ Explain your answer.   Completed ❑

_____

_____

14. Recheck your findings by checking the components that could cause the problems found in Steps 10 and 11. Be sure to recheck any components that appear to be damaged or out of adjustment. Also repeat all electrical checks.   Completed ❑

Name _____

15. List any defective components or systems located. Do they agree with your original diagnosis of the problem? _____ Explain your answer.

Completed ❑

_____

_____

16. Correct the defect by making necessary repairs or adjustments. Describe the services performed, including parts used, and related operations such as brake bleeding. Be brief, but mention every task performed.

Completed ❑

_____

_____

_____

17. Recheck ABS/traction control system operation by testing with the scan tool or other test equipment.

Completed ❑

18. Road test the vehicle and determine whether the ABS/traction control system is operating properly.

Completed ❑

19. Did the service operations in Step 16 correct the problem? _____ If not, what steps should you take now?

Completed ❑

_____

_____

20. Clean the work area and dispose of old parts.

Completed ❑

21. Return all service manuals and test equipment to storage.

Completed ❑

22. Have your instructor check your work and sign this worksheet.

Completed ❑

Instructor's signature: _____

Job 27

Replace MacPherson Strut Cartridge and Shock Absorber

Introduction

MacPherson struts are widely used on the front suspensions and are sometimes used on the rear suspensions of automobiles. At the same time, shock absorbers continue to be used on many trucks and SUVs, and on the rear suspensions of passenger cars. Both struts and shock absorbers are usually replaced in pairs. You should know how to replace both MacPherson strut cartridges and shock absorbers.

Objective

Given the needed tools and equipment, you will remove and replace MacPherson strut assemblies and shock absorbers.

Materials and Equipment

- Vehicle in need of MacPherson strut replacement.
- Vehicle in need of shock absorber replacement.
- Service information as needed.
- Basic hand tools.
- Lift or floor jack and jack stands.
- MacPherson strut compressor.
- Tire crayon.
- Replacement MacPherson strut cartridges.

Instructions

Study Chapter 36, *Suspension System Service*, in your workbook and textbook. Ask your instructor to assign vehicles for the job and to provide any additional information about the job. The following procedure is general in nature. Some MacPherson struts are attached by different methods, and some vehicles have the spring separate from the strut assembly. On other vehicles, the strut cartridge can be removed from under the hood. Always consult the proper service manual before proceeding with replacement. After strut replacement, the vehicle must be realigned. Shock absorbers are relatively easy to remove and replace, and do not require realignment when they are replaced.

As you read the job instructions, answer the questions and perform the tasks. Print your answers neatly and use complete sentences. Take your time and do not hesitate to ask for help.

Warning

Before performing this job, review all pertinent safety information in the text and discuss safety procedures with your instructor.

Name _____

Procedure

Replacing MacPherson Strut Cartridges

STOP **Warning: Note that the following procedure applies only to MacPherson strut assemblies that have the spring installed around the strut** cartridge. If the spring is separate from the strut cartridge, do not loosen the fasteners until the lower control arm is safely supported.

1. Raise the vehicle on a lift or raise the vehicle with a floor jack and support it with jack stands.   Completed ❑

2. Remove the wheel and tire. If necessary, remove the brake line from the strut body.   Completed ❑

3. Remove the fasteners holding the strut assembly to the vehicle.   Completed ❑

4. Do the upper or lower strut mountings have a provision for alignment adjustment? _____ If yes, describe it. _____   Completed ❑

_____

_____

What should you do to make sure that the alignment is kept as close as possible to the original settings?

_____

_____

5. Remove the strut assembly from the vehicle.   Completed ❑

6. Place the strut assembly in the strut compressor.   Completed ❑

7. Mark the spring, cartridge body, and cartridge top with a tire crayon before going on to Step 8.   Completed ❑

8. Compress the strut assembly spring according to the strut compressor directions. If you have any questions about how the strut compressor works, read the directions and consult your instructor.   Completed ❑

9. Remove the strut assembly shaft nut and remove the cartridge top from the spring.   Completed ❑

Name _____

⚠️ **Caution: Hold the strut assembly while removing the top nut. If the strut assembly is not held, it will fall when the nut is removed.** Make sure that the compressor has relieved spring tension from the strut assembly shaft nut before completely removing the nut.

---

10. Lower the strut body from the spring and place it in a vise.   Completed ❑

11. Remove the cartridge from the strut body. Most cartridges are held in place by a threaded top. A few cartridges must be cut from the body. In many designs, the cartridge and body are replaced as a single unit.   Completed ❑

12. Compare the old and new strut cartridges.   Completed ❑

Note: If the cartridges do not match, obtain the correct cartridges before going on to the next step.

---

13. Assemble the cartridge to the strut body if necessary.   Completed ❑

Note: Some strut bodies must be drained and refilled with oil before replacing the cartridge. Check the manufacturer's directions.

---

14. Place the new cartridge and strut body assembly in position inside of the spring. Line up the crayon marks made in Step 7, and make sure that all rubber insulators are in place.   Completed ❑

15. Install and tighten the strut assembly shaft nut.   Completed ❑

16. Decompress the spring, making sure that all parts are aligned.   Completed ❑

17. Remove the strut assembly from the compressor and place it in position on the vehicle.   Completed ❑

18. Install and tighten the strut fasteners.   Completed ❑

Name _____

19. Repeat Steps 2–18 to replace the other strut cartridge.   Completed ❑

20. Reinstall the wheels and tires.   Completed ❑

21. Lower the vehicle and check strut operation.   Completed ❑

22. Arrange to have the vehicle aligned. Your instructor may direct you to perform Job 32 at this time   Completed ❑

23. Return all tools and equipment to storage.   Completed ❑

24. Clean the work area and dispose of the old strut cartridges.   Completed ❑

Removing and Replacing Shock Absorbers

Note: This procedure applies to front or rear shock absorbers. Some steps apply to specific front axle or rear axle procedures.

25. Raise the vehicle on a lift or jack stands.   Completed ❑

   Why are the shock absorbers being replaced?

   _____

   _____

26. If rear shock absorbers are being replaced, ensure that the axle will not drop when the shock absorbers are removed. If necessary, support the rear axle with jack stands.   Completed ❑

27. Remove the fasteners holding the shock absorber to the vehicle.   Completed ❑

   What types of fasteners hold the shock absorber to the vehicle?

   • Upper mounting: _____

   • Lower mounting: _____

28. Remove the shock absorber from the vehicle.   Completed ❑

29. Compare the old and new shock absorbers.   Completed ❑

Note: If the shocks do not match, obtain the correct shocks before proceeding.

30. Place the new shock absorbers in position.   Completed ❑

Name _____

> **Note: It may be necessary to compress the new shock absorber slightly to place it in position.**

31. Install the shock absorber fasteners.   Completed ❑

32. Repeat Steps 26–31 for the other shock absorbers to be changed. How many shock absorbers were changed? _____   Completed ❑

33. Lower the vehicle and check shock absorber operation.   Completed ❑

34. Return all tools and equipment to storage.   Completed ❑

35. Clean the work area and dispose of the old shock absorbers.   Completed ❑

36. Have your instructor check your work and sign this worksheet.   Completed ❑

   Instructor's signature: _____

Job 28

Change Ball Joints and Bushings

Introduction

Eventually, all vehicles will need some kind of suspension service. The suspension is constantly in motion whenever the vehicle is being driven, and parts inevitably wear out. Accidents can cause suspension part damage. You should know how to remove and install suspension ball joints and bushings installed on late-model vehicles.

Objective

Given the needed tools and equipment, you will remove and replace a ball joint and suspension bushing.

Tools and Equipment

- Vehicle in need of ball joint or bushing service.
- Service information as needed.
- Lift or floor jack and jack stands.
- Basic hand tools.
- Air operated tools.
- Bushing drivers.
- MacPherson strut compressor.
- Spring compressor.
- Ball joint removal tools.
- Shop towels.
- Replacement parts as needed.

Instructions

Review Chapter 36, *Suspension System Service*, in your workbook and textbook. Ask your instructor to assign vehicles and provide any addition information for the job. Because of the vast number and variety of suspension parts, the following is a general procedure only. You should refer to the appropriate service manual for the specific procedure for your assigned vehicle. The vehicle must be realigned after any of these parts are replaced.

As you read the job instructions, answer the questions and perform the tasks. Print your answers neatly and use complete sentences. Ask your instructor for help as needed.

Warning

Before performing this job, review all pertinent safety information in the text and discuss safety procedures with your instructor.

Name _____

Procedure

 Note: After a control arm has been removed, the vehicle must be aligned. Check with your instructor to determine whether you should perform Job 33 after completing this job.

1. Raise the vehicle on a lift or raise the vehicle with a floor jack and support it with jack stands.

Completed ❑

Replacing Ball Joints

2. Remove the wheel and tire.

Completed ❑

3. If the control arm to be serviced is under spring tension, place a jack stand or floor jack under the control arm, and then lower the vehicle until the control arm rests on the jack stand. If the vehicle is supported by jack stands on the frame, it will be necessary to remove the jack stands and lower the vehicle using a floor jack. As an alternative, the floor jack can be raised under the lower control arm to compress the spring. No matter which method is used, the spring tension must be removed in a safe manner.

Completed ❑

4. If necessary, compress the coil spring to prevent injury or parts damage. See **Figure 28-1.**

Completed ❑

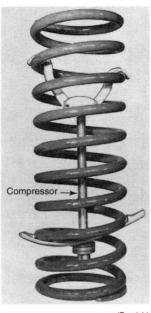

Compressor →

(Branich)

Figure 28-1: The proper positioning for one type of spring compressor is shown here. Always use care when compressing any spring.

Name _____

5. Remove the cotter pin from the ball joint nut.   Completed ❑

6. Is this ball joint a loaded or follower joint? _____ If it is loaded, it is tension or compression loaded? _____   Completed ❑

7. Loosen but do not remove the ball joint nut.   Completed ❑

8. Break the ball-joint-to-steering-knuckle taper using a special tool and a hammer. **Figure 28-2** shows a special tool being used to break the taper. Once the tool is placing tension on the taper, the spindle is struck with a hammer to loosen the taper.   Completed ❑

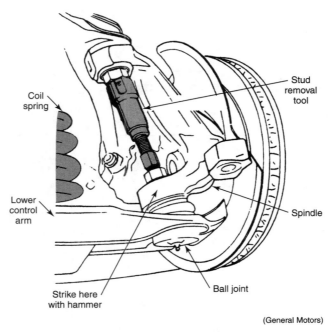

(General Motors)

Figure 28-2: Hit the spindle body with a hammer to remove a ball joint stud. Be careful not to damage any drive axles or anti-lock brake sensors that may be present.

9. Remove the nut from the ball joint stud.   Completed ❑

▽ **Caution: If the spring is installed on the lower control arm, support the arm before removing the nut. This should be done even when the upper ball joint is being replaced.**

10. Remove the steering knuckle from the ball joint stud.   Completed ❑

11. Remove the ball joint from the control arm.   Completed ❑

Name _____

Note: There are many ball joint attachment methods. Some ball joints are pressed into the control arm; others are attached by rivets, and a few are threaded into the control arm. Consult the service manual if you have any doubts as to the correct removal method.

- What was the attachment method? _____
- How did you remove the ball joint? _____

12. Compare the old and new ball joints.   Completed ❏

Caution: Do not attempt to install an incorrect ball joint.

13. Install the new ball joint in the control arm.   Completed ❏
    - What is the attachment method? _____
    - Does this differ from the original attachment method? _____
      If yes, explain?

      _____

      _____

14. Install the ball joint stud in the steering knuckle.   Completed ❏

15. Install the nut on the ball joint stud.   Completed ❏

16. Install a new cotter pin.   Completed ❏

17. Lubricate the new ball joint if it has a grease fitting.   Completed ❏

    Did the new ball joint have a grease fitting? _____

18. Repeat Steps 2–17 to replace the other ball joints.   Completed ❏

Note: If no other parts are being replaced, proceed to Step 33. Your instructor may ask you to perform the second part of this job on a different vehicle.

Replacing Bushings

Note: Control arm removal may not be necessary if you are replacing stabilizer bar or sway bar bushings.

Name _____

19. Remove the ball joint stud from the steering knuckle as covered in the previous section of this job.

Completed ❑

20. If necessary, remove the strut rod, stabilizer bar, alignment shims, and shock absorber/strut rod mounting. Which parts required removal?

Completed ❑

_____

_____

21. Remove the control arm or cross bar attaching bolts. Be sure to mark the position of alignment components to assist in the rough resetting of camber and caster.

Completed ❑

Are there eccentric cams or slotted holes on the control arm bolts?

_____

22. Remove the control arm from the vehicle.

Completed ❑

23. Place the control arm in a vise.

Completed ❑

24. Examine the bushings and determine the best method of removing them. Many bushings may be pressed out using an arrangement similar to the one shown in **Figure 28-3.** Threaded bushings may be removed by unscrewing.

Completed ❑

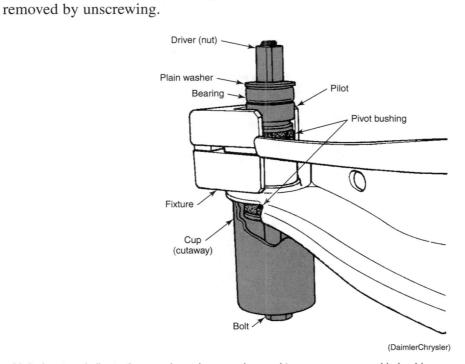

(DaimlerChrysler)

Figure 28-3: A setup similar to the one shown here can be used to remove a pressed-in bushing.

Name _____

Briefly describe how you removed the bushings.

_____

_____

25. Compare the old and new bushings.   Completed ❑

26. Install the new bushings using the proper   Completed ❑
    driver. See **Figure 28-4.**

    Briefly describe how you installed the bushings. _____

_____

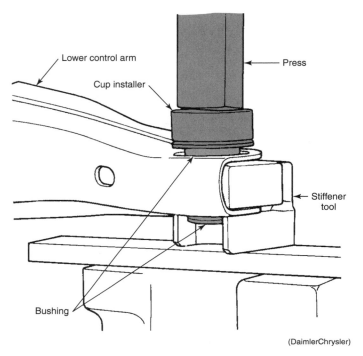

(DaimlerChrysler)

Figure 28-4: Use a press with the proper driver and a stiffener tool to press in new bushings.

27. Reinstall the control arm on the vehicle.   Completed ❑

28. Install the fasteners at the control arm bush-   Completed ❑
    ings or cross bar.

29. Install all related parts.   Completed ❑

30. Install the ball joint on the steering knuckle.   Completed ❑

31. Install the ball joint nut and a new cotter pin.   Completed ❑

32. Repeat Steps 18–30 to replace other bush-   Completed ❑
    ings as needed.

33. Install the wheel and tire.   Completed ❑

Name _____

34. Lower the vehicle.   Completed ❏

35. Arrange to have the vehicle aligned. Your   Completed ❏
    instructor may direct you to perform Job 33
    at this time.

36. Return all tools and equipment to storage.   Completed ❏

37. Clean the work area.   Completed ❏

38. Have your instructor check your work and   Completed ❏
    sign this worksheet.

    Instructor's signature: _____

Job 29

Replace Steering Linkage Parts

Introduction

All vehicles have steering linkage, and you must know how to replace linkage parts. Today most cars have rack-and-pinion steering, while large trucks and SUVs use parallelogram, sometimes called conventional, linkage. Smaller trucks and SUVs can be found with either type of steering. Parts replacement is relatively easy, since the steering linkage is not under spring attention.

Objective

Given the needed tools and equipment, you will remove and replace a steering linkage part.

Materials and Equipment

- Vehicle in need of steering linkage service.
- Service information as needed.
- Lift or floor jack and jack stands.
- Basic hand tools.
- Air-operated tools as needed.
- Rack-and-pinion tie-rod tools if necessary.
- Pittman-arm puller if necessary.
- Pickle fork if necessary.
- Shop towels.
- Replacement parts.

Instructions

Review Chapter 37, *Steering System Service*, in your workbook and textbook before beginning this job. Ask your instructor to assign vehicles and provide any further information about the job. Because of the vast number and variety of steering components, the following is a general procedure only. As you read the job instructions, answer the questions and perform the tasks. Print your answers neatly and use complete sentences.

Warning

Before performing this job, review all pertinent safety information in the text and discuss safety procedures with your instructor.

Most vehicles must be realigned whenever a steering linkage part is replaced. Check with your instructor to determine whether you should perform Job 33 after completing this job.

Name _____

Procedure

1. Raise the vehicle on a lift or raise the vehicle with a floor jack and support it with jack stands. List the steering part(s) to be replaced. _____

_____

_____

Completed ❑

2. If the steering system part contains an adjuster (such as the tie rods), measure the length of the old assembly from the center of each ball stud. Record this reading. _____

Completed ❑

3. Remove the cotter pin from the nut(s) holding ball socket(s) of the part to be replaced.

Completed ❑

4. Remove the nut(s) from the ball socket(s).

Completed ❑

5. If necessary, use a special tool, such as the one in **Figure 29-1,** or a hammer and pickle fork to break the stud tapers. How did you break the tapers? _____

Completed ❑

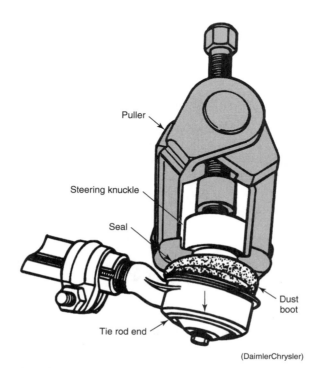

Puller

Steering knuckle

Seal

Dust boot

Tie rod end

(DaimlerChrysler)

Figure 29-1: A puller similar to the one shown here may be used to remove a tie rod end from a steering knuckle.

Name _____

6. Loosen and remove the fasteners holding the part to the frame or steering gear, as applicable. The attachment methods between rack-and-pinion and parallelogram systems are very different and you should consult the applicable service manual for removal and installation information.

   Describe how the part is fastened to the frame or gear.

   Completed ❑

   _____

   _____

7. Remove the part from the vehicle.

   Completed ❑

8. Compare the old and new parts. If the parts are incorrect, what should you do next?

   Completed ❑

   _____

9. If the steering system part contains an adjuster, set the new assembly to the same length as was recorded in Step 2.

   Completed ❑

10. Install the new part on the vehicle.

   Completed ❑

11. Install and tighten all fasteners and install new cotter pins as necessary.

   Completed ❑

12. Repeat Steps 2–11 for other linkage parts to be changed.

   Completed ❑

13. Lower the vehicle and check steering system operation.

   Completed ❑

14. Arrange to have the vehicle aligned. Your instructor may direct you to perform Job 33 at this time.

   Completed ❑

15. Return all tools and equipment to storage.

   Completed ❑

16. Clean the work area.

   Completed ❑

17. Have your instructor check your work and sign this worksheet.

   Completed ❑

   Instructor's signature: _____

Job **30**

Service Wheel Bearings

Introduction

Wheel bearings have the tough job of supporting the weight of the vehicle while spinning for thousands of miles. A worn bearing can become loose, causing tire wear, poor handling, erratic brake application, and lowered gas mileage. A badly worn and dry bearing can seize (lock), causing the wheel to shear off of spindle while the vehicle is being driven. Serviceable (or open) tapered roller bearings should be periodically disassembled, cleaned, and repacked with new grease. Sealed bearings must be replaced when worn and noisy.

Objective

Given the proper tools and equipment, you will service both open and sealed wheel bearings.

Materials and Equipment

- Vehicle in need of wheel bearing service.
- Service literature as needed.
- Basic hand tools.
- Torque wrench.
- Grease cap pliers.
- Parts cleaning tank.
- Wheel-bearing grease.
- Lug wrench.
- Air nozzle.
- Cotter pins.
- Ruler.
- Shop towels.
- Seal driver.
- Hydraulic press with appropriate adaptors.

Instructions

Read the information about wheel bearing service in Chapter 38, *Wheel and Tire Service*, of your workbook and textbook. Ask your instructor which vehicles and parts will be used during this job. You should have access to a vehicle with tapered roller bearings on the front wheels. You will also need a steering knuckle with a pressed-in sealed bearing. The first part of the job includes the disassembly and service of a serviceable bearing. The second section summarizes the service of a sealed bearing.

As you read the job instructions, answer the questions and perform the tasks. Print your answers neatly and use complete sentences. Take your time and do not hesitate to ask for help.

Name _____

Warning

Before performing this job, review all pertinent safety information in the text and discuss safety procedures with your instructor.

Procedure

1. Drum and disc brakes require different service procedures. Look on the inside of the front wheel and determine whether the hub uses drum or disc brakes.   Completed ❑

2. What type of brakes are used on the hub?   Completed ❑

   _____

Removing the Wheel

3. Loosen the wheel lug nuts one-half turn but do not remove them all the way.   Completed ❑

4. Raise the vehicle on a lift or a jack and jack stands. Block the wheels so that the vehicle cannot roll.   Completed ❑

5. Before removing the wheel and tire, mark one of the lug studs and the wheel with chalk or crayon. See **Figure 30-1.** This will let you install the wheel exactly as it was removed. If the wheel and tire were balanced on the vehicle, it could be thrown out of balance if installed in a different position.   Completed ❑

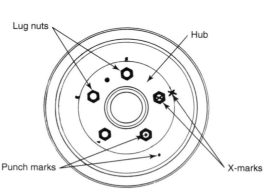

Figure 30-1: A lug stud and the wheel should be marked with chalk, crayon, a center punch.

Name _____

Removing the Hub

6. If the vehicle has disc brakes, remove the brake caliper before removing the rotor and hub. The caliper is usually fastened to the spindle with two or three bolts, **Figure 30-2.**

Completed ❑

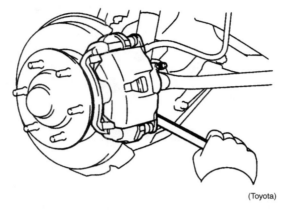

(Toyota)

Figure 30-2: If you have disc brakes, you will have to remove the fasteners securing the caliper before removing the hub.

7. Using the dust cover pliers or channel locks, remove the dust cap. Then, use the diagonal cutters to straighten and remove the cotter pin. Place all of the parts into a container.

Completed ❑

8. Unscrew and remove the adjusting nut and wiggle the drum or disc sideways. This will make the safety washer and outer bearing pop out for removal.

Completed ❑

 Note: Steps 9 and 10 explain one method of quickly removing the inner grease seal and bearing. If not done properly, this method could damage the threads on the spindle. Your instructor may direct you to remove the inner bearing and grease seal by removing the seal with a punch and hammer after the drum is removed.

9. Thread the adjusting nut onto the end of the spindle without the safety washer and outer bearing.

Completed ❑

10. Slide the hub assembly briskly off the spindle while placing a slight downward pressure on the hub.

Completed ❑

Name _____

11. Inspect the brake pads or linings for wear. Compare the thickness against the manufacturer's specifications.   Completed ❑

   • What is the minimum measured thickness of the brake pads or linings? _____

   • Are the pads or linings worn enough to need replacement? _____ Explain. _____

   _____

▽ **Caution: Brake dust may contain asbestos, a known carcinogen.**

12. Do not use air to blow the dust off of the brake parts. Use a vacuum dust removal machine if available, or wash the brake parts clean with a parts washer. Cleaning the parts in a brake parts washer is the preferred method of removing brake dust. Use a rag to wipe the spindle clean.   Completed ❑

13. If you are disassembling the wheel bearings on both sides of the vehicle, keep the bearings from right and left sides separated. Wheel bearings are matched to their cup. Mixing up used bearings and cups can cause problems.   Completed ❑

Cleaning and Inspecting Bearings

14. Submerse and clean the bearings in part cleaning solvent. Use a brush to remove hardened grease.   Completed ❑

15. Blow the bearings dry with compressed air, but do *not* allow them to spin. If spun, a wheel bearing can explode with tremendous force. The bearing rollers can shoot out as if shot from a gun. Remember to wear eye protection.   Completed ❑

16. Inspect the bearings and cup for faults, **Figure 30-3.** Look at them closely. Rotate and inspect each bearing roller. Run your finger over the surface of the bearing cups as you check for imperfections. The slightest amount of roughness can cause bearing noise and quick failure.   Completed ❑

Name _____

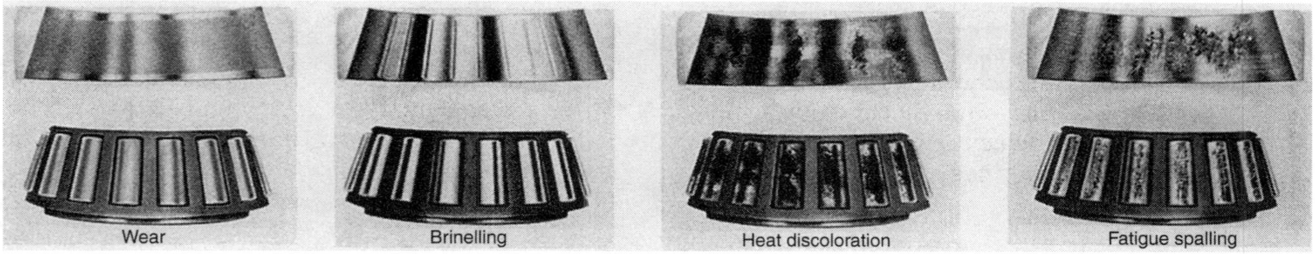

| Dulled cup and rollers | Shiny lines | Darkened and burned | Pitted surface |
| Wear | Brinelling | Heat discoloration | Fatigue spalling |

(General Motors)

Figure 30-3: Some typical bearing conditions are shown here. Study them.

17. Can you detect any problem with the bearings or cups? _____ Explain.   Completed ❑

_____

_____

18. Pack the wheel bearings with grease. Use a bearing packer if one is available. If a bearing packer is not available, place a tablespoon of grease on your palm and push the large side of the bearing into the grease. Repeat this process until the grease has been worked into all of the bearing rollers. Lay the packed bearings on a clean, lint-free cloth. Do not lay the bearing on shop towels or paper towels as the grease could pick up lint or paper fibers.   Completed ❑

19. Inspect the inner lip of the grease seal for splits or tears. See **Figure 30-4.** To check the seal for wear, slide it over the enlarged portion of the spindle. The seal should fit snugly over the spindle.   Completed ❑

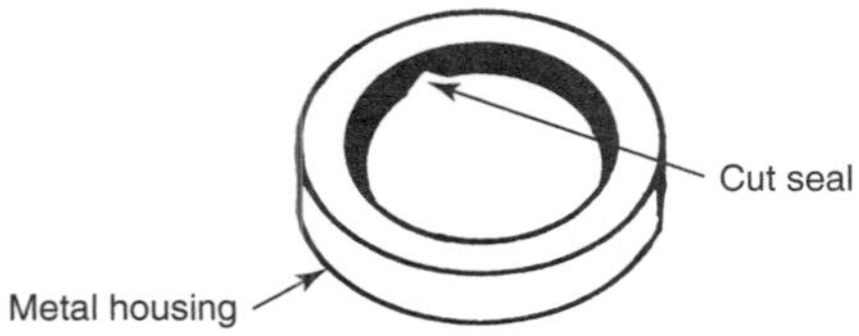

Cut seal

Metal housing

Figure 30-4: Inspect the inner rubber lip of the grease seal for damage or wear. During an actual service, the grease seal would be replaced regardless of its apparent condition.

20. Can you find any problem with the grease seal? _____ Explain.   Completed ❑

_____

_____

Name _____

 Note: If this job is being performed on a vehicle that will be driven on the road, replace the grease seals.

21. Wipe out the inside of the hub to remove all of the old grease. Coat the inside of the hub with new grease. Do not overfill the hub. If the hub is overfull, heat and expansion during operation may force grease out of the hub and onto the brake linings.

Completed ❑

Assembling Bearings

22. Lay the greased inner bearing onto the inner cup.

Completed ❑

23. If available, use a seal driver and a ball-peen hammer to install the inner grease seal, **Figure 30-5.** Lightly tap the seal squarely into place. Make sure you do not dent or bend the seat housing.

Completed ❑

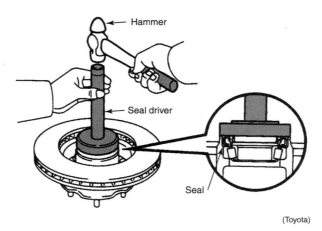

(Toyota)

Figure 30-5: Use a seal driver and a ball-peen hammer to install seals.

24. Slide the hub assembly onto the spindle. Be careful not to damage the grease seal on the threads of the spindle.

Completed ❑

25. Install the outer bearing, safety washer, and adjusting nut.

Completed ❑

26. Adjust the wheel bearings following the procedures in **Figure 30-6** or **Figure 30-7,** whichever is appropriate for your application. A torque wrench and socket will be needed to properly set the bearings.

Completed ❑

Name _____

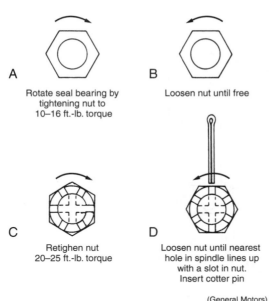

(General Motors)

Figure 30-6: This four-step adjustment procedure is recommended for cars with a slotted nut only.

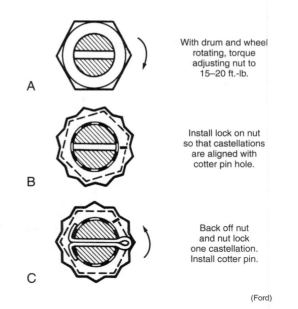

(Ford)

Figure 30-7: This three-step adjustment procedure is recommended for cars with a locknut.

27. Install a new cotter pin and bend it around the outside of the adjusting nut.   Completed ❑

28. Carefully recheck the installation of the cotter pin. If the cotter pin is left out or installed improperly, the front wheel of the vehicle may fall off.   Completed ❑

29. Being careful not to dent it, tap the dust cap into the hub with a ball peen hammer. If you dent the cap, straighten it by hammering on the inside of the cap with a blunt driver and a hammer.   Completed ❑

Removing and Replacing Sealed Bearings

30. Obtain a steering knuckle with a pressed-in sealed bearing from your instructor.   Completed ❑

31. Remove any dust covers or snap rings in the steering knuckle.   Completed ❑

32. Place the knuckle in an appropriate press.   Completed ❑

33. Using the proper adapters, press the bearing out of the steering knuckle. See **Figure 30-8.**   Completed ❑

34. Clean the steering knuckle and study it for cracks and wear. Does the knuckle show signs of wear or damage? _____   Completed ❑

Name _____

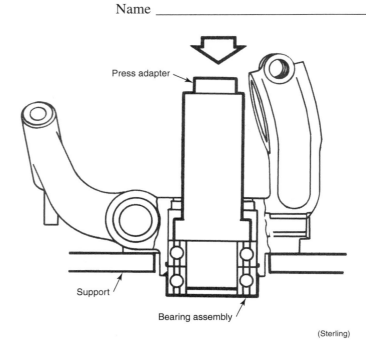

Press adapter

Support

Bearing assembly

(Sterling)

Figure 30-8: Use an arbor press and a special bearing adapter to remove the bearing from the steering knuckle.

If so, explain.

_____

_____

35. Inspect all parts removed from the steering   Completed ❑
    knuckle. Replace worn or dented dust
    covers and wear plates.

36. Position the bearing assembly on the   Completed ❑
    steering knuckle. If this were an actual
    service, the bearing would be replaced.

37. Using the proper adapters, press the bearing   Completed ❑
    into the steering knuckle.

38. Install snap rings or dust covers in the   Completed ❑
    correct positions in the steering knuckle.

Torquing Lug Nuts

39. If applicable, line up the marks on the   Completed ❑
    wheel and lug and slid the wheel into place.
    The tapered edge of the lug nuts should face
    the wheel. This centers the wheel on the hub.

40. Snug the lug nuts with the lug wrench and   Completed ❑
    lower the vehicle to the ground using the
    reverse order utilized when raising the
    vehicle.

Name _____

41. Use a torque wrench to tighten the lug nuts to specification. The chart in **Figure 30-9** shows average torque values for various vehicle makes. Torque the lug nuts in a crisscross pattern similar to the one shown in **Figure 30-10.** Go over all of the lug nuts two or three times to double-check their torque.

Completed ☐

Car make	Lug nut torques in ft.-lb.		
	Compact	Intermediate	Full size
Ford	55-65	75-95	80-110
GMC	60–70	75–85	90–130
Chrysler	50–55	55–60	60–80

Figure 30-9: This chart lists typical lug nut torques for three manufacturers. Use the values listed in the service manual during actual repairs.

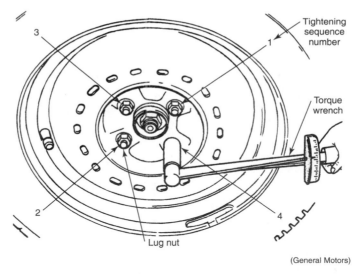

(General Motors)

Figure 30-10: Torque lug nuts in a crisscross pattern to help center and evenly tighten the wheel.

⚠️ **Caution: Proper lug nut torque is very critical, especially on vehicles using disc brakes or aluminum or composite wheels.** Over-tightened lug nuts can warp the disc brake rotor or wheel rim, causing the vehicle to vibrate when the brakes are applied.

42. How tight did you or would you torque the lug nuts? _____

Completed ☐

Name _____

43. What was the final torque value of the   Completed ❑
    bearing adjusting nut before loosening?
    _____

44. Return all tools and equipment to storage.   Completed ❑

45. Clean the work area and dispose of old   Completed ❑
    parts.

46. Have your instructor check your work and   Completed ❑
    sign this worksheet.

    Instructor's signature: _____

Name _____ Date _____

Score _____ Instructor _____

Job 31

Change and Repair Tires

Introduction

Changing, repairing, and balancing tires are fundamental service operations. Damaged or improperly maintained tires cause numerous drivability problems, such as wheel tramp, shimmy, vibration, hard steering, poor steering recovery, steering pull, steering wander, tire squeal, hard ride, tire wear, lowered gas mileage, and flats or blowouts. Anyone who plans on becoming a qualified automotive technician and troubleshooter should learn as much as possible about basic tire service.

Objective

Given a tubeless tire, a wheel rim, and the listed tools, you will properly service and repair a wheel and tire assembly.

Materials and Equipment

- Wheel rim.
- Tubeless tire.
- Tire changer.
- Valve core tool.
- Tire repair kit.
- Wheel-weight pliers.
- Small steel ruler.
- Tread depth gauge or Lincoln-head penny.
- Tire lubricant or soapy water.
- Safety glasses.
- Tire pressure gauge.
- Diagonal-cutting pliers.
- Shop towels.

Instructions

Review the wheel and tire repair information in Chapter 38, *Wheel and Tire Service*, in your workbook and textbook. Ask your instructor for the location of the tire and wheel to be used in the job and for any other details. You should have seen a demonstration on the safe and proper use of the tire changer. Be sure to wear safety glasses, especially while inflating the tire.

As you read the job instructions, answer the questions and perform the tasks. Print your answers neatly and use complete sentences. Take your time and do not hesitate to ask for help.

Name _____

Warning

Before performing this job, review all pertinent safety information in the text and discuss safety procedures with your instructor.

Procedure

1. Obtain the tools, equipment, tire, and wheel to be used for the job.   Completed ❑

2. If the tire is installed on a vehicle, raise the vehicle and carefully remove the wheel and tire.   Completed ❑

3. From the information printed on the sidewall of the tire, fill in the following tire data.   Completed ❑
   • Brand name:_____
   • Wheel diameter:_____
   • Ply information:_____
   • Construction type: _____
   • Load range: _____
   • Maximum air pressure: _____
   • Maximum load: _____
   • DOT number:_____

Checking Tire Wear

4. As previously mentioned in your textbook, tire wear patterns are indicators of suspension, inflation, alignment, and driving problems. Compare the tire wear pattern on your assigned vehicle to the patterns in **Figure 31-1.**   Completed ❑

5. What type of wear pattern does the tire have?_____   Completed ❑

6. What was the most likely cause for the wear pattern?_____   Completed ❑

   _____

7. Can you find any tire damage (for instance cuts, cracks, or tread separation) aside from wear? _____ If so, describe the damage.   Completed ❑

   _____

   _____

Name _____

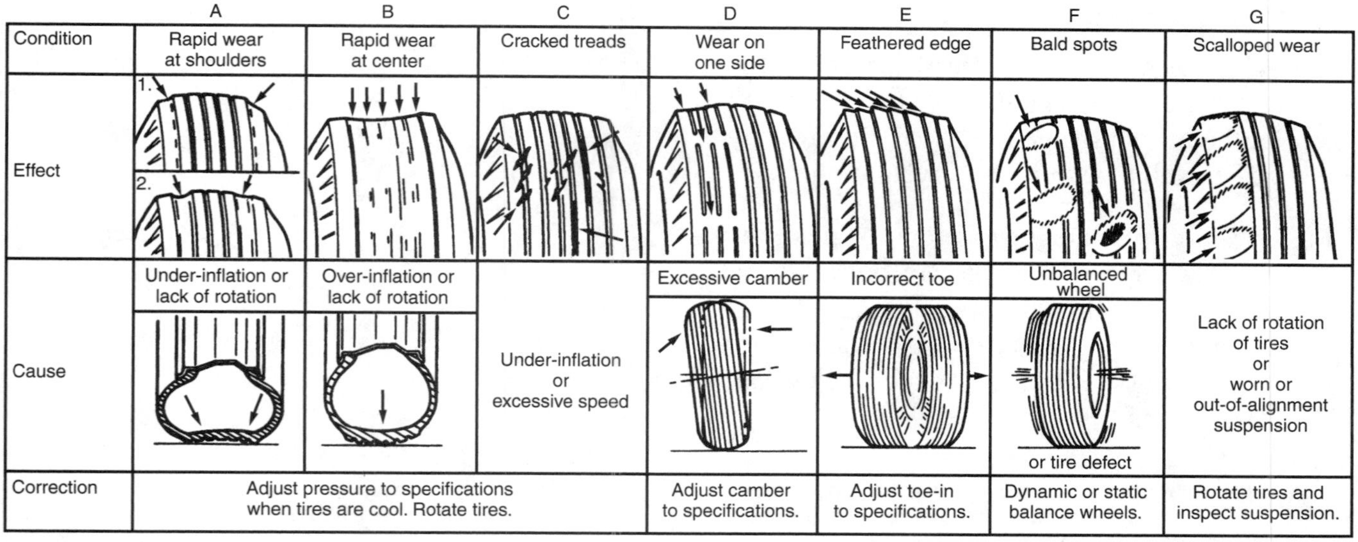

Condition	A Rapid wear at shoulders	B Rapid wear at center	C Cracked treads	D Wear on one side	E Feathered edge	F Bald spots	G Scalloped wear
Effect							
Cause	Under-inflation or lack of rotation	Over-inflation or lack of rotation	Under-inflation or excessive speed	Excessive camber	Incorrect toe	Unbalanced wheel or tire defect	Lack of rotation of tires or worn or out-of-alignment suspension
Correction	Adjust pressure to specifications when tires are cool. Rotate tires.			Adjust camber to specifications.	Adjust toe-in to specifications.	Dynamic or static balance wheels.	Rotate tires and inspect suspension.

(DaimlerChrysler)

Figure 31-1: The cause, effect, and correction of several abnormal tire wear patterns are shown here.

8. As a rule of thumb, if the tread on a tire, at any point, is less than 1/16″ (1.59 mm) deep, the tire is unsafe. It should be replaced. Many modern tires have tread ribs at the bottom of the tire tread. The tread ribs are at a right angle (90°) to the tread. When the tread wears down so that these ribs can be seen to extend across the tire tread, the tire is worn out and should be replaced.

Completed ❏

9. Use a special depth gauge, a steel rule, or a Lincoln penny to measure the depth of the tire tread. A common procedure is to insert a penny, top of the head pointing downward, into the tire tread at various locations. If the top of Lincoln's head shows, then the tread is less than 1/16″ (1.59 mm) deep and the tire is worn out. If the tread covers part of Lincoln's head, the tire is usually considered safe. Measure the tread depth at its deepest and shallowest points.

Completed ❏

10. What is the measured tread depth at its shallowest point? _____

Completed ❏

11. What is the measured tread depth at its deepest point? _____

Completed ❏

12. Describe the tread condition of the tire.

Completed ❏

Name _____

Removing the Tire

13. While wearing safety glasses, remove the valve stem core with the core tool. This will let the air out of the tire. Also, remove any wheel weights with wheel weight pliers.

Completed ❑

14. Use the tire changer to break or push the tire bead away from the lip or flange of the wheel. Follow the specific instructions provided with the tire changer. Keep your fingers out of the way and follow all safety rules.

Completed ❑

> Note: If you are using a power tire changer, do not catch the bead breaker on the edge of the wheel. It can bend a steel wheel or break an alloy wheel.

15. Rub some special rubber lubricant or soapy water on the tire bead and the wheel flange. This will ease tire removal.

Completed ❑

16. Next, use the proper end of the large steel bar of the tire changer to remove the tire from the wheel, **Figure 31-2.** You must use one hand to hold the tire down into the drop center of the wheel while prying off the opposite side of the tire. Be careful not to cut or split the tire bead. If you run into difficulty, ask your instructor for help.

Completed ❑

17. After the tire is off of the wheel, inspect the inside of the tire for splits, cracks, punctures, patches, or repairs.

Completed ❑

Repairing a Punctured Tire

> Note: In the past, many tires were repaired without dismounting the tire. Rubber plugs were inserted into the puncture. Today, many tire makers and service technicians do not consider this an acceptable tire repair. Many shops will repair tires from the inside only. The following procedure is for repairing a puncture with an internal patch. If available, a one-piece head-type internal plug can be used. This type of plug eliminates the need for a separate patch. Follow the manufacturer's directions.

18. After removing the tire from the rim, remove the puncturing object and note the angle of penetration. Clean the area to be repaired.

Completed ❑

Name _____

Safety goggles

Air hose

Air pressure gauge

Wheel spacer

Mount-demount tool

Protective sleeve

Bead lever

Tire and wheel

Mounting paste

Side-mounted bead breaker

Foot controls

(Hunter Engineering Co.)

Figure 31-2: Demounting a tire. Use caution and wear safety goggles.

19. From the inside of the tire, fill the puncture with a plug or a liquid sealer. After filling the hole, cut off the plug (if used) slightly above the tire's inside surface.

Completed ❑

20. Scuff the inside surface of the tire well beyond the repair area. Clean the scuffed area thoroughly.

Completed ❑

21. Apply cement to the scuffed area and place a patch over the damaged area. Use a stitcher to apply pressure to the patch. This bonds the patch to the inner surface of the tire. See **Figure 31-3.**

Completed ❑

22. Clean the outer pressure sealing edge of the wheel as needed. See **Figure 31-4.** Wipe it off with rags or towels. If the rim is rusted or dirty, clean it with steel wool.

Completed ❑

Name _____

(Goodyear Tire and Rubber Co.)

Figure 31-3: Installing a tire patch. A—Buff an area slightly larger than the patch. Clean the buffed area thoroughly. B—Apply the cement with a brush. Allow the cement to dry for the recommended period of time. C—Install the patch. Use a stitcher tool to firmly roll the patch into contact with the cement. Roll over the entire surface of the patch.

(Rubber Mfg. Association)

Figure 31-4: Clean the rim and check for cracks, dents, and old wheel weights before mounting the tire.

23. Check the condition of the valve stem. Bend it sideways and look for weather cracks or splits.   Completed ❑

24. What is the condition of the valve stem? Cracked or weathered valve stems should be replaced.   Completed ❑

 Note: If this is a practice job and the tire will not be installed on a vehicle, do not remove the valve stem unless instructed to do so.

Mounting a Tire

25. To remount the tire, wipe tire lubricant or soapy water on the tire bead and the flange of the wheel. Use the particular procedures for the tire changer to pry the tire back on the wheel.   Completed ❑

Name _____

> **Caution: Push the tire bead (bead opposite the pry bar) down in the drop center of the wheel or the tire will not go on the rim.**

26. With the tire on the rim, pull up on the tire while twisting it. Try to get the tire to catch on the upper safety ridge. You want the upper tire bead to catch and hold on the upper flange of the rim. Then, the tire can be filled with air.

Completed ❏

27. Fill the tire with air. If you can hear air rushing out of the tire, push in lightly on the leaking part of the tire. When the tire begins to take air, do not over inflate it and loosen the wheel hold-down cone. Be careful not to get your fingers caught between the tire and the wheel. If the tire inflates properly, go on to Step 31. If the tire will not take air and expand, continue on to Step 28.

Completed ❏

28. If the tire will not hold enough air to expand and seal on the wheel, you may need to use a bead expander. Clamp the expander around the outside of the tire. This will push the tire bead against the wheel flange. Again fill the tire with air.

Completed ❏

> **Warning: As soon as the tire begins to expand, release the bead expander. If not released immediately, the expander can break and fly off of the tire with dangerous force.**

29. While inflating the tire, lean away from the tire to prevent possible injury. The tire could blow out. Do not inflate the tire to over 40 psi. Check that the bead ridge on the sidewall of the tire is even or true with the wheel. If not, remount the tire on the wheel.

Completed ❏

30. After the bead has popped over the safety ridges, screw in the valve core and tighten it.

Completed ❏

Checking Tire Pressure

31. Use the pressure gauge to check tire pressure. The tire pressure should be a few pounds under the maximum pressure rating

Completed ❏

Name _____

labeled on the sidewall of the tire. As a tire is operated, it will heat up, causing the tire pressure to go up.

32. What is the inflation pressure for the tire?   Completed ❑
   _____

33. To check for leaks, pour rubber lubricant or a soap solution over the tire beads, puncture repair, and valve stem. Watch for bubbles.   Completed ❑

34. Is the tire holding air? _____ Why or why not?   Completed ❑

   _____

   _____

35. Have your instructor check your work and sign this worksheet.   Completed ❑

   Instructor's signature: _____

Job **32**

Replace a Wheel Stud

Introduction

Wheel studs are frequently damaged and require replacement. Replacing a wheel stud is relatively simple. Wheel studs are always pressed into the hub, brake rotor, or brake drum. A missing stud places extra strain on the remaining studs and nuts. Do not allow your customer to drive around with a broken wheel stud.

Objective

Given the proper tools and replacement stud, you will remove a damaged wheel stud and install a replacement stud.

Materials and Equipment

- Vehicle in need of wheel stud replacement.
- Correct service literature as needed.
- Basic hand tools.
- Lift or floor jack and jack stands.
- Replacement stud.

Instructions

Review the wheel stud replacement information in Chapter 38, *Wheel and Tire Service*, in your workbook and textbook. Ask your instructor for vehicle assignments and additional information about the job. Be sure to wear safety glasses when driving out the old stud.

As you read the job instructions, answer the questions and perform the tasks. Print your answers neatly and use complete sentences. If you have any questions or difficulties, ask your instructor for help.

Warning

Before performing this job, review all pertinent safety information in the text and discuss safety procedures with your instructor.

Procedure

---

1. Raise the vehicle on a lift or raise the vehicle with a floor jack and support it with jack stands. Which wheel contains the damaged stud? _____

   Completed ❑

2. Remove the wheel and tire.

   Completed ❑

Name _____

3. Knock out the damaged stud with a large hammer, or use a special stud removal tool, such as the one shown in **Figure 32-1.**

Completed ❑

 Note: If necessary, remove the rotor and splash shield to allow the stud to be removed from the rear of the wheel assembly.

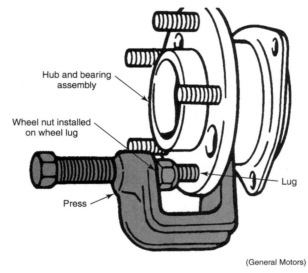

(General Motors)

Figure 32-1: You can use a special press to remove the lug stud.

4. Remove the damaged stud from the vehicle.

Completed ❑

5. Compare the old and new studs.

Completed ❑

 Note: If the studs do not match, obtain the correct stud before proceeding. Stud length, diameter and thread size should all match. Do not attempt to install an incorrect stud. You will end up with a ruined hub.

6. Place the new stud in position on the hub.

Completed ❑

7. Lubricate the stud threads.

Completed ❑

8. Install washers and a lug nut on the stud as shown in **Figure 32-2.**

Completed ❑

9. Slowly tighten the nut to draw the stud into the hub.

Completed ❑

10. Check that the stud is fully seated in the hub. The back of the lug stud should be flush with the other lug studs installed in the hub.

Completed ❑

Name _____

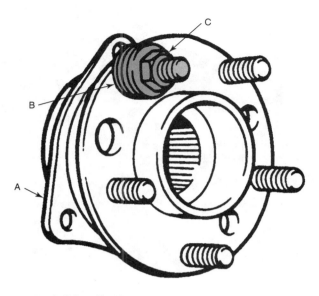

A–Hub and bearing assembly
B–Insert washers over wheel lug
C–Tighten nut to draw wheel lug into correct position

(General Motors)

Figure 32-2: Arrange washers and a lug nut over the new lug stud as shown. Tightening the lug nut will draw the new stud into position.

11. Remove the nut and washers from the stud.   Completed ❏

12. Install the wheel and tire.   Completed ❏

13. Return all tools and equipment to storage.   Completed ❏

14. Clean the work area and dispose of old parts.   Completed ❏

15. Have your instructor check your work and sign this worksheet.   Completed ❏

   Instructor's signature: _____

Job 33

Perform a Wheel Alignment

Introduction

Uneven tire wear is an obvious sign of improper wheel alignment. Misalignment can ruin a tire within a few miles. Incorrect wheel alignment can also cause poor handling and pulling to one side while accelerating or braking. Incorrect alignment cannot cause vibration directly, but it can wear a tire to the point where the tire causes vibration. Correct wheel alignment is critical to the driveablity and dependability of modern vehicles.

Objective

Given the needed tools and equipment, you will measure the alignment angles on all four wheels of a vehicle and correct alignment as necessary.

Materials and Equipment

- Vehicle requiring a four-wheel alignment.
- Appropriate service literature.
- Basic hand tools.
- Specialized alignment tools as needed.
- Alignment machine.

Instructions

Review the alignment information in Chapter 39, Wheel Alignment, of your workbook and text-book. After obtaining all needed tools and access to the shop alignment machine, and getting your instructor's approval, proceed to align the four wheels on the vehicle. If the vehicle has a nonadjustable rear axle, take measurements only.

As you read the job instructions, answer the questions and perform the tasks. Print your answers neatly and use complete sentences. Take your time and follow each step carefully. Ask your instructor for help as needed.

Warning

Before performing this job, review all pertinent safety information in the text and discuss safety procedures with your instructor.

Procedure

Service Manual Information

1. In a service manual for the assigned vehicle, read the service information about four-wheel alignment.

Completed ❑

Name _____

2. What are the alignment specifications for your vehicle?
   Completed ❑
   - Front-right wheel caster: _____
   - Front-left wheel caster: _____
   - Front-right wheel camber: _____
   - Front-left wheel camber: _____
   - Front toe: _____
   - Rear-right wheel camber: _____
   - Rear-left wheel camber: _____
   - Rear toe: _____

Steering and Suspension Inspection

3. Position the vehicle on the alignment rack. Have someone guide you as you drive onto the rack.
   Completed ❑

4. Once in position, place the transmission in Park or place it in gear and block the rear wheels.
   Completed ❑

5. What kind of alignment equipment do you have in your shop?
   Completed ❑

   _____

   _____

6. Have someone turn the steering wheel to the right and then to the left while you check for worn steering parts.
   Completed ❑

7. Summarize the condition of the following steering parts:
   Completed ❑
   - Tie rod ends: _____
   - Rack-and-pinion assembly: _____
   - Power steering belt and pump: _____
   - Other parts: _____

8. Check the suspension system for wear. Summarize the condition of the following suspension system parts:
   Completed ❑
   - Ball joints: _____
   - Shock absorbers: _____
   - Control arm bushings: _____
   - Other parts: _____

Name _____

9. Inspect the tires for signs of wear. What does the tire wear pattern tell you about the vehicle's wheel alignment?

Completed ❑

_____

_____

10. Since the steering and suspension systems must be in good condition before adjusting wheel alignment, replace worn or damaged parts as necessary. Consult you instructor about what to do next.

Completed ❑

Measuring the Alignment Angles

11. How do you prepare to measure alignment angles?

Completed ❑

_____

_____

12. Many of today's alignment machines are computerized, **Figure 33-1.** They will give instructions, specifications, and even pictures of what should be done to align the wheels. Is your alignment machine computerized? _____

Completed ❑

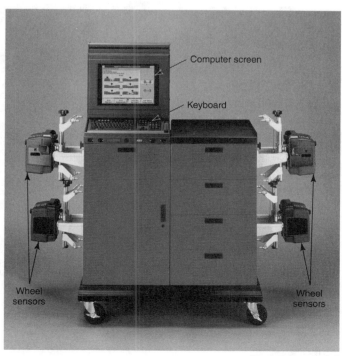

(Hunter Engineering Co.)

Figure 33-1: This is a typical computerized alignment machine. The sensors are mounted on the vehicle's wheels and the alignment readings are displayed on the screen.

Name _____

13. Read through the operating manual for your
    alignment machine.

Completed ❑

14. Mount the alignment machine's wheel
    attachments (usually called heads) on the
    vehicle as outlined in the operating manual.
    See **Figure 33-2.**

Completed ❑

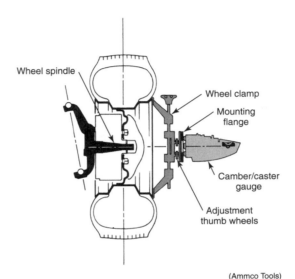

Wheel spindle

Wheel clamp

Mounting
flange

Camber/caster
gauge

Adjustment
thumb wheels

(Ammco Tools)

Figure 33-2: This alignment tool is properly attached to the wheel. Handle these tools with care.

15. Follow the directions in the alignment
    machine's operating manual to adjust the
    heads. This is sometimes called compen-
    sating the heads and allows the alignment
    machine to correct for bent wheel rims.

Completed ❑

⚠ **Caution: It may be necessary to place the
transmission in Neutral and release the
parking brakes while compensating the
heads. Do not forget to place the transmission in Park
and reset the parking brake after the heads are
compensated.**

16. Take readings for four-wheel alignment and
    record them.

Completed ❑

    • Front-right wheel caster: _____
    • Front-left wheel caster: _____
    • Front-right wheel camber: _____
    • Front-left wheel camber: _____
    • Front toe: _____

Name _____

- Rear-right wheel camber _____
- Rear-left wheel camber: _____
- Rear toe: _____

17. How much do these readings vary from specs?   Completed ❑
    - Front-right caster difference: _____
    - Front-left caster difference: _____
    - Front-right camber difference: _____
    - Front-left camber difference: _____
    - Front toe difference: _____
    - Rear-right camber difference: _____
    - Rear-left camber difference: _____
    - Rear toe difference: _____

Adjusting the Wheel Alignment

Note: Make the following adjustments only if the alignment measurements taken previously are not within specifications. In the following steps, if a particular alignment value is not changed, indicate how it would be changed. Remember that toe must be adjusted last, as it is affected by the other alignment angles. Refer to the appropriate service manual for information about making adjustments to your assigned vehicle.

18. Adjust the rear camber. Explain how this is accomplished on this vehicle, and what types of adjusters are used.   Completed ❑

    _____

    _____

19. Adjust the rear toe. **Figure 33-3** shows one method of adjusting rear toe. Describe the adjusting method used on your assigned vehicle.   Completed ❑

    _____

    _____

20. Adjust the front caster. Identify the caster-adjusting device and explain how to adjust caster on this vehicle.   Completed ❑

    _____

    _____

Name _____

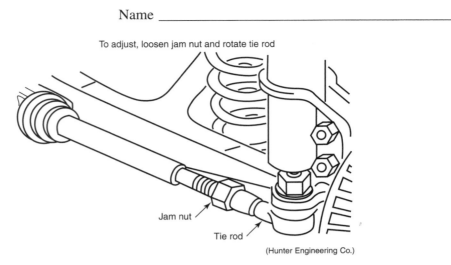

To adjust, loosen jam nut and rotate tie rod

Jam nut
Tie rod

(Hunter Engineering Co.)

Figure 33-3: Rotate the strut rod to set the rear toe to specifications.

21. Adjust the front camber. **Figure 33-4** shows two camber adjustment methods. Explain the adjustment method used on your assigned vehicle.

Completed ❑

_____

_____

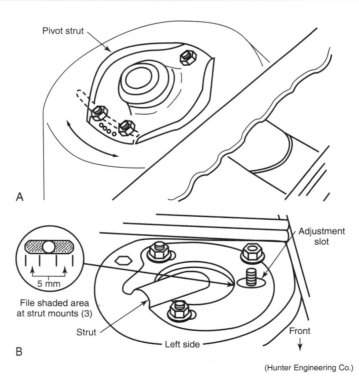

Pivot strut

A

Adjustment slot

5 mm

File shaded area at strut mounts (3)

Strut

Front

Left side

B

(Hunter Engineering Co.)

Figure 33-4: These camber adjustment methods use slotted holes in the body. A—Camber is adjusted by pivoting the strut. B—Camber is adjusted by sliding the strut sideways.

22. Finally, adjust front toe, **Figure 33-5.** Identify the toe-adjusting device and explain how to adjust the toe on this vehicle.

Completed ❑

Name _____

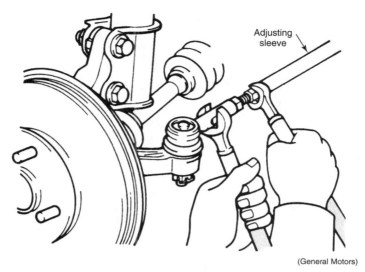

(General Motors)

Figure 33-5: This method of front toe adjustment is common on vehicles with rack-and-pinion steering gears. Turn the adjusting sleeve as specified in the service manual.

23. Is the steering wheel still centered when the front wheels are straight ahead? _____ See **Figure 33-6.** If not the wheel is not centered, how would you correct this problem?

Completed ❏

_____

_____

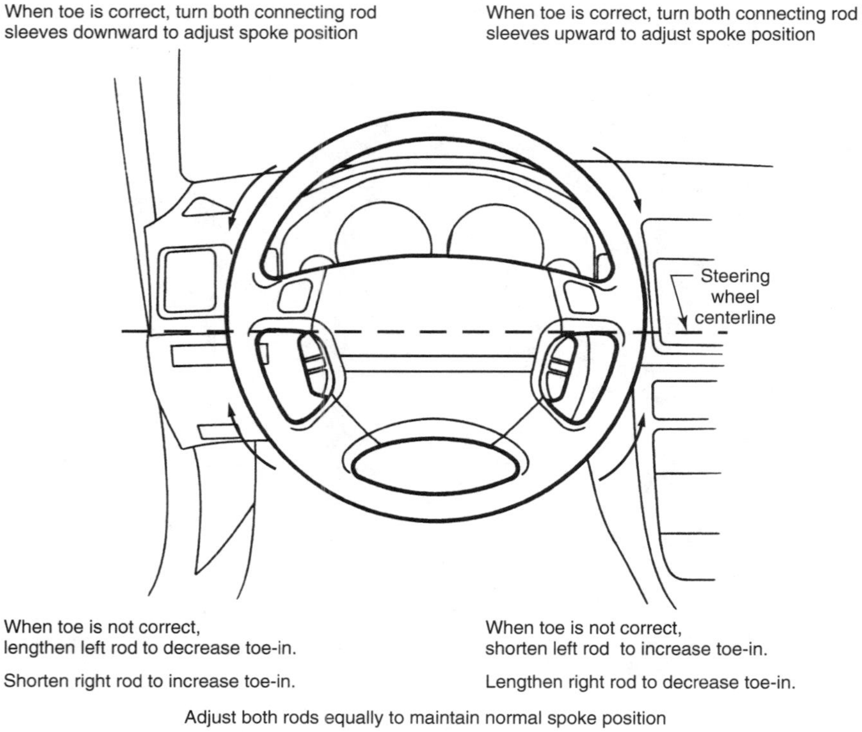

When toe is correct, turn both connecting rod sleeves downward to adjust spoke position

When toe is correct, turn both connecting rod sleeves upward to adjust spoke position

Steering wheel centerline

When toe is not correct, lengthen left rod to decrease toe-in.

Shorten right rod to increase toe-in.

When toe is not correct, shorten left rod to increase toe-in.

Lengthen right rod to decrease toe-in.

Adjust both rods equally to maintain normal spoke position

(Honda)

Figure 33-6: Correct the steering position by adjusting the tie rods. The steering wheel should be centered when the wheels are pointing straight ahead.

Name _____

24. Make sure all steering and suspension sys-   Completed ❑
    tem fasteners are retorqued to specifications.

25. If the alignment machine has a printer, print   Completed ❑
    out the final alignment readings and attach
    the printout to this worksheet.

26. Have your instructor check your work and   Completed ❑
    sign this worksheet.

    Instructor's signature: _____

Name _____ Date _____

Score _____ Instructor _____

Job 34

Service an Air Conditioning System

Introduction

In an effort to reduce depletion of the earth's ozone layer, R-134a has completely replaced R-12 in factory-installed air conditioning systems. Be aware that there are two types of refrigerants now in wide use in automotive air conditioning. You may also encounter any of a number of lesser-used refrigerants or refrigerant blends. It is important to remember that different refrigerants cannot be interchanged. You must know how to work with different refrigerants, and know how to recover the system refrigerant to keep it from entering the atmosphere.

Objective

Given the needed tools and equipment, you will recover, evacuate, and recharge a vehicle air conditioning system.

Materials and Equipment

- Basic hand tools.
- Gauge manifold.
- Refrigerant service center.
- Vacuum pump.
- Thermometer.
- Leak detector.
- Safety glasses.
- Gloves.

Instructions

Review the information in Chapter 40, *Air Conditioning and Heating*, of your workbook and textbook. Ask your instructor to assign a vehicle for the job and provide any additional information about the job or the equipment being used.

As you read the job instructions, answer the questions and perform the tasks. Print your answers neatly and use complete sentences. Take your time and follow each step carefully. Refer to the appropriate service literature for added instructions if needed. Also, feel free to ask your instructor for help if you have difficulty.

Warning

Before performing this job, review all pertinent safety information in the text and discuss safety procedures with your instructor.

Name _____

Procedure

1. In a service manual, look up the informa-   Completed ❑
tion on servicing your vehicle's air condi-
tioning system. What type refrigerant does
the vehicle use? _____

Inspecting the A/C System

 **Warning: Before doing any work to the air
conditioning system, put on safety glasses
and other protective equipment.**

2. Attach a gauge manifold to the appropriate   Completed ❑
service fittings, **Figure 34-1.** Make sure
that all shutoff valves on the manifold and
in the hoses are closed before making the
connections.

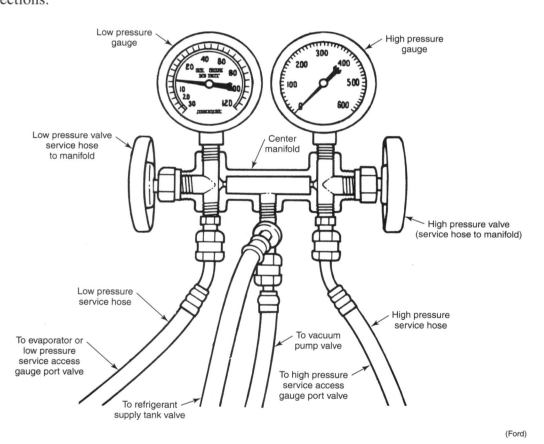

(Ford)

Figure 34-1: This is a typical manifold gauge set.

**Note: R-134a systems have quick disconnect
service fittings. R-12 service fittings are
threaded. Other types of refrigerants have
different service fitting designs. This is to prevent
mixing of refrigerants.**

Name _____

3. Once the gauge manifold is installed, check system static pressures. Normal static pressures are usually in the 70–125 psi (345 to 690 kPa) range. If static pressure is very low or zero, proceed to the leak-checking section of this job.

Completed ❑

4. Start the engine. Set the air conditioning controls for maximum cooling. Allow the system to run for about 10 minutes with the windows up.

Completed ❑

Note: If the compressor does not engage, skip Steps 5–7 and turn off the engine.

---

5. Using a thermometer, measure the air temperature leaving the vent nearest the evaporator. What was the vent temperature reading? _____

Completed ❑

6. Is the temperature measured in the previous step within specifications? _____

Completed ❑

7. Check system pressures. System pressures will vary with type of refrigerant, temperature, and engine speed. Consult the appropriate service literature for exact specifications.

Completed ❑

- High-side reading: _____
- Low-side reading: _____

8. Stop the engine and inspect the air conditioning system for signs of obvious troubles. Summarize the condition of the following:

Completed ❑

- Compressor drive belt: _____
- Compressor and clutch: _____
- Wiring and vacuum hoses: _____
- Condenser: _____
- Dash controls: _____

9. Which parts do you think might be in need of service or replacement?

Completed ❑

_____

_____

How did you determine this? _____

_____

Name _____

Recovering Refrigerant from the Air Conditioning System

> **Note: When discharging an air conditioning system, a refrigerant recovery unit or service center must be used. See Figure 34-2. Always follow the manufacturer's instructions when using this equipment.**

(Robinair)

Figure 34-2: This refrigerant service center is acceptable for use with either R-134a or R-12.

10. Connect the center hose of the manifold gauge to the refrigerant service center as outlined in the manufacturer's instructions.   Completed ❑

11. Turn on the refrigerant service center.   Completed ❑

12. Open both manifold gauge valves slightly to allow refrigerant to slowly enter the refrigerant service center.   Completed ❑

13. After all refrigerant has been recovered, the air conditioning system will go into a slight vacuum. This can be verified by the readings on the manifold gauges. At this point, shut off the service center.   Completed ❑

14. Close the manifold gauge valves and allow the system to remain closed for approximately two minutes.   Completed ❑

Name _____

15. If vacuum remains constant, disconnect the manifold gauge from the refrigerant service center. If vacuum drops, repeat Steps 12–14 until vacuum remains constant.

Completed ❑

16. Using the information in your textbook, describe what could happen if refrigerant squirted into your face and eyes.

Completed ❑

_____

_____

Evacuating the System

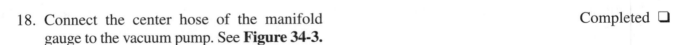 **Note: This section covers evacuating an air conditioning system with a vacuum pump. However, most refrigerant service center units have the capability to evacuate the system, making a separate vacuum pump unnecessary. When using a service center unit to evacuate an air conditioning system, follow the manufacturer's instructions.**

17. Before evacuating the air conditioning system, make sure that both gauge valves are off. Also make sure that there is no pressure in the system.

Completed ❑

18. Connect the center hose of the manifold gauge to the vacuum pump. See **Figure 34-3.**

Completed ❑

19. Slowly open the high and low side manifold gauge valves.

Completed ❑

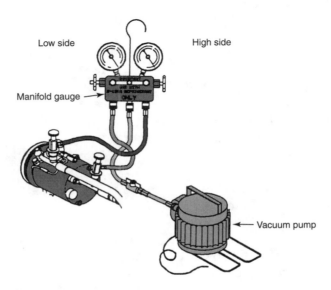

Low side   High side

Manifold gauge

Vacuum pump

Figure 34-3: This is a typical setup for evacuating a system. Always follow the procedure recommended by the manufacturer.

Name _____

20. Start the vacuum pump.   Completed ❑

21. While the pump is running, slowly open the pump shutoff valve to prevent oil from being drawn from the pump.   Completed ❑

22. Watch the vacuum reading on the low-side gauge. When the gauge reads 29″ Hg, allow the pump to run for an additional 15 minutes.   Completed ❑

23. Close the gauge shutoff valves; and then turn off the vacuum pump.   Completed ❑

24. After turning off the vacuum pump, the system vacuum should not drop more than 2″ Hg in 5 minutes. If it does, there is a leak in the system.   Completed ❑

Charging the System

Note: This section covers charging systems with a refrigerant cylinder. If using a pound can or a refrigerant service center unit, discuss the charging procedure with your instructor.

25. Install a safety valve on the refrigerant cylinder, if necessary.   Completed ❑

26. Disconnect the center hose of the manifold gauge set from the vacuum pump and attach it to the refrigerant cylinder.   Completed ❑

27. Open the high-side manifold gauge valve. This will allow the refrigerant vapor to enter the high side of the system.   Completed ❑

28. When refrigerant stops flowing into the system, shut off the high-pressure gauge valve.   Completed ❑

29. Start the engine and run it at 1500 rpm. Set the air conditioner for maximum cooling with the blower at high speed.   Completed ❑

30. Open the tank valve and briefly loosen the center hose at the gauge manifold to purge the center hose.   Completed ❑

31. Open the low-side manifold gauge valve to draw refrigerant into the suction side of the system.   Completed ❑

Name _____

32. While charging the system, watch the gauges to determine when the system is properly charged. On this vehicle, what should the gauges read with a properly charged and functioning system? _____

Completed ❑

Warning: Do not overcharge the system. If high-side pressure rises above 350 psi, (2412 kPa), stop adding refrigerant and allow the system to operate for a few minutes to stabilize. If pressures stabilize above 350 psi, remove some refrigerant and check for condenser air flow or high side restriction problems.

---

33. When the system is properly charged, turn the blower speed to Low and check the thermometer reading at the air duct nearest the evaporator. Normal reading should be between 35°F and 45°F. What does the thermometer read? _____

Completed ❑

Using an Electronic Leak Detector

34. Turn the engine off and allow the vehicle to sit undisturbed for a few minutes.

Completed ❑

35. Turn the detector on and allow it to warm up for about one minute away from the refrigeration system components.

Completed ❑

36. Turn the sensitivity adjustment knob on the detector face. A common initial detector sensitivity setting would be to detect a leak rate of about 1 1/2 ounce (45 ml) per year. In some cases, the electronic detector's sensitivity must be reduced when a large leak is present or when other engine or shop fumes trigger the detector.

Completed ❑

37. After setting sensitivity, slowly pass the sensing tip closely around possible leak areas and check for an increase in the ticking noise. See **Figure 34-4.** Also pass the tip under suspected leak areas. Since refrigerant is heavier than air, it will flow downward from a leak, **Figure 34-5.** Most leak detectors will make a ticking noise that increases when the probe encounters refrigerant. Large leaks raise the ticking to a high-pitched squeal. Newer leak detectors have a display that indicates the leak rate.

Completed ❑

Name _____

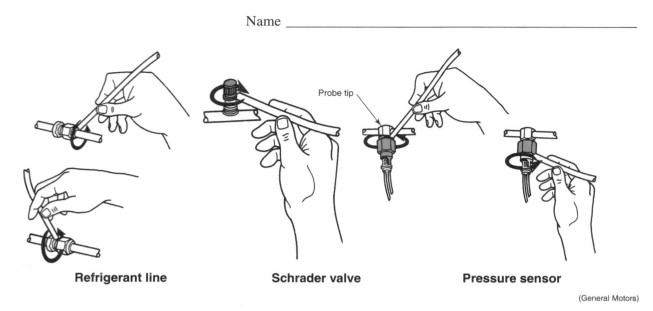

Refrigerant line   **Schrader valve**   **Pressure sensor**

(General Motors)

Figure 34-4: Move the tip of the detector all the way around suspected leaks.

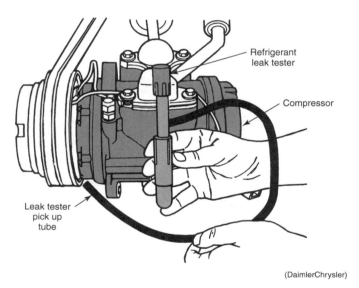

(DaimlerChrysler)

Figure 34-5: Because refrigerant is heavier than air, you should always pass the tip of the detector underneath suspected leaks.

38. List the location of any leaks found.   Completed ❑

_____

What must be done to correct them? _____

_____

39. Disconnect the manifold gauge set.   Completed ❑

40. Return all tools and equipment to their storage areas.   Completed ❑

41. Have your instructor check your work and sign this worksheet.   Completed ❑

Instructor's signature: _____

Job **35**

Complete a Repair Order

Introduction
Completing paperwork is as important as performing the actual repair jobs. Most shops use a repair order, sometimes abbreviated RO. Repair orders are used to bill the customer, and often used to calculate the technician's pay for the work performed. While some repair orders are computer generated, the technician will usually fill them out by hand. Every repair order should list the *condition* (the customer's complaint), *cause* (what is causing the problem), and *correction* (what you did to fix the problem). Listing the condition, cause, and correction is especially important when doing warranty repairs. You must understand the purpose of repair orders and know how to correctly fill them out.

Objective
Upon completion of this activity sheet, you will be able to complete a repair order listing condition, cause, and correction of a problem, and labor and parts costs.

Materials and Equipment
- Pencil or pen.

Instructions
As you read the job instructions, answer the questions and perform the tasks. Print your answers neatly and use complete sentences. If you have any problems or questions, ask your instructor for help.

Procedure

Note: When calculating costs on a repair order, write out each equation so the customer can quickly see how the costs were determined.

1. Pick a job that you have performed earlier in this class. It is suggested that you select a job that involves diagnosis and repair instead of preventive maintenance. Your instructor may select a job for you.

Completed ❑

2. Fill out the customer information section of the repair order using the name of the vehicle's owner from the previous job or your own name if the owner's name is not known.

Completed ❑

Name _____

List the make, model, year, mileage, vehicle identification number, and other information about the vehicle that was worked on in the previous job.

3. Review the previous job material.　　　　　　　　　　　　　　Completed ❑

4. In the Labor Instructions section of the repair order, write down the condition (problem) that caused the vehicle to be brought in for service.　　　　　　　　　　　　　　　　　Completed ❑

   • Write down what you determined to be the cause of the problem.

   • Write down what was done to correct the problem.

5. In the Part No. or Description section, list all parts used in the job.　　　　　　　　　　　　　　　　　Completed ❑

   • List the retail prices of all parts. To find parts prices, refer to the original receipts, look in catalogs or call local parts suppliers. Remember that the customer pays retail (not discount) prices.

   • Add the cost of all parts and list the total on the repair order.

6. If an outside shop performed any work, list the type and cost in the Sublet Repairs section. The Sublet Repairs section is in the lower left-hand corner of the repair order. List each sublet repair separately.　　　　　　　　　　Completed ❑

7. Next to the corrective actions listed in the Labor Instructions, list the time used to make each repair. List the repair time for each task separately. Ask your instructor whether you should use a labor rate manual to find repair times or use the times that it actually took you to perform the work.　　　　　　Completed ❑

8. Ask your instructor to assign a labor rate for this job. To determine local labor rates, call several repair facilities and average the results.　　　　　　　　　　　　　　　　　　　　Completed ❑

   • Determine the labor cost for each task by multiplying the labor time for the task by the labor rate. Write this cost next to the time.

9. Add all of the labor costs and list the total in the Service Dept. Sales column, in the lower right-hand corner of the repair order.　　　　　　Completed ❑

Name _____

10. Determine a reasonable charge for shop supplies, such as cleaning fluids and shop towels. Ask your instructor to suggest an amount for this job. Add the shop supply fee to Service Dept. Sales section on the repair order.

Completed ❑

11. Write the cost of parts and sublet repairs in Service Dept. Sales section. Add the cost of parts, labor, shop supplies, and sublet repairs together to determine the total pretax price of the service. Record the price in the Service Dept. Sales section of the repair order.

Completed ❑

12. Determine the sales tax rate in your area.

Completed ❑

13. Multiply the total pretax price of the service by the tax rate. For example, if the tax in your area is 5%, multiply the total pretax price (step 11) by .05. This is the tax for the service.

Completed ❑

14. Add the tax (Step 13) to the total pretax price (Step 11) and list the grand total at the bottom of the Service Dept. Sales section of the repair order.

Completed ❑

15. If your instructor directs you, fill out other sections of the repair order.

Completed ❑

16. Were there any sections of the form that could not be completed? What sections could not be completed and why?

Completed ❑

17. Have your instructor check your work and sign this worksheet.

Completed ❑

Instructor's signature: _____

Name _____

TOM DUNCAN PONTIAC, GMC TRUCK, INC.

201 South Buncombe Rd. - P.O. Box 1907

GREER, S. C. 29652

R.O. NUMBER 38505

DATE

VEHICLE IDENTIFICATION NO

MILEAGE   YEAR   MAKE - MODEL

38505

TERMS: STRICTLY CASH UNLESS ARRANGEMENTS MADE

I hereby authorize the repair work hereinafter set forth to be done along with the necessary material and agree that you are not responsible for loss or damage to vehicle or articles left in vehicle in case of fire, theft or any other cause beyond your control or for any delays caused by unavailability of parts or delays in parts shipments by the supplier or transporter. I hereby grant you and/or your employees permission to operate the vehicle herein described on streets, highways or elsewhere for the purpose of testing and/or inspection. An express mechanic's lien is hereby acknowledged on vehicle to secure the amount of repairs thereto.

X _____

The only warranties applying to this part(s) are those which may be offered by the manufacturer. The selling dealer hereby expressly disclaims all warranties, either express or implied, including any implied warranties of merchantability or fitness for a particular purpose, and neither assumes nor authorizes any other person to assume for it any liability in connection with the sale of this part(s) and/or service. Buyer shall not be entitled to recover from the selling dealer any consequential damages, damages to property, damages for loss of use, loss of time, loss of profits, or income, or any other incidental damages.

"This dealership utilizes the hours published in Pontiac-GMC Truck's Labor Time Guide, which reflects an average time requirement for the performance of specific vehicle repairs, and which may therefore, be either more or less than the actual clock time in any given instance."

CUSTOMER LABOR CHARGES ARE BASED ON A RATE OF $ _____ PER HOUR

	OK'D BY	CASH	CHARGE	INTERNAL

DELIVERY DATE   OPERATION   LABOR

CUST. P.O. NO

TIME PROMISED   AM / PM

NAME _____

ADDRESS _____

CITY _____ STATE _____

PHONE WHEN READY   YES / NO

BUS. PHONE

RES. PHONE

	OK BY	CASH	CHARGE	INTERNAL

OPERATION list: LUBRICATE, CHANGE OIL, CHANGE OIL FILTER CART, TUNE ENGINE, CHARGE TRANS. MISSION FLUID & FILTER, ENG. FUEL SYSTEM FILTER, STATE INSPECTION, FLUSH RADIATOR & ADD COOLANT, PACK WHEEL BRGS., ALIGN WHEELS, ROTATE TIRES

INT - WC

REPAIR ORDER - LABOR INSTRUCTIONS

	TC	OPERATION	TIME	MECH NO
O				
P				
E				
R				
A				
T				
I				
O				
N				
S				

WARRANTY CLAIMS					INTERNAL SALES				SERVICE DEPT. SALES				
ACCT	K	AMOUNT	K	COST	ACCT.	K	AMOUNT	K	DESC.	ACCT	K	COST	AMOUNT	K
462	-		..	463	-		..	LABOR- MECH.	460	-			
480	-		..	481	-		..	PTS./ACC. MECH.	467	-			
466	-		..	466	-		..	SUBLET- MECH.	466	-			
472	-		..	473	-		..	LABOR- BODY	470	-			
458	-		..	476	-		..	SUBLET- BODY	476	-			
480	-		..	481	-		..	PTS./ACC. BODY	477	-			
479	-		..	479	-		..	P & B MATL.	479	-			
490	-		..	490	-		..	TIRES	490	-			
491	-		..	491	-		..	GAS OIL GREASE	491	-			

POLICY WORK- VEHICLES

POLICY WORK- MECH. & BODY

	TAX	324	-	CUST NO
SUBLET REPAIRS		261A	-	67 E	CLM	CHARGE	220	-
TOTAL SUBLET REPAIRS	263	+		STK	CASH	225	+

	PO NO	SUBLET REPAIRS			CLM			

W.I.	COST	QTY	PART NO. OR DESCRIPTION	SALE		K	AMOUNT	K	COST
			A						
			B						
			C						
			D						
			E						
			F						
			G						
			H						
			I						
			J						
			K						

Job 36

Complete an Employment Application

Introduction

You will not be able to do any automotive repairs if you don't get a job in the automotive repair industry. The first step in getting a job is to fill out an employment application. This is not something to be done carelessly. A legible, well-thought-out employment application is your first chance to present yourself in a positive way, a way that will get you an interview and a job.

Objective

After working through this activity sheet, you should be able to properly complete an employment application form, listing your education, former employers, and other relevant information.

Tools and Equipment

- Pencil or pen.

Instructions

As you read the job instructions, answer the questions and perform the tasks. Print your answers neatly and use complete sentences. Take your time and follow each step carefully. Also, feel free to ask your instructor for help if you have difficulty.

Procedure

1. Gather the personal information needed to fill out an employment application. This usually includes your social security and driver's license numbers, the names and addresses of schools you have attended, the names and addresses of former employers, your salary history, and personal references.

Completed ❑

Note: When you are looking for a job, it is a good idea to carry a copy of your employment and salary histories, your personal references, and a list of the names and addresses of the schools you have attended with you at all times. By doing so, you will be prepared if a surprise opportunity presents itself.

Name _____

2. Go to the employment application attached to this job. Write today's date and put "Automotive Technician" as the position applied for. Use your best penmanship.

Completed ❑

3. List your personal information as required on the application. Do not list your social security or driver's license number on this sample application.

Completed ❑

4. Add the information requested below the personal information section. Note that some information may not apply to you. Ask your instructor if you are unclear about what information is desired.

Completed ❑

5. List your educational experience, including the schools attended, the dates of attendance, and diplomas or degrees earned. If you attended more than one high school, list the school you are currently attending or from which you graduated. If you are less than one full semester away from graduation at the school you are currently attending, write down the number of years you plan to complete, the degree you intend to earn, and the date you plan to graduate. If you are more than one full semester away from graduating, write down the number of years you have already completed.

Completed ❑

6. List any special abilities or training you have in the Special Skills section. Concentrate on skills that would interest a potential employer. This would be a good place to list any automotive areas in which you are particularly interested or have special training.

Completed ❑

7. List your work experience, starting with your current or last employer and working backward. List company names, dates of employment, job title and type of work performed, your supervisor's name, and your reason for leaving the job. If you have not yet held a job, list any volunteer work or other experience that could be substituted for employment. Be honest but present your job information in a positive manner. Do

Completed ❑

Name _____

not take this opportunity to discuss the shortcomings of your former employers. Use an extra sheet of paper if necessary.

8. If there are any gaps in your employment history of more than 90 days, account for them in the space provided. For example, list schooling, military service, and illnesses. If there are no gaps in your employment record, skip this section.

Completed ❑

9. List the names, addresses, and phone numbers of two professional references. These should be people who have worked with you and are familiar with the quality of your work and your overall attitude toward the job. If you have not held any jobs, list fellow students or neighbors.

Completed ❑

10. Carefully read the section labeled, "Important," and be sure that you comprehend it before signing your name at the bottom.

Completed ❑

11. Add any other information needed to complete the employment application form.

Completed ❑

12. Give the completed employment application to your instructor.

Completed ❑

13. Discuss the information that you presented on the employment application form. Your instructor may suggest methods of improving your presentation on future employment application forms.

Completed ❑

14. Have your instructor check your work and sign this worksheet.

Completed ❑

Instructor's signature: _____

Name _____

EMPLOYMENT APPLICATION

Date _____

Position Applied For _____

PERSONAL INFORMATION

Name _____

Address _____

Telephone _____ Social Security Number _____

Work Telephone _____ Cellular Telephone _____

E-mail Address _____

If employed and you are under 18, can you furnish a work permit?   ❑ Yes   ❑ No
  ❑ Not Applicable (Current age is 18 or over)

Have you ever served in the U.S. Armed Forces?   ❑ Yes   ❑ No

Are you prevented from lawfully becoming employed in this country because of VISA or Immigration Status?   Yes   No
(Proof of citizenship or immigratation status will be required upon employment, as required by Federal Law)

Have you previously applied at _____ ❑ Yes ❑ No   If yes, when? _____ Where? _____

Have you previously been employed by _____ ❑ Yes ❑ No

If yes, when and where? _____

Interested in working: ❑ Full-time ❑ Part-time ❑ Temporary   Willing to work: ❑ Days ❑ Evenings ❑ Weekends ❑ Holidays

Desired wage $ _____ per _____ Date available for work? _____

Have you been convicted of a crime within the last seven years? (Do not include convictions prior to age 18)   ❑ Yes ❑ No

If yes, please explain _____
(Conviction will not necessarily disqualify applicant from employment)

How were you referred to us

    ❑ Newspaper   ❑ TV/Radio   ❑ Internet   ❑ Walk-In

❑ Employee _____ ❑ Job fair   ❑ Please specify _____

EDUCATIONAL BACKGROUND

Please list all educational or specialized experience which relates to the position(s) applied for, such as high school, college, business, technical, or graduate school.

Schools	Name & Location of School	No. of Years Attended	Did You Graduate?	Highest Degree Earned
High School				
College				
Business/ Technical				
Graduate				

Summarize special skills which you want us to consider in evaluating your qualifications:

SPECIAL SKILLS

_____

_____

_____

_____

_____

_____